Engineering Properties of Rocks

Engineering Properties of Rocks

Second Edition

Lianyang Zhang
University of Arizona
Tucson, Arizona, United States

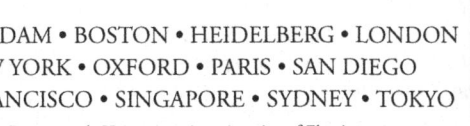

AMSTERDAM • BOSTON • HEIDELBERG • LONDON
NEW YORK • OXFORD • PARIS • SAN DIEGO
SAN FRANCISCO • SINGAPORE • SYDNEY • TOKYO

Butterworth-Heinemann is an imprint of Elsevier

Butterworth-Heinemann is an imprint of Elsevier
The Boulevard, Langford Lane, Kidlington, Oxford OX5 1GB, United Kingdom
50 Hampshire Street, 5th Floor, Cambridge, MA 02139, United States

Notices
Knowledge and best practice in this field are constantly changing. As new research and experience
broaden our understanding, changes in research methods, professional practices, or medical
treatment may become necessary.

Practitioners and researchers must always rely on their own experience and knowledge in evaluating
and using any information, methods, compounds, or experiments described herein. In using
such information or methods they should be mindful of their own safety and the safety of others,
including parties for whom they have a professional responsibility.

To the fullest extent of the law, neither the Publisher nor the authors, contributors, or editors,
assume any liability for any injury and/or damage to persons or property as a matter of products
liability, negligence or otherwise, or from any use or operation of any methods, products,
instructions, or ideas contained in the material herein.

Library of Congress Cataloging-in-Publication Data
A catalog record for this book is available from the Library of Congress

British Library Cataloguing-in-Publication Data
A catalogue record for this book is available from the British Library

ISBN: 978-0-12-802833-9

For information on all Butterworth Heinemann publications
visit our website at https://www.elsevier.com/

Working together
to grow libraries in
developing countries

www.elsevier.com • www.bookaid.org

Publisher: Joe Hayton
Acquisition Editor: Andre Gerhard Wolff
Editorial Project Manager: Mariana Kühl Leme
Production Project Manager: Vijayaraj purushothaman
Cover Designer: Maria Ines Cruz

Typeset by SPi Global, India

Contents

7. Strength 251

About the Author

Lianyang Zhang is a professor in Civil Engineering at the University of Arizona. He received his B.S. degree in Naval Architecture and Ocean Engineering from Shanghai JiaoTong University, his M.S. degree in Geotechnical Engineering from Tongji University, and his Ph.D. degree in Geotechnical Engineering from Massachusetts Institute of Technology. His areas of expertise include rock characterization, pile foundations, and sustainable construction materials. Dr. Zhang has published more than 100 technical papers in international journals and is also the author of the book *Drilled Shafts in Rock—Analysis and Design*.

Preface to the Second Edition

It has been more than a decade since the first edition of this book was published. During this time, much progress has been made in the field of rock mechanics and rock engineering, including the 2008 release of the World Stress Map (WSM) and the development of the three-dimensional Hoek-Brown strength criteria. This, along with the popular reception of the first edition of the book, has motivated the author to update the book with a second edition. The purpose of preparing the second edition is to expand the various topics presented in the first edition and add new topics that have either gained in importance or been developed since the publication of the first edition.

Because many readers are familiar with the layout of the first edition, the same format with the same number of chapters is followed in the second edition. Yet all chapters have been revised to include the latest development in rock mechanics and rock engineering. The following is a summary of the main additions and changes:

- Chapter 1: A new section has been added to describe briefly the rock expert system for evaluation of rock engineering properties.
- Chapter 2: The ISRM suggested method for establishing the final rock stress model has been added in the section on the strategy for determining in situ rock stresses. The 2004 WSM has been replaced by the 2008 one. Also, new empirical relations between in situ stresses and depth have been added.
- Chapter 3: Many new typical values and empirical correlations for the index properties of intact rock have been added.
- Chapter 4: The effect of direction on rock quality designation (RQD) has been briefly discussed. New methods for determining discontinuity frequency and trace length using planar and nonplanar sampling windows have been included. New information has been added on determination of block size for rock masses containing nonpersistent discontinuities. The discussion on discontinuity shape has also been expanded.
- Chapter 5: New methods for estimating the geological strength index (GSI) have been included. A new section has been added to describe the rock mass index (RMi). A number of new empirical correlations between different classification indices have been added. The section on classification of weathering of rock is also expanded.
- Chapter 6: New data and empirical relations for the elastic modulus of intact rock have been added. New subsections have been added on the Poisson's

ratio of intact rock and rock masses, respectively. The section on evaluating the deformation modulus of rock masses has been expanded by including new methods. The discussion on scale effect and anisotropy for rock deformability has also been updated.

- Chapter 7: New data and empirical relations for the unconfined compressive strength (UCS) and tensile strength of intact rock have been added. The section on evaluating the UCS of rock masses has been expanded by including new methods especially those based on RQD. Two new sections, one on the three-dimensional Hoek-Brown strength criteria and the other on the residual strength, have been added. The discussion on scale effect and anisotropy for rock strength has also been updated.
- Chapter 8: The section on the permeability of rock masses has been expanded by including new empirical relations between permeability and porosity and those between permeability and RQD and RMR. The discussion on the effect of different factors and the anisotropy of rock permeability has also been updated by including new data.

It is believed that this is still one of the very few books devoted to the evaluation of rock engineering properties. As with the first edition, it is still intended to be a book that provides a single source of information and serves as a valuable tool for practitioners to determine the engineering properties of rocks required for particular projects. It is also a useful reference for researchers and students to look into the typical values of different rock properties and the factors affecting them.

I would like to add some acknowledgments to those I made in the preface to the first edition. Working with the Elsevier staff was once again a pleasure. I thank, in particular, André Gerhard Wolff, the Publisher of Elsevier S&T Books, and Mariana Kühl Leme, the Editorial Project Manager of Elsevier S&T Books. I moved from industry back to academia in 2007. My gratitude extends to my colleagues in the Department of Civil Engineering and Engineering Mechanics at the University of Arizona, Tucson, AZ, as well as my students, from whom I learn new things all the time. Finally, I would like to thank my wife and two children for their understanding and support during the preparation of this book.

Lianyang Zhang
Tucson, AZ

Series Preface to the First Edition

The objective of the Elsevier Geo-Engineering Book Series is to provide high-quality books on subjects within the broad geo-engineering subject area—eg, on engineering geology, soil mechanics, rock mechanics, civil/mining/environmental/petroleum engineering, etc. The first three books in the series have already been published:

- "Stability Analysis and Modelling of Underground Excavations in Fractured Rocks" by Weishen Zhu and Jian Zhao.
- "Coupled Thermo-Hydro-Mechanical-Chemical Processes in Geo-Systems" edited by Ove Stephansson, John A. Hudson and Lanru Jing.
- "Ground Improvement—Case Histories" edited by Buddhima Indraratna and Jian Chu.

These three books already represent an admirable, high-quality start to the series.

Now, I am delighted to introduce the fourth book in the series:

- "Engineering Properties of Rocks" by Lianyang Zhang.

This book provides expert, up-to-date information on rock mechanics and rock engineering for both the engineering and academic communities. It is a particularly logical and helpful book because it sequentially outlines the key aspects of the rock mechanics problem: the rock stress, and then the intact rock, the discontinuities and rock masses, followed by the deformability, strength and permeability of these components.

The author, in his own preface, states that "The typical values of and correlations between rock properties come in many forms and are scattered in different textbooks, reference manuals, reports, and articles published in technical journals and conference proceedings. It is often difficult, time-consuming, or even impossible for a practitioner to find appropriate information to determine the rock properties required for a particular project." Not only is this true, but the rock property values are the key to rock engineering design, whether it be by an empirical approach or by numerical modeling—as is evident from the content of the first two books in the Geo-Engineering Series.

The rock engineer must be able to predict the consequences of a particular excavation design. This can only be done via an adequate model, and the model can only be adequate if it is supported by the appropriate rock property values. Thus, the content of this book has a value which transcends the direct explanations of the rock mechanics and the rock properties: it represents one of the fundamental and essential keys to good rock engineering design. I am more than pleased to recommend that you read the book from cover to cover.

We hope that you enjoy the book and we welcome proposals for new books. Please send these to me at the email address as follows.

Professor John A. Hudson FREng
Geo-Engineering Series Editor
jah@rockeng.co.uk

Preface to the First Edition

For different reasons, it is often difficult for rock engineers to obtain directly the specific design parameter(s) of interest. As an alternative, they use the typical values or empirical correlations of similar rocks to estimate the specific parameter(s) of interest indirectly. For example, the unconfined compressive strength (UCS) of intact rock is widely used in designing surface and underground structures. The procedure for measuring UCS has been standardized by both the American Society for Testing and Materials (ASTM) and the International Society for Rock Mechanics (ISRM). Although the method is relatively simple, it is time-consuming and expensive; also, it requires well-prepared rock cores, which is often difficult or even impossible for weak rocks. Therefore, indirect tests are often conducted to estimate the UCS by using empirical correlations, such as point load, Schmidt hammer, sound velocity, and impact strength tests. Another example is the determination of the deformation modulus of rock masses. Rock masses usually contain discontinuities. To obtain realistic values of rock mass deformation modulus, in situ tests, such as plate bearing, flat jack, pressure chamber, borehole jacking and dilatometer tests, need be conducted. The in situ tests, however, are time-consuming, expensive and, in some cases, even impossible to carry out. Therefore, the deformation modulus of rock masses is often estimated indirectly from correlations with classification indices such as rock quality designation, rock mass rating, Q-system, and geological strength index.

The typical values of and correlations between rock properties come in many forms and are scattered in different textbooks, reference manuals, reports, and articles published in technical journals and conference proceedings. It is often difficult, time-consuming, or even impossible for a practitioner to find appropriate information to determine the rock properties required for a particular project. The main purpose of this book is to summarize and present, in one volume, the correlations between different rock properties, together with the typical values of rock properties.

This book contains eight chapters which are presented in a logical order. Chapter 1 provides a general introduction to rock engineering problems and methods for determining rock properties, and presents examples on using empirical correlations to estimate rock properties. Chapter 2 describes in situ rock stresses and presents different empirical correlations for estimating them. Chapters 3–5 describe the classification of intact rock and rock masses and the characterization of rock discontinuities. Chapters 6–8 present the typical values and correlations of deformability, strength and permeability of intact rock, rock discontinuities, and rock masses.

It must be noted that the typical values and correlations should never be used as a substitute for a proper testing program, but rather to complement and verify specific project-related information.

This book is intended for people involved in rock mechanics and rock engineering. It can be used by practicing engineers to determine the engineering properties of rocks required for particular projects. It will be useful for teaching to look into the typical values of different rock properties and the factors affecting them. It will also be useful for people engaged in numerical modeling to choose appropriate values for the properties included in the model.

Prof. Harun Sönmez of Hacettepe University, Turkey provided the deformation modulus data that was included in Fig. 6.14. The author sincerely thanks him.

Dr. Evert Hoek, Evert Hoek Consulting Engineer Inc., Canada sent the author his discussion papers and provided valuable information on the rock mass strength data included in Fig. 7.11. The author is grateful to him.

Portions of Chapters 4, 6, and 7 are based on the author's doctoral research conducted at the Massachusetts Institute of Technology. The author acknowledges the support and advice given by Prof. Herbert Einstein.

Finally, the author wants to thank Dr. Francisco Silva and Mr. Ralph Grismala of ICF Consulting for their support during the preparation of this book.

Lianyang Zhang
Lexington, MA

Chapter 1

Introduction

1.1 ROCK ENGINEERING PROBLEMS

Rock has been used as a construction material since the down of civilization. Different structures have been built on, in or of rock, including houses, bridges, dams, tunnels, and caverns (Bieniawski, 1984; Goodman, 1989; Brown, 1993; Fairhurst, 1993; Hudson, 1993; Hudson and Harrison, 1997; Hoek, 2000; Zhang, 2004). Table 1.1 lists different types of structures built on, in or of rock and the fields of their applications. Brown (1993) produced this table by adding surface civil engineering structures to that given by Bieniawski (1984) in his book on rock mechanics design in mining and tunneling.

Rock differs from most other engineering materials in that it contains discontinuities such as joints, bedding planes, folds, sheared zones, and faults which render its structure discontinuous. A clear distinction must be made between the intact rock or rock material and the rock mass. The intact rock may be considered a continuum or polycrystalline solid between discontinuities consisting of an aggregate of minerals or grains. The rock mass is the in situ medium comprised of intact rock blocks separated by discontinuities such as joints, bedding planes, folds, sheared zones, and faults. The properties of the intact rock are governed by the physical characteristics of the materials of which it is composed and the manner in which they are bonded to each other. Rock masses are discontinuous and often have heterogeneous and anisotropic properties.

Because rock masses are discontinuous and variable in space, it is important to choose the right domain that is representative of the rock mass affected by the structure being analyzed. Fig. 1.1 shows a simplified representation of the influence exerted on the selection of a rock mass behavior model by the relation of the discontinuity spacing and the size of the problem domain. When the problem domain is much smaller than the blocks of rock formed by the discontinuities, such as the excavation of rock by drilling, the behavior of the intact rock material will be of concern. When the block size is of the same order of the structure being analyzed or when one of the discontinuity sets is significantly weaker than the others, the stability of the structure should be analyzed by considering failure mechanisms involving sliding or rotation of blocks and wedges defined by intersecting structural features. When the structure being analyzed is

Engineering Properties of Rocks. http://dx.doi.org/10.1016/B978-0-12-802833-9.00001-8

TABLE 1.1 Different Types of Structures on, in or of Rock

Field of Application	Types of Structures on, in or of Rock
Mining	Surface mining: slope stability; rock mass diggability; drilling and blasting; fragmentation
	Underground mining: shaft, pillar, draft, and stope design; drilling and blasting; fragmentation; cavability of rock and ore; amelioration of rockbursts; mechanized excavation; in situ recovery
Energy development	Underground power stations (hydroelectric and nuclear); underground storage of oil and gas; energy storage (pumped storage or compressed air storage); dam foundations; pressure tunnels; underground repositories for nuclear waste disposal; geothermal energy exploitation; petroleum development including drilling, hydraulic fracturing, wellbore stability
Transportation	Highway and railway slopes, tunnels, and bridge foundations; canals and waterways; urban rapid transport tunnels and stations; pipelines
Utilities	Dam foundations; stability of reservoir slopes; water supply tunnels; sanitation tunnels; industrial and municipal waste treatment plants; underground storages and sporting and cultural facilities; foundations of surface power stations
Building construction	Foundations; stability of deep open excavations; underground or earth-sheltered homes and offices
Military	Large underground chambers for civil defense and military installations; uses of nuclear explosives; deep basing of strategic missiles

Based on Brown, E.T., 1993. The nature and fundamentals of rock engineering. In: Hudson, J.A. (Ed.), Comprehensive Rock Engineering—Principle, Practice and Projects, vol. 1. Pergamon, Oxford, UK, pp. 1–23.

much larger than the blocks of rock formed by the discontinuities, the rock mass can be simply treated as an equivalent continuum (Brady and Brown, 1985; Brown, 1993; Hoek, 2000).

Hudson (1993) developed a general three-tier approach to all rock engineering problems as represented by the three borders shown in Fig. 1.2. The main project subjects concerned, such as foundations, rock slopes, shafts, tunnels, and caverns, are illustrated within the three borders of the diagram. The words in the borders at the top of the diagram represent the entry into the design problem: the whole project complete with its specific objective in the outer border, the inter-relation between various components of the total problem in the middle border, and the individual aspects of each project in the central border.

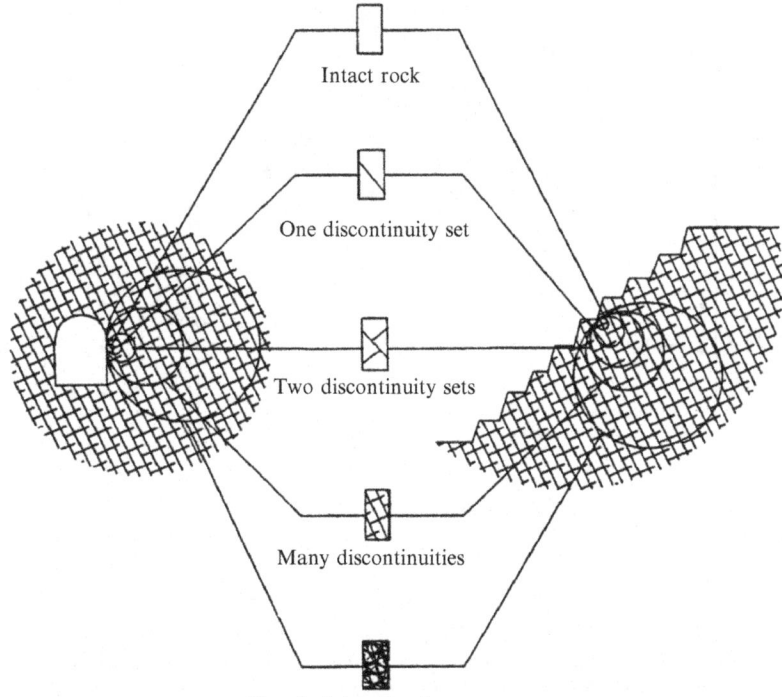

Intact rock

One discontinuity set

Two discontinuity sets

Many discontinuities

Heavily jointed rock mass

FIG. 1.1 Simplified representation of the influence of scale on the type of rock mass behavior. *(Based on Hoek, E., Kaiser, P.K., Bawden, W.F., 1995. Support of Underground Excavations in Hard Rock. Balkema, Rotterdam.)*

The words in the borders at the lower part of the diagram illustrate how the different components of the design might be executed. Different methods, such as the knowledge-based expert system, the rock mechanics interaction matrix analysis and the numerical analysis, can be used to consider the problem. It is noted that, for any project problem considered and any deign method used, the material properties (highlighted as intact rock, discontinuities, and permeability in the figure) and the boundary conditions (highlighted as in situ stress and the hydrogeological regime in the figure) should be known. Therefore, determination of the engineering properties of rocks (including the boundary conditions) is an essential part of all rock engineering problems.

Fig. 1.3 shows the components of a general rock mechanics program for predicting the responses of rock masses associated with rock engineering projects (Brady and Brown, 1985). Determination of the engineering (or geotechnical) properties of rock masses is an important part of the general rock mechanics program. Brown (1986) clearly stated the importance of site characterization for determining the engineering properties of rock masses: "Inadequacies in site characterization of geological data probably present the major impediment

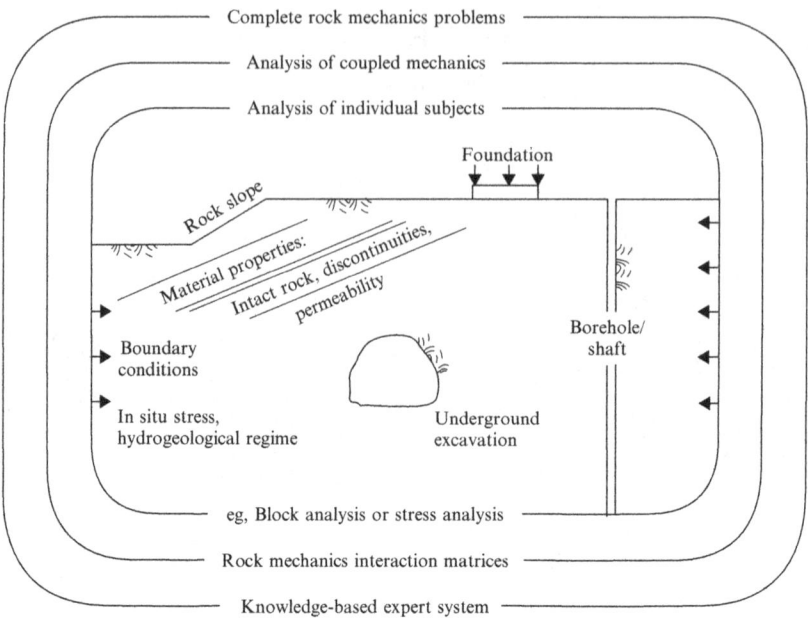

FIG. 1.2   Three-tier approach to all rock engineering problems. *(Based on Hudson, J.A., 1993. Rock properties, testing methods and site characterization. In: Hudson, J.A. (Ed.), Rock Engineering—Principle, Practice & Projects, vol. 3. Pergamon, Oxford, UK, pp. 1–39.)*

to the design, construction, and operation of excavations in rock. Improvements in site characterization methodology and techniques, and in the interpretation of the data are of primary research requirements, not only for large rock caverns, but for all forms of rock engineering."

1.2   DETERMINATION OF ENGINEERING PROPERTIES OF ROCKS

As stated earlier, determination of the engineering properties of rocks is an important part of all rock engineering problems. Because of the discontinuous and variable nature of rock masses, however, it is a complex and difficult task to determine the engineering properties of rocks. As Hudson (1992) noted, "The subject of rock characterization is far more complex and intractable than might appear at first sight. The subject does not merely concern the optimal length-to-diameter ratio for a compression test specimen and other similar tactical aspects of testing procedures; it concerns the whole strategic concept of how to characterize naturally occurring rock masses, which have been in existence for millions of years, have been operating as natural process-response systems for all that time and are about to be perturbed by engineers in order to achieve particular objectives."

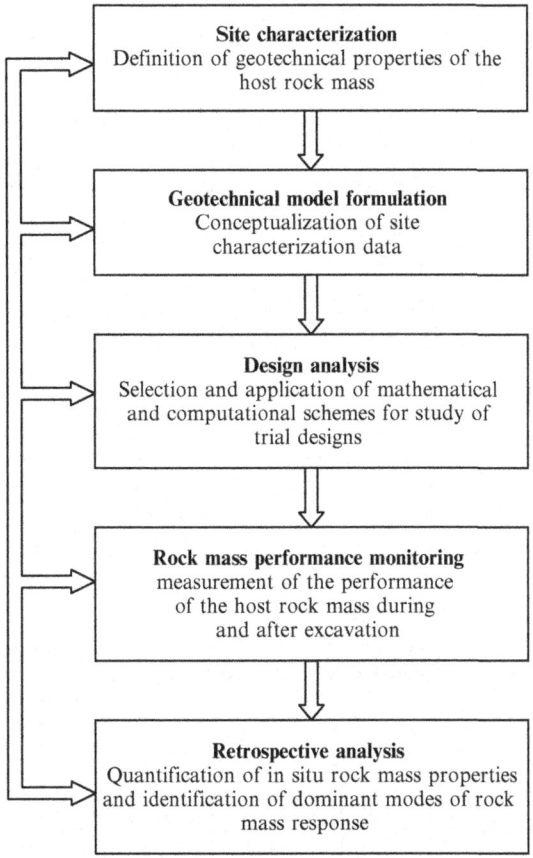

FIG. 1.3 Components of a general rock mechanics program. *(Based on Brady, B.H.G., Brown, E. T., 1985. Rock Mechanics for Underground Mining. George Allen & Unwin, London.)*

Despite the complexity and difficulty in determining the engineering properties of rocks, we still need to do the best we can to decide the specific rock properties required for a particular rock engineering problem and assign reliable values to them. Table 1.2 lists some of the typical rock engineering applications and the required accuracy for corresponding rock properties suggested by Pine and Harrison (2003).

There are different methods for determining the engineering properties of rocks, which can be divided in two general categories: direct and indirect methods (Table 1.3). The direct methods include laboratory and in situ tests. Many rock mechanics and rock engineering textbooks provide information on conducting laboratory and in situ tests to determine the engineering properties of rocks. In addition, the American Society for Testing and Materials (ASTM) and International Society for Rock Mechanics (ISRM) standards

TABLE 1.2 Suggested Levels of Accuracy Required for Rock Mass Properties in Different Applications

Application	Strength	Deformability	Permeability/ Hydraulic Conductivity
Mining	Pillars (10%) Walls (10%) Roofs (10%)	Shafts (25%)	Total inflow rates (50%)
Civil excavations	Tunnels (25%)	Tunnels (25%)	Total inflow rates (50%)
	Caverns (10%)	Caverns (25%)	
		Pressure tunnels and shafts (10%)	Total leakage rates (25%)
Nuclear/radioactive waste			Mass transport (factor of 10^{-2} to 10^{2})
Oil and gas	Borehole stability (10%)	Reservoir subsidence (25%)	Connectivity/ transmissivity (50%)
Civil foundations/ pile sockets		Settlement (25%)	

Based on Pine, R.J., Harrison, J.P., 2003. Rock mass properties for engineering design. Q. J. Eng. Geol. Hydrogeol. 36, 5–16.

TABLE 1.3 Methods for Determining Rock Mass Properties

Direct Methods	Indirect Methods
Laboratory tests	Empirical or theoretical correlations
In situ tests	Combination of intact rock and discontinuity properties using analytical or numerical methods
	Back-analysis using field observations of prototype observations

Based on Brown, E.T., 1993. The nature and fundamentals of rock engineering. In: Hudson, J.A. (Ed.), Comprehensive Rock Engineering—Principle, Practice and Projects, vol. 1. Pergamon, Oxford, UK, pp. 1–23; Pine, R.J., Harrison, J.P., 2003. Rock mass properties for engineering design. Q. J. Eng. Geol. Hydrogeol. 36, 5–16; Zhang, L., 2004. Drilled Shafts in Rock—Analysis and Design. Balkema, Leiden.

provide guidance related to the specific procedures for performing the different types of laboratory and in situ tests. Table 1.4 lists the categories of the suggested test methods by ISRM.

The direct methods have different limitations. To obtain realistic results of rock mass properties, rock specimens of different volumes having a number of different known discontinuity configurations should be tested at relevant stress levels under different stress paths. Such an experimental program is almost impossible to carry out in the laboratory. With in situ tests, such an experimental program would be very difficult, time-consuming, and expensive.

TABLE 1.4 Categories of Test Methods Suggested by ISRM

1. Laboratory Tests
   (a) Characterization
       (i) Porosity, density, water content
       (ii) Absorption
       (iii) Hardness—Schmidt rebound, shore scleroscope
       (iv) Resistance to abrasion
       (v) Point load strength index
       (vi) Uniaxial compressive strength and deformability
       (vii) Swelling and slake-durability
       (viii) Sound velocity
       (ix) Petrographic description
   (b) Engineering Design
       (i) Triaxial strength and deformability test
       (ii) Direct shear test
       (iii) Tensile strength test
       (iv) Permeability
       (v) Time dependent and plastic properties
2. In Situ Tests
   (a) Characterization
       (i) Discontinuity orientation, spacing, persistence, roughness, wall strength, aperture, filling, seepage, number of sets, and block size
       (ii) Drill core recovery/RQD
       (iii) Geophysical borehole logging
       (iv) In situ sound velocity
   (b) Engineering Design
       (i) Plate and borehole deformability tests
       (ii) In situ uniaxial and triaxial strength and deformability test
       (iii) Shear strength—direct shear, torsional shear
       (iv) Field permeability measurement
       (v) In situ stress determination

Based on Brown, E.T., 1981. Rock Characterization, Testing and Monitoring—ISRM Suggested Methods. Pergamon, Oxford, UK.

The indirect methods include empirical or theoretical correlations, combination of intact rock and discontinuity properties using analytical or numerical methods, and back-analysis using field observations of prototype structures. Because of the limitations of the direct methods, current practice relies heavily on the indirect methods. The indirect methods can be used not only for determining the rock properties but also for checking the test results. The data resulting from laboratory and in situ tests are often not completely consistent with other data obtained for a particular project. The indirect methods such as the empirical or theoretical correlations can be used to check the data from the tests and investigate the reasons for the inconsistency. The two examples presented in next section also show the applications of existing data and empirical correlations in the determination of the engineering properties of rocks.

1.3   EXAMPLES ON DETERMINING ENGINEERING PROPERTIES OF ROCKS

This section presents two examples to show how the existing data and empirical correlations are used to determine the engineering properties of rocks.

(a) Estimation of rock discontinuity shear strength (Wines and Lilly, 2003)

This example shows the estimation of rock discontinuity shear strength in part of the Fimiston open pit operation in Western Australia (Wines and Lilly, 2003). There are four major discontinuity sets at the pit site:

- The discontinuities in Set 1 are generally rough, planar and clean, with occasional quartz infill and have an average dip/dip direction of 65°/271°.
- The discontinuities in Set 2 are generally rough and planar to undulating, with regular quartz infill and have an average dip/dip direction of 2°/306°.
- The discontinuities in Set 3 are generally rough and planar, with regular quartz infill and have an average dip/dip direction of 82°/323°.
- The discontinuities in Set 4 include tightly healed, rough and undulating quartz veins and have an average dip/dip direction of 86°/001°.

The shear strength data of the discontinuities were required in order to design a major part of the eastern wall in the Fimiston open pit.

The empirical shear strength criterion proposed by Barton (1976) (Eq. (7.37) in Chapter 7) was used to describe the shear strength of discontinuities at the site. To use this criterion, the following three input parameters need to be determined:

- JRC—Discontinuity roughness coefficient
- JCS—Discontinuity wall compressive strength
- ϕ_r—Residual friction angle of the discontinuity.

The JRC values for the four discontinuity sets were recorded during scanline mapping and diamond core logging using the profiles presented in Fig. 6.10. Since the discontinuities in the study area generally exhibited no wall softening due to weathering, JCS was assumed to be equal to the unconfined compressive strength of the intact rock. The estimated values of JRC and JCS are summarized in Table 1.5.

Since the discontinuities in the study area generally exhibited no wall softening due to weathering, the residual friction angle ϕ_r could be simply taken equal to the basic friction angle ϕ_b (Eq. 6.38). The basic friction angle was determined using the following three different methods:

(1) Conducting direct shear test along smooth and clean saw cut samples.
(2) Conducting tilt test on split core samples and using Eq. (6.39) to calculate the basic friction angle.
(3) Using typical values available in the literature (Table 6.11).

The estimated values of the basic friction angle using the above three methods are shown in Table 1.6. It is noticed that the estimated values using the three methods are in good agreement except that the values of Paringa basalt from the tilt test are a bit higher.

(b) Estimation of strength and deformability of rock masses (Ozsan and Akin, 2002)

This example describes the estimation of strength and deformability of rock masses at the proposed Urus Dam site in Turkey (Ozsan and Akin,

TABLE 1.5 Estimated Values of JRC and JCS for the Four Main Discontinuity Sets

Parameter	Statistic	Paringa Basalt			Golden Mine Dolerite			
		Set 1	Set 2	Set 4	Set 1	Set 2	Set 3	Set 4
JRC	Mean	6.4	7.1	4.7	7.8	7.3	5.9	7.0
	SD	3.0	2.9	2.9	2.7	2.3	2.2	3.1
	Min	2.0	2.0	2.0	2.0	2.0	2.0	2.0
	Max	14.0	12.0	10.0	16.0	14.0	10.0	14.0
JCS (MPa)	Mean	86.9	86.9	86.9	95.9	95.9	95.9	95.9
	SD	28.9	28.9	28.9	34.4	34.4	34.4	34.4
	Min	43.7	43.7	43.7	34.3	34.3	34.3	34.3
	Max	156.5	156.5	156.5	156.3	156.3	156.3	156.3

From Wines, D.T., Lilly, P.A., 2003. Estimates of rock joint shear strength in part of the Fimiston open pit operation in Western Australia. Int. J. Rock Mech. Min. Sci. 40, 929–937.

TABLE 1.6 Estimated Values of the Basic Friction Angle ϕ_b

	Direct Shear Testing				Tilt Testing				Values From
Rock	Mean	SD	Min	Max	Mean	SD	Min	Max	Table 6.11
Paringa basalt	36.9	2.0	32.9	39.4	39.6	1.1	37.4	42.1	35–38
Golden Mine dolerite	34.2	1.5	32.0	36.0	36.3	1.4	32.6	40	36

From Wines, D.T., Lilly, P.A., 2003. Estimates of rock joint shear strength in part of the Fimiston open pit operation in Western Australia. Int. J. Rock Mech. Min. Sci. 40, 929–937.)

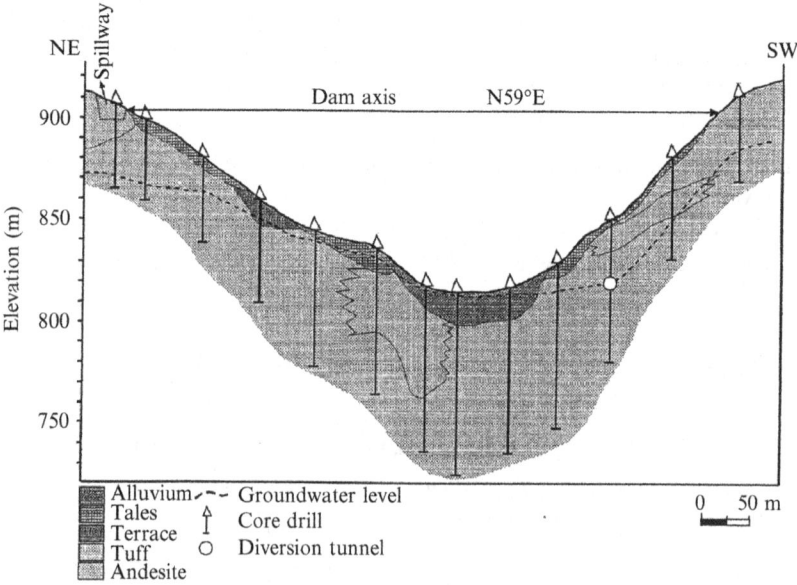

FIG. 1.4 Geological cross-section along dam axis at the Urus Dam site. (*From Ozsan, A., Akin, M., 2002. Engineering geological assessment of the proposed Urus Dam, Turkey. Eng. Geol. 66, 271–281.*)

2002). The site is located on volcanic rocks of the Neogene Age and on Quaternary deposits. The volcanic rocks consist of andesite, basalt, and tuff (see Figs. 1.4 and 1.5).

Characterization of discontinuities was carried out by exposure logging following ISRM (1978). A total of 399 discontinuities, 372 on the left bank and 27 on the right bank, were measured. Three major discontinuity sets on the left bank (29°/352°, 87°/333°, 30°/079°) and two (85°/146°, 60°/297°)

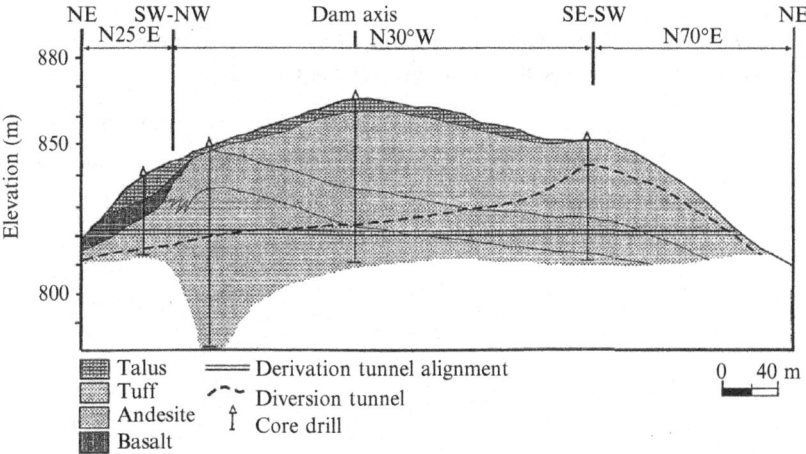

FIG. 1.5   Geological cross-section of the diversion tunnel alignment at the Urus Dam site. *(From Ozsan, A., Akin, M., 2002. Engineering geological assessment of the proposed Urus Dam, Turkey. Eng. Geol. 66, 271–281.)*

on the right bank were determined. Table 1.7 shows the quantitative descriptions and statistical distributions of the discontinuities of basalt and andesite at the site. Since tuff is moderately-highly weathered, no discontinuity was observed during exposure logging.

Borings were made at the site to verify foundation conditions and to obtain rock samples for laboratory tests. Rock quality designation (RQD) and total core recovery (TCR) values were determined for different structural areas of the dam site. Table 1.8 lists the average values.

Laboratory tests were carried out to determine physical and mechanical properties including unit weight, porosity, unconfined compressive and tensile strength, cohesion, internal friction angle, and deformation parameters. The results are summarized in Table 1.9.

To estimate the strength and deformation properties of the rock masses, the geological strength index (GSI) was estimated using the quantified GSI chart proposed by Sonmez and Ulusay (1999, 2002) (Fig. 5.3 in Chapter 5). Table 1.10 lists the estimated GSI values. The strength of the rock masses at this site was expressed using the Hoek-Brown criterion (Eq. 7.50 in Chapter 7). The rock mass constants m_b, s, and a for the Hoek-Brown criterion were estimated by using Eq. (7.55). The intact rock constants were selected from Hoek and Brown (1997). The unconfined compressive strength σ_{cm} of rock masses was determined by inserting the minor principal stress σ'_3 of zero into Eq. (7.50). The results are shown in Table 1.10.

The deformation modulus of the rock masses at this site was estimated by using Eq. (6.63) in Chapter 6. The results are also shown in Table 1.10.

TABLE 1.7 Quantitative Descriptions and Statistical Distributions of Discontinuities of Basalt and Andesite at the Urus Dam Site

	Range	Description	Distribution (%)	
			Basalt	*Andesite*
Spacing (mm)	<20	Extremely close	4	–
	20–60	Very close	47	26
	60–200	Close	32	51
	200–600	Moderate	17	23
Persistence (m)	1–3	Low	17	13
	3–10	Medium	59	47
	10–20	High	24	40
Aperture[a] (mm)	0.25–0.5	Partly open	26	31
	0.5–2.5	Open	55	54
	2.5–10	Moderately wide	19	15
Roughness	1[b]	0–2[c]	25	23
	2	2–4	33	36
	3	4–6	19	25
	4	6–8	13	8
	5	8–10	10	8

[a]*Aperture of discontinuities contains mostly limonite, hematite, and clay infilling materials.*
[b]*Roughness profile numbers.*
[c]*JRC values.*
From Ozsan, A., Akin, M., 2002. Engineering geological assessment of the proposed Urus Dam, Turkey. Eng. Geol. 66, 271–281.

1.4   ROCK EXPERT SYSTEM FOR EVALUATION OF ROCK ENGINEERING PROPERTIES

The two examples presented in the previous section clearly show the importance of utilizing existing information for estimating the engineering properties of rocks. The past decades have seen sustained research and development efforts to improve the methods for determining the engineering properties of rocks. Although much has been learned, all the major findings are scattered in different sources such as textbooks, reference manuals, reports, and articles published in technical journals and conference proceedings. It is often difficult, time-consuming, or even impossible for a practitioner to find appropriate information for determining the rock properties required for a particular project. It is

TABLE 1.8 Average RQD and TCR Values Obtained From Core Drilling at the Urus Dam Site

Location	Andesite		Basalt		Tuff	
	RQD (%)	TCR (%)	RQD (%)	TCR (%)	RQD (%)	TCR (%)
Left bank	59	93	–	–	34	79
Dam axis	46	85	–	–	10	94
Right bank	35	100	–	–	34	87
Diversion tunnel	52	100	15	58	8	75
Spillway	38	83	–	–	0	75

From Ozsan, A., Akin, M., 2002. Engineering geological assessment of the proposed Urus Dam, Turkey. Eng. Geol. 66, 271–281.

therefore important to develop a central database of the existing information and an easy-to-use and effective tool for engineers to evaluate rock properties based on the existing well determined values and empirical correlations.

The ISRM Commission on Design Methodology has been developing the procedure for implementing the "corporate memory" system outlined in Fig. 1.6. The system includes tables of intact rock and rock mass properties, libraries of standard and case example modeling solutions and libraries of design and construction case examples, and can be used for determining rock engineering properties (Hudson, 2012).

Separately to the developments of the ISRM Commission on Design Methodology, Zhang et al. (2012) have been developing a rock expert system (RES) for the evaluation of rock properties. The RES has three main components: RES database, web application platform, and data application tools (Fig. 1.7). The RES intends to be an easy-to-use and effective tool for engineers to evaluate rock properties. The RES will be multi-functional and allow a user to: (1) search for typical values of interested rock properties; (2) calculate rock properties based on empirical correlations; (3) check the measured values of rock properties against the typical values; and (4) integrate regional and/or local rock property databases into the system for re-use and reducing the uncertainty of predictions from the empirical correlations.

1.5   CONTENT OF THE BOOK

This book focuses on the determination of the engineering properties of rocks. The emphasis is on the indirect methods for determining the rock properties, including empirical or theoretical correlations and combination of intact rock

TABLE 1.9 Laboratory Test Results of Rocks at the Urus Dam Site

Parameter	Andesite		Basalt		Tuff	
	Range	Average	Range	Average	Range	Average
Unit weight (kN/m³)	21.6–25.5	23.7	22.1–2.57	24.0	18.8–21.5	19.9
Porosity (%)	3.26–4.13	3.73	3.03–3.54	3.29	12.5–18.6	16.1
Unconfined compressive strength (MPa)	40–148	93	64–249	142	17–33	24
Tensile strength (MPa)	7.55–9.60	8.58	6.20–8.30	7.25	0.75–2.94	1.97
Cohesion (MPa)		9.72		10.81		9.29
Internal friction angle (°)		53.21		43.18		36.77
Elastic modulus (GPa)		41.9		40.0		11.6
Poisson's ratio		0.22		0.30		0.21

From Ozsan, A., Akin, M., 2002. Engineering geological assessment of the proposed Urus Dam, Turkey. Eng. Geol. 66, 271–281.

TABLE 1.10 GSI Values, Rock Mass Constants and Deformations Modulus of Rock Masses at the Urus Dam Site

Parameter	Andesite	Basalt	Tuff
Unconfined compressive strength of intact rock σ_c (MPa)	93	142	24
Geological strength index (GSI)	41	42.5	31
Hoek-Brown intact rock constant m_i	19[a]	17[a]	15[a]
Hoek-Brown rock mass constant m_b	2.31	2.18	1.28
Hoek-Brown rock mass constant s	0.00142	0.00168	0.00047
Hoek-Brown rock mass constant a	0.5	0.5	0.5
Unconfined compressive strength of rock masses σ_{cm} (MPa)	3.51	5.82	0.52
Deformation modulus of rock masses E_m (GPa)	5.74	7.74	1.64

[a]Marinos, P., Hoek, E., 2001. Estimating the geotechnical properties of heterogeneous rock masses such as flysch. Bull. Eng. Geol. Environ. 60, 85–92 updated the table for m_i (see Table 7.10 in Chapter 7). If the updated table were used, the corresponding values of m_i would be, respectively, 25, 25, and 13.
From Ozsan, A., Akin, M., 2002. Engineering geological assessment of the proposed Urus Dam, Turkey. Eng. Geol. 66, 271–281.

and discontinuity properties using analytical or numerical methods. The reader can refer to many rock mechanics and rock engineering textbooks about the direct methods: laboratory and in situ tests. The ASTM and ISRM standards also provide guidance related to the specific procedures for performing the laboratory and in situ tests.

The main purpose of this book is to provide the reader a single source of information required for determining rock properties by summarizing and presenting the latest information in one volume. The eight chapters in this book are presented in a logical order starting with this chapter that provides a general introduction to rock engineering problems and methods for determining rock properties, presents examples on determining rock properties, briefly describes the RES, and outlines the various topics covered by the main chapters of this book.

Chapter 2 describes in situ rock stresses, presents the world stress map, empirical correlations, and analytical solutions for estimating the in situ rock stresses, and briefly discusses the main factors that affect the in situ rock stress state.

Chapter 3 discusses the classification and index properties of intact rocks. The typical values of and empirical or theoretical correlations between different index properties are also presented.

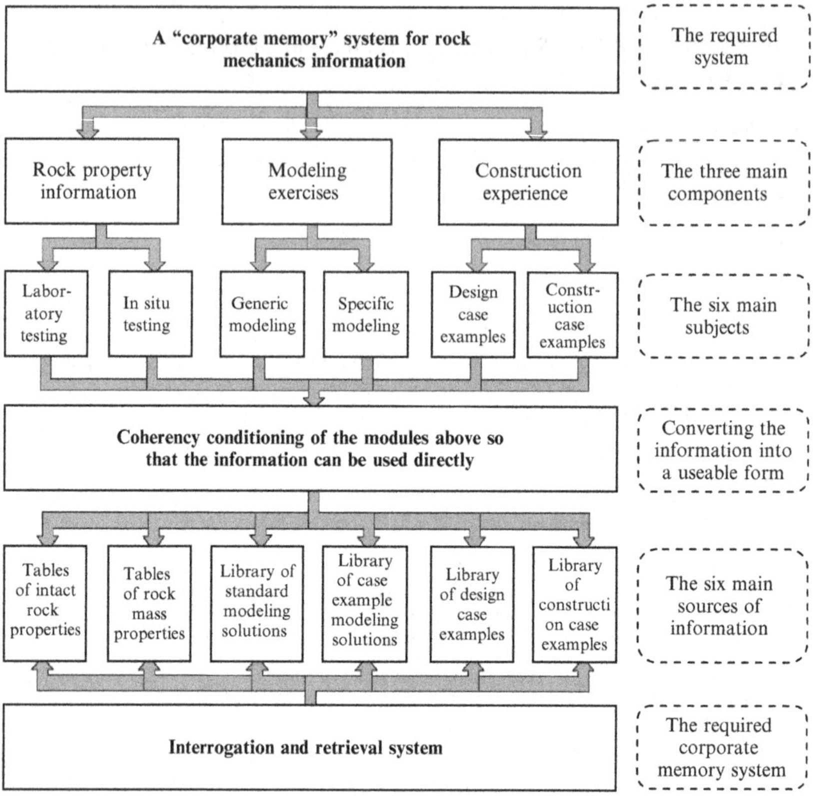

```
┌─────────────────────────────────────────────────┐   ┌ ─ ─ ─ ─ ─ ─ ┐
│     A "corporate memory" system for rock          │     The required
│          mechanics information                    │   │   system     │
└─────────────────────────────────────────────────┘   └ ─ ─ ─ ─ ─ ─ ┘
```

┌──────────────┐ ┌──────────────┐ ┌──────────────┐   ┌ ─ ─ ─ ─ ─ ─ ┐
│ Rock property │ │ Modeling   │ │ Construction │   The three main
│ information   │ │ exercises   │ │ experience   │   │ components   │
└──────────────┘ └──────────────┘ └──────────────┘   └ ─ ─ ─ ─ ─ ─ ┘

┌──────┐┌──────┐ ┌──────┐┌──────┐ ┌──────┐┌──────┐   ┌ ─ ─ ─ ─ ─ ─ ┐
│Labor-││In situ│ │Generic││Specific│ │Design ││Constr-│   The six main
│atory ││testing│ │modeling││modeling│ │case   ││uction │ │ subjects   │
│testing││   │ │   ││   │ │examples││case   │
└──────┘└──────┘ └──────┘└──────┘ └──────┘│examples│   └ ─ ─ ─ ─ ─ ─ ┘

┌───┐   ┌ ─ ─ ─ ─ ─ ─ ┐
│ Coherency conditioning of the modules above so   │   Converting the
│ that the information can be used directly   │   │ information into │
└───┘   a useable form
                                                        └ ─ ─ ─ ─ ─ ─ ┘

┌──────┐┌──────┐┌────────┐┌────────┐┌────────┐┌────────┐   ┌ ─ ─ ─ ─ ─ ─ ┐
│Tables ││Tables ││Library of││Library ││Library ││Library │   The six main
│of intact││of rock││standard ││of case ││of design││of   │ │ sources of   │
│rock   ││mass   ││modeling ││example ││case   ││constructi│ information
│properties││properties││solutions││modeling││examples││on case │
│   ││   ││   ││solutions││   ││examples│   └ ─ ─ ─ ─ ─ ─ ┘
└──────┘└──────┘└────────┘└────────┘└────────┘└────────┘

                                                        ┌ ─ ─ ─ ─ ─ ─ ┐
┌───┐   The required
│   Interrogation and retrieval system   │   │   corporate   │
└───┘   memory system
                                                        └ ─ ─ ─ ─ ─ ─ ┘

FIG. 1.6 The structure for a corporate memory system for rock mechanics and rock engineering information *(Based on Hudson, J.A., 2012. Design methodology for the safety of underground rock engineering. J. Rock Mech. Geotech. Eng. 4 (3), 205–214.)*

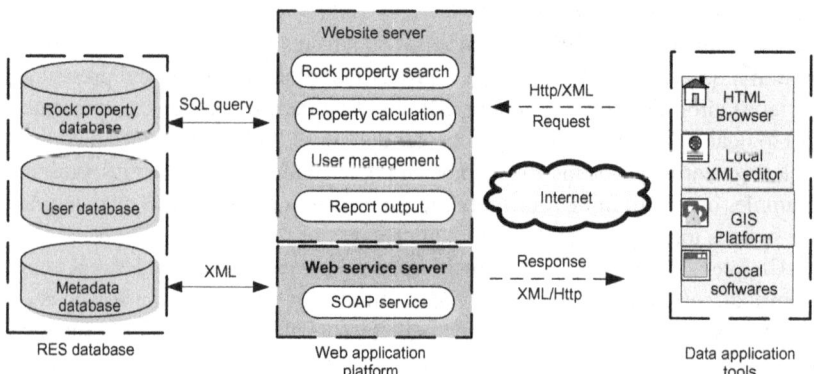

FIG. 1.7 Architecture of the rock expert system (RES). *(From Zhang, L., Ding, X., Budhu, M., 2012. A rock expert system for the evaluation of rock properties. Int. J. Rock Mech. Min. Sci. 50, 124–132.)*

The characterization of rock discontinuities is presented in Chapter 4. This chapter focuses on the geometric properties (orientation, intensity, persistence, trace length, shape, size, roughness, aperture, and fillings) of discontinuities. The mechanical and hydraulic properties of discontinuities are discussed in later chapters.

Chapter 5 describes the classification of rock masses using different rock mass classification systems and presents the correlations between the different classification indices. The classification of weathering of rocks is also discussed.

The deformability of intact rocks, rock discontinuities, and rock masses is discussed in Chapter 6. The typical values of the deformation parameters of different rocks are summarized in tables and figures. The different methods for determining the deformability of intact rocks, rock discontinuities, and rock masses are presented and the factors the affect the deformability of rocks are discussed.

Chapter 7 deals with the strength of intact rocks, rock discontinuities, and rock masses. The typical values of the strength parameters of different rocks are summarized in tables and figures. The different methods for determining the strength of intact rocks, rock discontinuities, and rock masses are presented and the factors that affect the strength of rocks are discussed.

Finally, Chapter 8 discusses the permeability of rocks. The typical values of the permeability of intact rocks and rock masses are presented. The various methods for estimating the permeability of intact rocks, discontinuities, and rock masses are described. The effects of different factors such as stress, temperature, and scale on rock permeability and the anisotropy of rock permeability are also discussed.

This book is intended for people involved in rock mechanics and rock engineering. It can be used by practicing engineers to determine the engineering properties of rocks required for particular projects. It will be useful for teaching to look into the typical values of different rock properties and the factors affecting them. It will also be useful for people engaged in numerical modeling to choose appropriate values for the properties included in the model.

This book focuses on the indirect methods with emphasis on empirical or theoretical correlations and combination of intact rock and discontinuity properties using analytical or numerical methods. It does not mean that the direct methods are not important. In practice, a project should always include some types of laboratory and/or in situ tests. The indirect methods can only be used to supplement the direct methods. When describing the use of correlations, Sabatini et al. (2002) states:

> *Correlations in general should never be used as a substitute for an adequate subsurface investigation program, but rather to complement and verify specific project-related information.*

The above statement about correlations also applies to the indirect methods covered in this book.

REFERENCES

Barton, N., 1976. The shear strength of rock and rock joints. Int. J. Rock Mech. Min. Sci. Geomech. Abstr. 13, 255–279.

Bieniawski, Z.T., 1984. Rock Mechanics Design in Mining and Tunneling. Balkema, Rotterdam.

Brady, B.H.G., Brown, E.T., 1985. Rock Mechanics for Underground Mining. George Allen & Unwin, London.

Brown, E.T., 1986. Research and development for design and construction of large rock caverns. In: Proceedings International Symposium on Large Rock Caverns. Helsinki, Finland, 1937–1948.

Brown, E.T., 1993. The nature and fundamentals of rock engineering. In: Hudson, J.A. (Ed.), In: Comprehensive Rock Engineering—Principle, Practice and Projects, vol. 1. Pergamon, Oxford, UK, pp. 1–23.

Fairhurst, C., 1993. Analysis and design in rock mechanics—the general context. In: Hudson, J.A. (Ed.), In: Comprehensive Rock Engineering—Principle, Practice and Projects, vol. 2. Pergamon, Oxford, UK, pp. 1–29.

Goodman, R.E., 1989. Introduction to Rock Mechanics. Wiley, New York.

Hoek, E., 2000. Rock Engineering. http://www.rocscience.com.

Hoek, E., Brown, E.T., 1997. Practical estimates of rock mass strength. Int. J. Rock Mech. Min. Sci. 34, 1165–1186.

Hudson, J.A., 1992. Rock characterization. In: Proceedings EUROCK'92 Symposium, Chester, UK, Telford, London.

Hudson, J.A., 1993. Rock properties, testing methods and site characterization. In: Hudson, J.A. (Ed.), In: Rock Engineering—Principle, Practice & Projects, vol. 3. Pergamon, Oxford, UK, pp. 1–39.

Hudson, J.A., 2012. Design methodology for the safety of underground rock engineering. J. Rock Mech. Geotech. Eng. 4 (3), 205–214.

Hudson, J.A., Harrison, J.P., 1997. Rock Engineering Mechanics—An Introduction to the Principles. Elsevier, Oxford.

ISRM, 1978. Suggested methods for the quantitative description of discontinuities in rock masses. International society for rock mechanics, commission on standardization of laboratory and field tests. Int. J. Rock Mech. Min. Sci. Geomech. Abstr. 15, 319–368.

Ozsan, A., Akin, M., 2002. Engineering geological assessment of the proposed Urus Dam, Turkey. Eng. Geol. 66, 271–281.

Pine, R.J., Harrison, J.P., 2003. Rock mass properties for engineering design. Q. J. Eng. Geol. Hydrogeol. 36, 5–16.

Sabatini, P.J., Bachus, R.C., Mayne, P.W., Schneider, J.A., Zettler, T.E., 2002. Evaluation of Soil and Rock Properties. Geotechnical Engineering Circular No. 5, Report No.: FHWA-IF-034, FHWA, US.

Sonmez, H., Ulusay, R., 1999. Modifications to the geological strength index (GSI) and their applicability to stability of slopes. Int. J. Rock Mech. Min. Sci. 36, 743–760.

Sonmez, H., Ulusay, R., 2002. A discussion on the Hoek-Brown failure criterion and suggested modification to the criterion verified by slope stability case studies. Yerbilimleri (Earthsciences) 26, 77–99.

Wines, D.T., Lilly, P.A., 2003. Estimates of rock joint shear strength in part of the Fimiston open pit operation in Western Australia. Int. J. Rock Mech. Min. Sci. 40, 929–937.

Zhang, L., 2004. Drilled Shafts in Rock—Analysis and Design. Balkema, Leiden.

Zhang, L., Ding, X., Budhu, M., 2012. A rock expert system for the evaluation of rock properties. Int. J. Rock Mech. Min. Sci. 50, 124–132.

Chapter 2

In Situ Stresses

2.1 INTRODUCTION

The distribution of in situ rock stresses is a major concern of rock mechanics and rock engineering, both with respect to understanding basic geological process such as plate tectonics and earthquakes, and the design of structures in and on rock masses (Amadei and Stephanson, 1997; Hudson and Harrison, 1997; Fairhurst, 2003; Zang and Stephansson, 2010; Hudson, 2012; Stephansson and Zang, 2012). A list of activities requiring knowledge of in situ rock stresses is given in Table 2.1, which was produced by Amadei and Stephanson (1997). As stated by Hudson and Harrison (1997),

"The basic motivations for in situ stress determination are two-fold:

1. To have a basic knowledge of the stress state for engineering, eg, in what direction and with what magnitude is the major principal stress acting? What stress effects are we defending ourselves and our structures against? In what direction is the rock most likely to break? All other things being equal, in what direction will the groundwater flow? Even for such basic and direct engineering questions, a knowledge of the stress state is essential.
2. To have a specific and "formal" knowledge of the boundary conditions for stress analyses conducted in the design phase of rock engineering projects."

Stress is a tensor quantity containing nine components: three normal stress components and six shear stress components (see Fig. 2.1A). With the complementary pairs of shear stresses being equal, the stress tensor has six independent components. Hence, to specify the in situ rock stress at a point, six independent pieces of information must be known. When the cube shown in Fig. 2.1A is rotated, the stress components on the faces change in value. At one, and only one, cube orientation, all the shear stress components on the faces will be zero (see Fig. 2.1B). When this occurs, the cube faces represents the principal stress planes and the corresponding normal stresses are the principal stresses (Hudson et al., 2003). So the in situ rock stress at a point can also be specified if we know the orientations and magnitudes of the principal stresses.

There are different methods for measuring in situ rock stresses. These methods can be classified into two main categories (Ljunggren et al., 2003). The first consists of methods that disturb the in situ rock conditions, ie, by inducing strains, deformations or crack opening pressures, including hydraulic

Engineering Properties of Rocks. http://dx.doi.org/10.1016/B978-0-12-802833-9.00002-X

TABLE 2.1 Activities Requiring Knowledge of In Situ Rock Stresses

Civil and mining engineering

Stability of underground excavations (tunnels, mines, caverns, shafts, stopes, haulages)

Drilling and blasting

Pillar design

Design of support systems

Prediction of rock bursts

Fluid flow and contaminant transport

Dams

Slope stability

Energy development

Borehole stability and deviation

Borehole deformation and failure

Fracturing and fracture propagation

Fluid flow and geothermal problems

Reservoir production management

Energy extraction and storage

Geology/geophysics

Orogeny

Earthquake prediction

Plate tectonics

Neotectonics

Structural geology

Volcanology

Glaciation

Based on Amadei, B., Stephanson, O., 1997. Rock Stress and Its Measurement. Chapman & Hall, London, UK.

fracturing and/or hydraulic testing of pre-existing fractures (HTPE) methods, borehole relief methods and surface relief methods. The second consists of methods based on observation of rock behavior without any major influence from the measuring method, including core discing, borehole breakouts, relief of large rock volumes (back analysis), acoustic methods (Kaiser effect), strain recovery methods, geological observational methods and earthquake focal mechanisms. Description of the various methods for measuring in situ rock stresses is out of the scope of this book. The reader can refer to Amadei and Stephanson (1997), Hudson and Harrison (1997), Sjöberg et al. (2003),

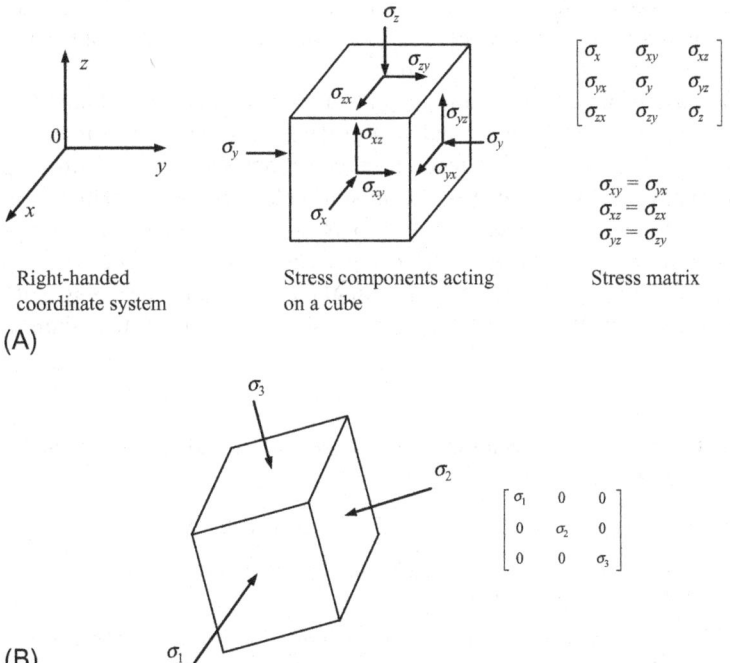

FIG. 2.1 (A) The components of the stress tensor acting on an infinitesimal cube within the rock mass; and (B) Principal stress cube and principal stress matrix.

Haimson and Cornet (2003), and Zang and Stephansson (2010) for detailed description of the various methods. Instead, this book will concentrate on the various methods for estimating the in situ rock stresses, including empirical correlations or observations obtained from stress measurements made in the past and different analytical models. "Estimating in situ stresses can be useful in the early stage of engineering design, for the planning process and when selecting stress measuring methods and the location of those measurements" (Amadei and Stephanson, 1997).

2.2 STRATEGY FOR DETERMINING IN SITU ROCK STRESSES

An exact determination of in situ rock stresses is very difficult and for all practical purposes impossible because "the current stress state is the end product of a series of past geological events and is the superposition of stress components of several diverse types. Further, since rock masses are rarely homogeneous and continuous, stresses can be expected to vary from place to place in a rock mass. In situ stresses not only vary in space but also with time due to tectonic events, erosion, glaciation, etc. The problem is further complicated in that the present rock fabric may or may not be correlated at all with the current stress field

(Terzaghi, 1962)" (Amadei and Stephanson, 1997). Hudson (2012) provided a good summary of the effects of different scales and natural and anthropogenic perturbations on the in situ rock stress state.

Because of the special nature of in situ rock stresses, they should be estimated using several methods and in a progressive process. The International Society of Rock Mechanics recommended an approach strategy to progressively build up a knowledge of the in situ rock stress tensor (Hudson et al., 2003). Table 2.2 lists the steps in the progression. Combining various methods based on their respective attributes can help in obtaining a more reliable estimate of the in situ rock stresses. It is important to integrate the stress estimates from various methods. The integration should check if the simplifying

TABLE 2.2 Steps in Developing a Knowledge of the Rock Stress Tensor Components

Use pre-existing information on the rock stress state at the site

Consider whether the vertical direction is a principal stress direction (from topography, geological evidence, and other information)

Estimate the vertical stress component magnitude (from the rock density and overburden depth)

Consider indications of the principal stress directions and the ratio of stress differences (from focal plane solutions inversion or seismic shear wave anisotropy)

Establish the minimum principal stress orientation (whether actual or minimum horizontal stress) from hydraulic or drilling induced fractures and borehole breakout orientations

Find components of the stress tensor using indirect methods on borehole core (such as the Kaiser effect and differential strain analysis)

Establish the complete stress state at one or more locations by overcoring tests	Establish the minimum principal stress (from hydraulic fracturing tests in boreholes)
	Establish the maximum principal stress magnitude (from hydraulic fracturing tests in boreholes and from borehole failure analysis)
	Establish the complete state of stress at one or more locations (by hydraulic testing of pre-existing fractures, HTPF)

Establish the variation of the stress state across the site due to different geological strata and fractures (as estimated through numerical analyses and further measurements)

Based on Hudson, J.A., Cornet, F.H., Christiansson, R., 2003. ISRM suggested methods for rock stress estimation—part 1: strategy for rock stress estimation. Int. J. Rock Mech. Min. Sci. 40, 991–998.

assumptions associated with each method are met and take into account uncertainties in each estimate. The number of estimates for each corresponding method should also be considered with care to avoid giving any inappropriate weight to the more numerous data set.

The ISRM further suggested a method for establishing the final rock stress model (FRSM) at a site or an area as shown in Fig. 2.2 (Stephansson and Zang, 2012). The FRSM is derived based on a combination of available stress data from the best estimate stress model (BESM), new stress data from stress measurement methods on site (SMM) and integrated stress determination (ISD) using previous data plus numerical modeling. The BESM is established through data collection and analysis, including (a) data extraction (assessment of stress types, estimation of rock stresses using stress data and World Stress Map (WSM), etc.), (b) analysis of field data on morphology, topography and geology, and (c) compilation and analysis of borehole and drill core data. After the establishment of the BESM, SMM can be selected for in situ stress measurement(s) and/or core-based stress measurement(s). With the established BESM and the results of various stress measurements, an ISD and numerical analyses are done to develop the FRSM.

As can be seen above, an important task for establishing the FRSM is to estimate the in situ rock stresses in order to develop the BESM. In the following sections, the various methods for making preliminary estimation of in situ rock stresses will be presented, including the empirical correlations or observations obtained from stress measurements made in the past and the different analytical models. The reader can refer to Amadei and Stephanson (1997), Hudson and Harrison (1997), Sjöberg et al. (2003), Haimson and Cornet (2003), and Zang and Stephansson (2010) for detailed description of the various measuring methods. It needs to be noted that the preliminary estimation of in situ rock stresses should not be considered a substitute for their measurements.

Best estimate stress model (BESM)	Stress measurement methods (SMM)	Integrated stress determination (ISD)	Final rock stress model (FRSM)
• Data extraction • Analysis of field data on morphology, topography, and geology • Compilation and analysis of borehole and drill core data	• Borehole methods • Core-based methods • Earthquakes	• ISD model • Numerical modeling	
Existing data	**New data**	**Integrated data**	**Final data**

FIG. 2.2   Establishment of the final rock stress model (FRSM) by combination of the best estimate stress model (BESM), new stress data from stress measurement methods (SMM) and integrated stress determination (ISD). *(Based on Stephansson, O., Zang, A., 2012. ISRM suggested methods for rock stress estimation—part 5: establishing a model for the in situ stress at a given site. Rock Mech. Rock Eng. 45, 955–969.)*

2.3   WORLD STRESS MAP (WSM)

The WSM is a global database of contemporary tectonic stress of the Earth's crust (Zobak et al., 1989; Reinecker et al., 2004; Heidbach et al., 2010). The 2008 release of WSM contains 21,750 data points and they are grouped into four major categories with the following percentage (Heidbach et al., 2010):

- earthquake focal mechanisms (72%);
- wellbore breakouts and drilling induced fractures (20%);
- in situ stress measurements (overcoring, hydraulic fracturing, borehole slotter) (4%); and
- young geologic data (from fault slip analysis and volcanic vent alignments) (4%).

The uniformity and quality of the WSM is guaranteed through quality ranking of the data according to international standards, standardized regime assignment, and guidelines for borehole breakout analysis and other methods. Each stress data record is assigned a quality between A and E, with A being the highest quality and E the lowest. A-quality indicates that the orientation of the maximum horizontal compressional stress is accurate to within $\pm 15°$, B-quality to within $\pm 20°$, C-quality to within $\pm 25°$, and D-quality to within $\pm 40°$. E-quality marks data records with insufficient or widely scattered stress information.

Fig. 2.3 shows the 2008 version of the WSM displaying the orientations of the maximum horizontal compressional stress, produced using the 11,346 stress data records of A–C quality but excluding all possible plate boundary events. The length of the stress symbols represents the data quality. The stress regimes are: NF for normal faulting, SS for strike-slip faulting, TF for thrust faulting and U for an unknown regime.

Stress maps of major continents and different countries can also be provided by the WSM (Heidbach et al., 2010). These maps can be used for a first estimate of the orientations of the maximum horizontal stress. In using stress data from the WSM, it is important to consider the depth for which the stress data are relevant.

2.4   VARIATION OF IN SITU STRESSES WITH DEPTH

As a first estimation, it is often assumed that the three principal stresses of an in situ rock stress state are acting vertically and horizontally. The validity of this assumption has been checked by many researchers based on in situ stress measurements, including Bulin (1971), Worotnicki and Walton (1976), Klein and Brown (1983), Li (1986), Zobak et al. (1989), Myrvang (1993), Stephansson (1993), and Kang et al. (2010). With this assumption concerning orientations, the magnitudes of these principal stresses can be estimated using

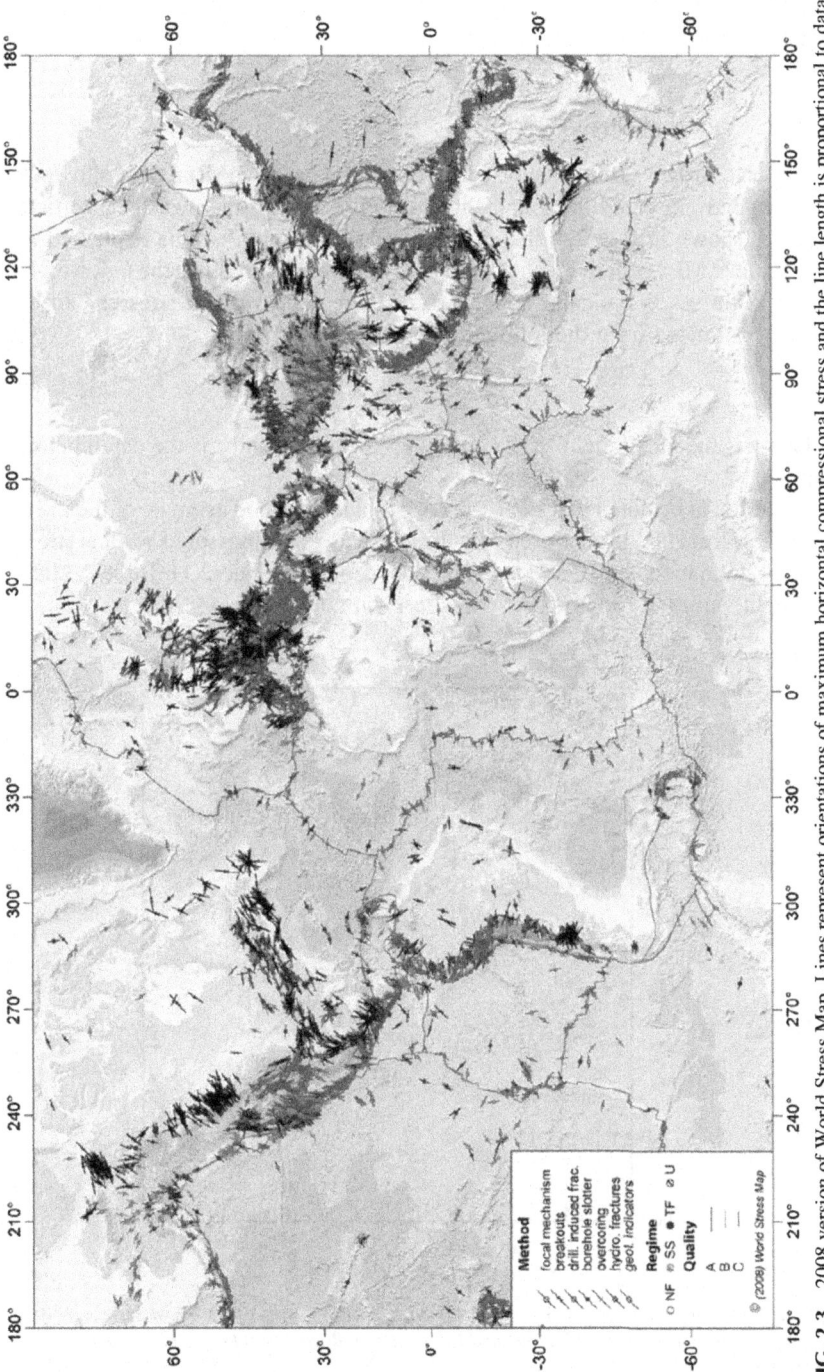

FIG. 2.3 2008 version of World Stress Map. Lines represent orientations of maximum horizontal compressional stress and the line length is proportional to data quality. The stress regimes are: normal faulting (NF), strike-slip faulting (SS), thrust faulting (TF), and unknown regime (U). (*From Heidbach, O., Tingay, M., Barth, A., Reinecker, J., Kurfeβ, D., Müller, B., 2010. Global crustal stress pattern based on the World Stress Map database release 2008. Tectonophysics 482, 3–15.*)

Method
focal mechanism
breakouts
drill. induced frac.
borehole slotter
overcoring
hydro. fractures
geol. indicators

Regime
○ NF ⊕ SS ● TF ⊘ U

Quality
A
B
C

© (2008) World Stress Map

the correlations between vertical and horizontal stresses and depth presented in the following subsections.

2.4.1   Vertical Stress

Hoek and Brown (1980) analyzed worldwide data on measured in situ rock stresses and presented the graph shown in Fig. 2.4. For the measured vertical stresses shown in Fig. 2.4, the average trend with depth can be expressed as $\sigma_v = 0.027z$ MPa, where z is the depth below surface in m. Since the unit weight of rock masses is typically about 0.027 MN/m^3, the vertical stresses can be simply estimated from the following relationship:

$$\sigma_v = \gamma z \qquad (2.1)$$

where γ is the unit weight of the overlying rock mass and z is the depth below surface.

It needs to be noted that Eq. (2.1) only provides a good estimate of the average stress from all the measured data. In some cases, the measured vertical stress may be dramatically different to the predicted one from Eq. (2.1). Table 2.3 lists different variation forms of vertical stress with depth.

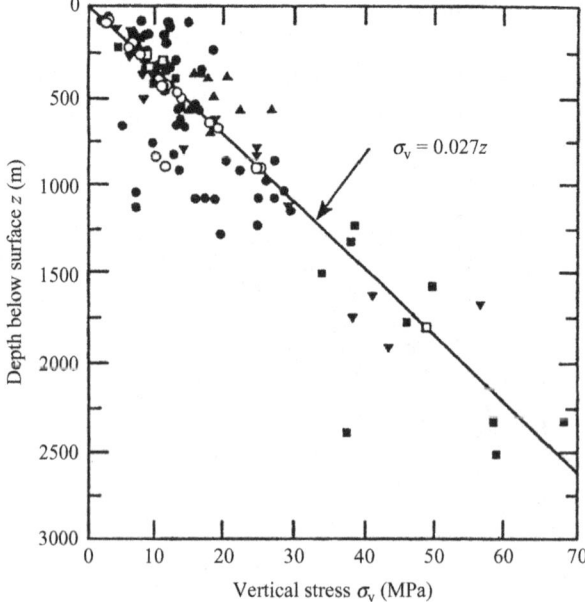

FIG. 2.4   Worldwide in situ rock stress data: vertical stress versus depth below surface. *(From Hoek, E., Brown, E.T., 1980. Underground Excavation in Rock. Institution of Mining and Metallurgy, London, UK.)*

TABLE 2.3 Variation of Vertical Stress With Depth

Reference	Variation of Vertical Stress σ_v (MPa) With Depth z (m)	Location and Depth Range (m)
Herget (1974)	$(1.9 \pm 1.26) + (0.0266 \pm 0.0028)z$	World data (0–2400)
Lindner and Halpern (1977)	$(0.942 \pm 1.1.31) + (0.0339 \pm 0.0067)z$	North American (0–1500)
McGarr and Gay (1978)	$0.0265z$	World data (100–3000)
Hoek and Brown (1980)	$0.027z$	World data (0–3000)
Herget (1987)	$(0.026 - 0.0324)z$	Canadian Shield (0–2200)
Arjang (1989)	$(0.0266 \pm 0.008)z$	Canadian Shield (0–2000)
Baumgärtner et al. (1993)	$(0.0275 - 0.0284)z$	KTP pilot hole (800–3000)
Herget (1993)	$0.0285z$	Canadian Shield (0–2300)
Sugawara and Obara (1993)	$0.027z$	Japanese Islands (0–1200)
Te Kamp et al. (1995)	$(0.0275 - 0.0284)z$	KTP hole (0–9000)
Lim and Lee (1995)	$0.233 + 0.024z$	South Korea (0–850)
Yokoyama et al. (2003)	$0.0255z$ (Crystalline rock) $0.0249z$ (Sedimentary rock)	Japan (0–1600)

2.4.2 Horizontal Stresses

The horizontal stresses in rock are much more difficult to estimate than the vertical stress. In many cases, the horizontal stresses at the same location in a rock mass are different in different directions. The maximum horizontal stress σ_{hmax} and the minimum horizontal stress σ_{hmin} can be related to the vertical stress σ_v as follows (see Fig. 2.5) (Anderson, 1951):

- $\sigma_v > \sigma_{hmax} > \sigma_{hmin}$ in normal fault area;
- $\sigma_{hmax} > \sigma_{hmin} > \sigma_v$ in thrust fault area; and
- $\sigma_{hmax} > \sigma_v > \sigma_{hmin}$ in strike-slip fault area.

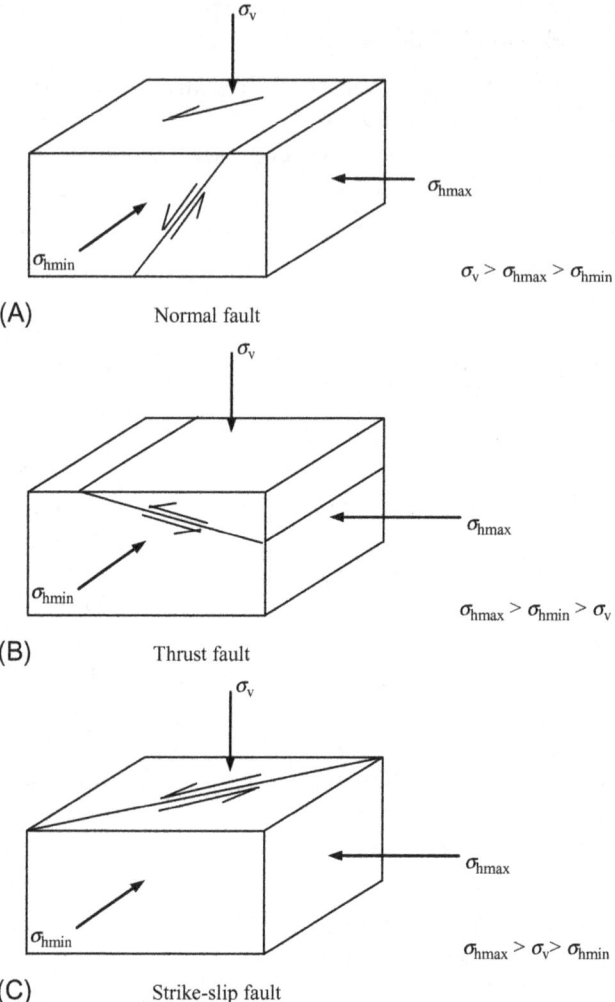

σ_v

σ_{hmax}

σ_{hmin}

$\sigma_v > \sigma_{hmax} > \sigma_{hmin}$

(A)   Normal fault

σ_v

σ_{hmax}

σ_{hmin}

$\sigma_{hmax} > \sigma_{hmin} > \sigma_v$

(B)   Thrust fault

σ_v

σ_{hmax}

σ_{hmin}

$\sigma_{hmax} > \sigma_v > \sigma_{hmin}$

(C)   Strike-slip fault

FIG. 2.5 Types of faults and state of in situ rock stresses. *(Modified from Anderson, E.M., 1951. The Dynamics of Faulting and Dyke Formation with Applications to Britain. Oliver and Boyd, Edinburgh, UK.)*

Table 2.4 lists different variation forms of σ_{hmax}, σ_{hmin} and average horizontal stress σ_{have} with depth.

Normally, the average horizontal stress is related to the vertical stress by the coefficient k such that:

$$\sigma_{have} = k\sigma_v = k\gamma z \qquad (2.2)$$

TABLE 2.4 Variation of Horizontal Stress With Depth

Reference	Variation of σ_{have}, σ_{hmax}, σ_{hamin} (MPa) or k, k_{max}, k_{min} With Depth z (m)	Location and Depth Range (m)
Voight (1966)	$\sigma_{have} = 8.0 + 0.043z$	World data (0–1000)
Herget (1974)	$\sigma_{have} = (8.3 \pm 0.5) + (0.0407 \pm 0.0023)z$	World data (0–800)
Van Heerden (1976)	$k = 0.448 + 248/z \, (r = 0.85)$	South Africa (0–2500)
Worotnicki and Denham (1976)	$\sigma_{have} = 7.7 + (0.021 \pm 0.002)z \, (r = 0.85)$	Australia (0–1500)
Haimson (1977)	$\sigma_{hmax} = 4.6 + 0.025z$ $\sigma_{hmin} = 1.4 + 0.018z \, (r = 0.95)$	Michigan Basin (0–5000)
Lindner and Halpern (1977)	$\sigma_{have} = (4.36 \pm 0.815) + (0.039 \pm 0.0072)z$	North America (0–1500)
Hoek and Brown (1980)	$0.3 + 100/z < k < 0.5 + 1500/z$	World data (0–3000)
Aytmatov (1986)	$5.0 + 0.058z < (\sigma_{hmax} + \sigma_{hmin}) < 9.5 + 0.075z$	World data (mostly former USSR) (0–1000)
Li (1986)	$\sigma_{have} = 0.72 + 0.041z$ $0.3 + 100/z < k < 0.5 + 440/z$	China (0–500)
Rummel (1986)	$k_{max} = 0.98 + 250/z$ $k_{min} = 0.5 + 150/z$	World data (500–3000)
Herget (1987)	$\sigma_{have} = 9.86 + 0.0371z$ $\sigma_{have} = 33.41 + 0.0111z$ $k = 1.25 + 267/z$ $k_{max} = 1.46 + 357/z$ $k_{min} = 1.10 + 167/z$	Canadian Shield (0–900) (990–2200) (0–2200)
Pine and Kwakwa (1989)	$\sigma_{hmax} = 15 + 0.028z$ $\sigma_{hmin} = 6 + 0.012z$	Carnmenellis granite Cornwall, UK (0–2000)
Arjang (1989)	$\sigma_{hmax} = 8.8 + 0.0422z$ $\sigma_{hmin} = 3.64 + 0.0276z$ $\sigma_{have} = 5.91 + 0.0349z$	Canadian Shield (0–2000)

Continued

TABLE 2.4 Variation of Horizontal Stress With Depth—cont'd

Reference	Variation of σ_{have}, σ_{hmax}, σ_{hamin} (MPa) or k, k_{max}, k_{min} With Depth z (m)	Location and Depth Range (m)
Baumgärtner et al. (1993)	$\sigma_{hmax} = 30.4 + 0.023z$ $\sigma_{hmin} = 16.0 + 0.011z$	KTP pilot hole (800–3000)
	$\sigma_{hmin} = 1.75 + 0.0133z$	Cajon pass hole (800–3000)
Sugawara and Obara (1993)	$\sigma_{have} = 2.5 + 0.013z$	Japanese Islands (0–1200)
Hast (in Stephansson, 1993)	$\sigma_{hmax} = 9.1 + 0.0724z$ $(r = 0.78)$ $\sigma_{hmin} = 5.3 + 0.0542z$ $(r = 0.83)$	Fennoscandia, overcoring (0–1000)
Stephansson (1993)	$\sigma_{hmax} = 10.4 + 0.0446z$ $(r = 0.61)$ $\sigma_{hmin} = 5.0 + 0.0286z$ $(r = 0.58)$	Fennoscandia: Leeman-Hiltscher overcoring (0–700)
	$\sigma_{hmax} = 6.7 + 0.0444z$ $(r = 0.61)$ $\sigma_{hmin} = 0.8 + 0.0329z$ $(r = 0.91)$	Leeman-type overcoring (0–1000)
	$\sigma_{hmax} = 2.8 + 0.0399z$ $(r = 0.79)$ $\sigma_{hmin} = 2.2 + 0.0240z$ $(r = 0.81)$	Hydraulic fracturing (0–1000)
Te Kamp et al. (1995)	$\sigma_{hmax} = 15.83 + 0.0303z$ $\sigma_{hmin} = 6.52 + 0.0157z$	KTP hole (0–9000)
Lim and Lee (1995)	$\sigma_{have} = 1.858 + 0.018z$ $(r = 0.869)$	South Korea: Overcoring (0–850)
	$\sigma_{have} = 2.657 + 0.032z$ $(r = 0.606)$	Hydraulic fracturing (0–500)
Rummel (2002)	$k_{max} = 1.30 + 110/z$ $k_{min} = 0.66 + 72/z$	Hong Kong (0–200)
Yokoyama et al. (2003)	Crystalline rock: $\sigma_{hmax} = -21.9 + 0.0301z$ $\sigma_{hmin} = 33.7 + 0.0219z$ Sedimentary rock: $\sigma_{hmax} = 23.5 + 0.0340z$ $\sigma_{hmin} = 47.5 + 0.0281z$	Japan (0–1600)

TABLE 2.4 Variation of Horizontal Stress With Depth—cont'd

Reference	Variation of σ_{have}, σ_{hmax}, σ_{hamin} (MPa) or k, k_{max}, k_{min} With Depth z (m)	Location and Depth Range (m)
Kang et al. (2010)	$k = 0.7 + 116.5/z$	Coal mines in China (70–1280)
Xu et al. (2015)	$\sigma_{hmax} = 2.0\gamma z$ $\sigma_{hmin} = 1.3\gamma z$	Guangdong, China (0–250)

Notes: $k = \sigma_{have}/\sigma_v$; $k_{max} = \sigma_{hmax}/\sigma_v$; $k_{min} = \sigma_{have}/\sigma_v$; γ = unit weight of rock (MN/m^3); and r is the correlation coefficient.

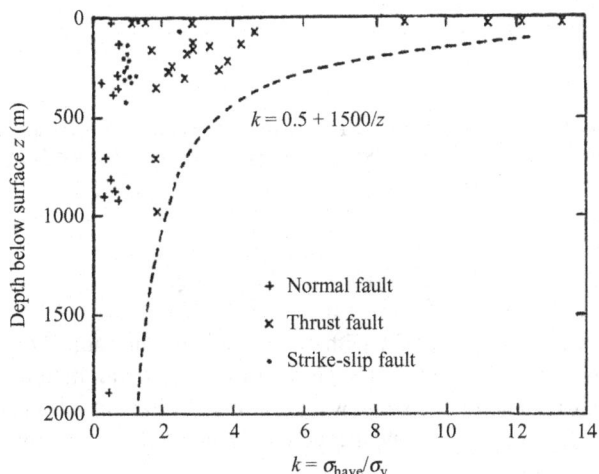

FIG. 2.6 Variation of average horizontal stress to vertical stress ratio versus depth with interpretation of faulting conditions. *(From Jamison, D.B., Cook, N.G.W., 1979. An analysis of the measured values for the state of stress in the earth's crust. In: Jaeger, J.C., Cook, N.G.W. (Eds.), Fundamentals of Rock Mechanics, third ed. Chapman and Hall, London.)*

Fig. 2.6 shows the variation of in situ k values with depth from Jamison and Cook (1979) with interpretation of faulting conditions. As would be expected, the values of σ_{have} in the normal fault areas are relatively low, the values of σ_{have} in the thrust fault areas are relatively high, and the values of σ_{have} in the strike-slip fault areas are intermediate.

Fig. 2.7 shows the worldwide in situ rock stress data compiled by Hoek and Brown (1980). All the data can be enveloped by the following formula:

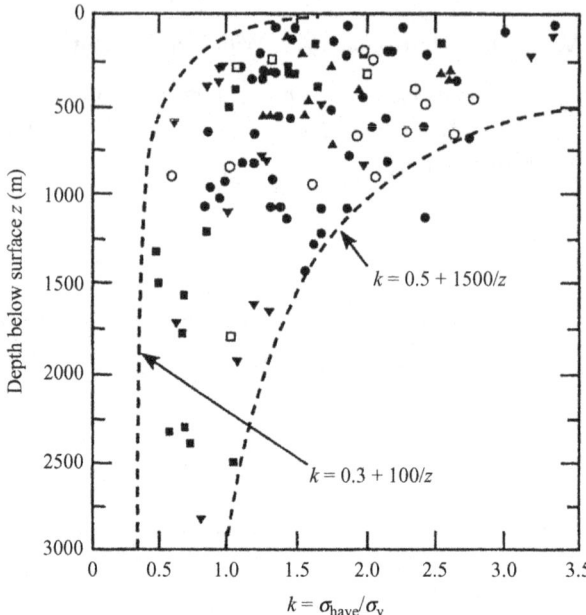

FIG. 2.7 Worldwide in situ rock stress data: Average horizontal stress to vertical stress ratio versus depth. *(From Hoek, E., Brown, E.T., 1980. Underground Excavation in Rock. Institution of Mining and Metallurgy, London, UK.)*

$$\frac{100}{z} + 0.3 < k < \frac{1500}{z} + 0.5 \tag{2.3}$$

Some other variation forms of k with depth are listed in Table 2.4.

Terzaghi and Richard (1952) suggested that, for a gravitationally loaded rock mass in which no lateral strain was permitted during formation of the overlying strata, the value of k is independent of depth and is given by

$$k = \frac{\nu}{1 - \nu} \tag{2.4}$$

where ν is the Poisson's ratio of the rock mass. For typical values of $\nu = 0.1$–0.4 for rock masses, Eq. (2.4) gives $k = 0.11$–0.67. Since the envelope formulae (2.3) tend towards $0.3 < k < 0.5$ as depth increases, Eq. (2.4) may provide a rough estimate of k at significant depths.

Sheorey (1994) developed an elasto-static thermal stress model of the earth. This model considers curvature of the crust and variation of elastic constants, density and thermal expansion coefficients through the crust and mantle. Sheorey (1994) presented the following simplified equation, which can be used for estimating the average horizontal stress:

$$\sigma_{\text{have}} = \frac{\nu}{1 - \nu}\gamma z + \frac{\beta E_h G}{1 - \nu}(z + 1000) \tag{2.5}$$

where ν is the Poisson's ratio of the rock; γ is the unit weight of the rock, in N/m^3; z is the depth below surface, in m; E_h is the average deformation modulus of the rock measured in the horizontal direction, in Pa; β is the coefficient of linear thermal expansion of the rock, in $1/°C$; and G is the geothermal gradient of the rock, in $°C/m$.

Table 2.5 lists the coefficient of linear thermal expansion of different rocks compiled by Sheorey et al. (2001). From this table, $\beta = 8 \times 10^{-6}/°C$ can be chosen as a reasonable representative value for different rocks except for coal. The thermal gradient for crustal rocks can be taken as $0.024°C/m$. Assuming the vertical stress $\sigma_v = \gamma z$ and taking the representative values of $\nu = 0.25$ and $\gamma = 2.7 \times 10^3 \, N/m^3$, the average horizontal to vertical stress ratio k can be derived from Eq. (2.5) as

$$k = 0.33 + 9.5E_h \left(0.001 + \frac{1}{z} \right) \tag{2.6}$$

where E_h is in the unit of GPa.

A plot of Eq. (2.6) is given in Fig. 2.8 for a range of deformation moduli. The curves relating k with depth below surface z are similar to those based on in situ stress data shown in Fig. 2.7. Hence Eq. (2.6) provides a reasonable basis for estimating the value of k.

The average horizontal stress to vertical stress ratio k is, in general, greater than 1. High horizontal stresses are caused by different factors, including erosion, tectonics, rock anisotropy and rock discontinuities (Amadei and

TABLE 2.5 Coefficient of Linear Thermal Expansion β of Some Rocks

Rock	$\beta \, (\times 10^{-6}/°C)$
Granite	6–9
Limestone	3.7–10.3
Marble	3–15
Sandstone	5–12
Schist	6–12
Dolomite	8.1
Conglomerate	9.1
Breccia	4.1–9.1
Coal	30

From Sheorey, P.R., Murali Mohan, G., Sinha, A., 2001. Influence of elastic constants on the horizontal in situ stress. Int. J. Rock Mech. Min. Sci. 38, 1211–1216.

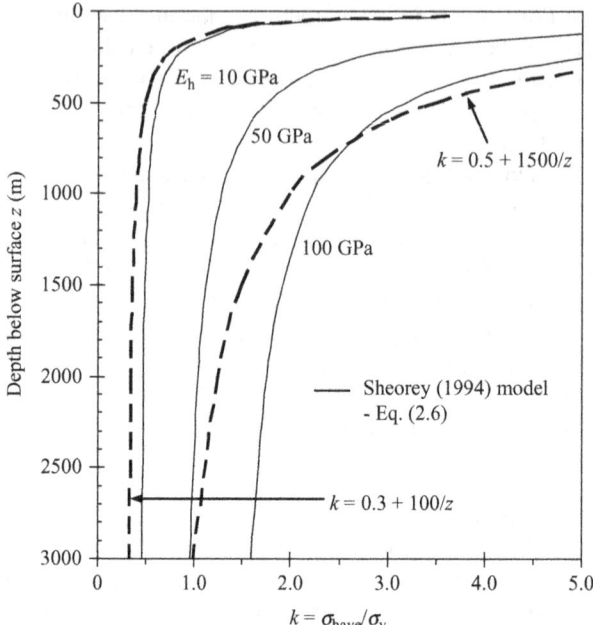

FIG. 2.8 Average horizontal stress to vertical stress ratio versus depth for different deformation moduli based on Sheorey (1994) model.

Stephanson, 1997; Hudson and Harrison, 1997). For detailed description of these factors, the reader can refer to the two references.

2.5   COMMENTS

The WSMs and the stress versus depth relationships presented in the above two sections can be useful in estimating the in situ rock stresses. However, these maps and relationships should be used with caution and the important issues presented in the following should be considered.

The assumption that the principal stresses are vertical and horizontal with depth breaks down when the ground surface is not horizontal (Amadei and Stephanson, 1997). This can be clearly seen from Fig. 2.9, which shows a semi-infinite isotropic, homogeneous rock mass with a complex topography consisting of a series of hills and valleys and no surface loads. The rock mass is under gravity alone with no lateral displacements. Because of the traction-free boundary conditions, the principal stresses are parallel and normal to the ground surface. As depth increases, the principal stresses approach the same directions as when the ground surface is horizontal.

Open discontinuities in the rock mass change the directions and magnitudes of the principal stresses (see Fig. 2.10). Since no normal or shear stress can be sustained, respectively, perpendicular and parallel to the discontinuity surface,

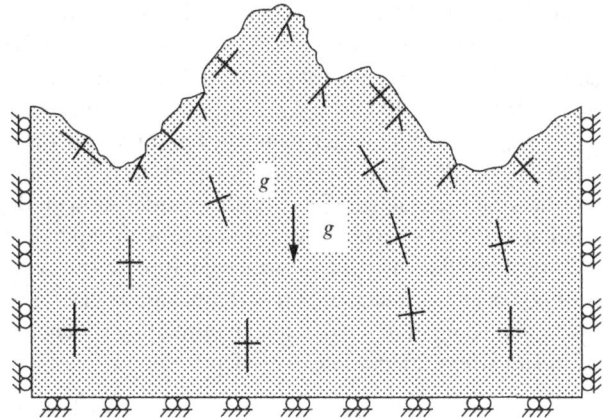

FIG. 2.9 Direction of principal stresses in a rock mass with a complex topography consisting of a series of hills and valleys. *(From Amadei, B., Stephanson, O., 1997. Rock Stress and Its Measurement. Chapman & Hall, London, UK.)*

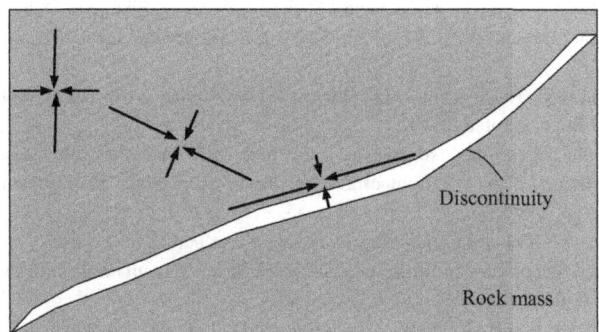

FIG. 2.10 An open discontinuity changes the stress field and causes the principal stresses to be locally parallel and perpendicular to the discontinuity plane. *(Based on Hudson, J.A., Cornet, F.H., Christiansson, R., 2003. ISRM suggested methods for rock stress estimation—part 1: strategy for rock stress estimation. Int. J. Rock Mech. Min. Sci. 40, 991–998.)*

the discontinuity surface becomes a principal stress plane with zero principal stress value (Hudson et al., 2003).

The anisotropy of the rock mass also affects the directions and magnitudes of the principal stresses (Zang and Stephansson, 2010). For example, consider an elastic homogeneous orthotropic rock mass with Poisson's ratio $\nu_{xy} \neq \nu_{yz} \neq \nu_{zx}$. The lateral-stress coefficient is related to the relative magnitudes of ν_{xy}, ν_{yz}, and ν_{zx}:

$$k_{h,x} = \frac{\nu_{xz} + \nu_{yz}\nu_{xy}}{1 - \nu_{xy}\nu_{yx}} \tag{2.7a}$$

$$k_{h,y} = \frac{\nu_{yz} + \nu_{yx}\nu_{xz}}{1 - \nu_{xy}\nu_{yx}} \tag{2.7b}$$

The ordering of in situ stresses may also change with depth. The hydraulic fracturing tests conducted by Haimson (1977) in an oil well near the center of the Michigan Basin revealed a change of the in situ stress ordering: $\sigma_{hmax} > \sigma_{hmin} > \sigma_v$ from 0 to 200 m, $\sigma_{hmax} > \sigma_v > \sigma_{hmin}$ from 200 to 4500 m, and $\sigma_v > \sigma_{hmax} > \sigma_{hmin}$ at depths larger than 4500 m. Dey and Brown (1986), Adams and Bell (1991), Plumb (1994), and Kang et al. (2010) also reported measurements showing the change of in situ stress ordering with depth.

The orientation of in situ stresses may also change with depth. Haimson and Rummel (1982) reported the variations in the orientation of the maximum horizontal stress of 60° over a distance of 500 m. Stephansson (1993), Martin and Chandler (1993), and Kang et al. (2010) also reported measurements showing the change of in situ stress orientations with depth.

REFERENCES

Adams, J., Bell, J.S., 1991. Crustal stresses in Canada. In: Slemmons, D.B., Engdahl, E.R., Zoback, M.D., Blackwell, D.D. (Eds.), Neotectonics of North America, The Geology of North America, Decade Map, vol. 1. Geological Society of America, Boulder, CO, pp. 367–386.

Amadei, B., Stephanson, O., 1997. Rock Stress and Its Measurement. Chapman & Hall, London, UK.

Anderson, E.M., 1951. The Dynamics of Faulting and Dyke Formation with Applications to Britain. Oliver and Boyd, Edinburgh, UK.

Arjang, B., 1989. Pre-mining stresses at some hard rock mines in the Canadian shield. In: Proceedings 30th US Symposium on Rock Mechanics, Morgantown. Balkema, Rotterdam, pp. 545–551.

Aytmatov, I.T., 1986. On virgin stress state of a rock mass in mobile folded areas. In: Proceedings International Symposium on Rock Stress and Rock Stress Measurements, Stockholm. Centek Publishers, Lulea, pp. 55–59.

Baumgärtner, J., Healy, J.H., Rummel, F.H., Zoback, M.D., 1993. Deep hydraulic fracturing stress measurements in the KTB (Germany) and Cajon Pass (USA) scientific drilling projects—a summary. In: Proceedings 7th Congress International Society Rock Mechanics, Aachen, vol. 3. Balkema, Rotterdam, pp. 1685–1690.

Bulin, N.K., 1971. The present stress field in the upper parts of the crust. Geotectonics 3, 133–139.

Dey, T.N., Brown, D.W., 1986. Stress measurements in a deep granitic rock mass using hydraulic fracturing and differential strain curve analysis. In: Proceedings International Symposium on Rock Stress and Rock Stress Measurements, Stockholm. Centek Publishers, Lulea, pp. 351–357.

Fairhurst, C., 2003. Stress estimation in rock: a brief history and review. Int. J. Rock Mech. Min. Sci. 40, 957–973.

Haimson, B.C., 1977. Recent in-situ stress measurements using the hydrofracturing technique. In: Proceedings 18th US Symposium Rock Mechanics. Johnson Publisher, Golden. 4C2/1–6.

Haimson, B.C., Cornet, F.H., 2003. ISRM suggested methods for rock stress estimation—part 3: hydraulic fracturing (HF) and/or hydraulic testing of pre-existing fractures (HTPF). Int. J. Rock Mech. Min. Sci. 40, 1011–1020.

Haimson, B.C., Rummel, F., 1982. Hydrofracturing stress measurements in the Iceland drilling project drillhole at Reydasfjordur, Iceland. J. Geophys. Res. 87, 6631–6649.

Heidbach, O., Tingay, M., Barth, A., Reinecker, J., Kurfeß, D., Müller, B., 2010. Global crustal stress pattern based on the World Stress Map database release 2008. Tectonophysics 482, 3–15.

Herget, G., 1974. Ground stress determinations in Canada. Rock Mech. 6, 53–74.

Herget, G., 1987. Stress assumptions for underground excavations in the Canadian shield. Int. J. Rock Mech. Min. Sci. Geomech. Abstr. 24, 95–97.

Herget, G., 1993. Rock stresses and rock stress monitoring in Canada. In: Hudson, J.A. (Ed.), In: Comprehensive Rock Engineering—Principle, Practice and Projects, vol. 3. Pergamon, Oxford, UK, pp. 473–496.

Hoek, E., Brown, E.T., 1980. Underground Excavation in Rock. Institution of Mining and Metallurgy, London, UK.

Hudson, J.A., 2012. Design methodology for the safety of underground rock engineering. J. Rock Mech. Geotech. Eng. 4 (3), 205–214.

Hudson, J.A., Cornet, F.H., Christiansson, R., 2003. ISRM suggested methods for rock stress estimation—part 1: strategy for rock stress estimation. Int. J. Rock Mech. Min. Sci. 40, 991–998.

Hudson, J.A., Harrison, J.P., 1997. Rock Engineering Mechanics—An Introduction to the Principles. Elsevier, Oxford.

Jamison, D.B., Cook, N.G.W., 1979. An analysis of the measured values for the state of stress in the earth's crust. In: Jaeger, J.C., Cook, N.G.W. (Eds.), Fundamentals of Rock Mechanics, third ed. Chapman and Hall, London.

Kang, H., Zhang, X., Si, L., Wu, Y., Gao, F., 2010. In-situ stress measurements and stress distribution characteristics in underground coal mines in China. Eng. Geol. 116, 333–345.

Klein, R.J., Brown, E.T., 1983. The state of stress in British rocks. Report DOE/RW/83.8.

Li, F., 1986. In-situ stress measurements, stress state in the upper crust and their application to rock engineering. In: Proceedings International Symposium on Rock Stress and Rock Stress Measurements, Stockholm. Centek Publishers, Lulea, pp. 69–77.

Lim, H.-U., Lee, C.-I., 1995. Fifteen years' experience on rock stress measurements in South Korea. In: Proceedings International Workshop on Rock Stress Measurements at Great Depth, Tokyo, Japan, pp. 7–12.

Lindner, E.N., Halpern, E.N., 1977. In-situ stress: an analysis. In: Proceedings 18th US Symposium Rock Mechanics. Johnson Publisher, Golden. 4C1/1–7.

Ljunggren, C., Chang, Y., Janson, T., Christiansson, R., 2003. An overview of rock stress measurement methods. Int. J. Rock Mech. Min. Sci. 40, 975–989.

Martin, C.D., Chandler, N.A., 1993. Stress heterogeneity and geological structure. Int. J. Rock Mech. Min. Sci. Geomech. Abstr. 30, 993–999.

McGarr, A., Gay, N.C., 1978. State of stress in the Earth's crust. Annu Rev. Earth Planet. Sci. 6, 405–436.

Myrvang, A.M., 1993. Rock stress and rock stress problem in Norway. In: Hudson, J.A. (Ed.), In: Comprehensive Rock Engineering—Principle, Practice & Projects, vol. 3. Pergamon, Oxford, UK, pp. 461–471.

Pine, R.J., Kwakwa, K.A., 1989. Experience with hydrofracture stress measurements to depths of 2.6 km and implications for measurements to 6 km in the Carnmenellis granite. Int. J. Rock Mech. Min. Sci. Geomech. Abstr. 26, 565–571.

Plumb, R.A., 1994. Variation of the least horizontal stress magnitude in sedimentary rocks. In: Proceedings 1st North America Rock Mechanics Symposium, Austin. Balkema, Rotterdam, pp. 71–78.

Reinecker, J., Heidbach, O., Tingay, M., Connolly, P., Müller, B., 2004. The 2004 release of the World Stress Map. Available online at: www.world-stress-map.org.

Rummel, F., 1986. Stresses and tectonics of the upper continental crust—a review. In: Proceedings International Symp. on Rock Stress and Rock Stress Measurements, Stockholm. Centek Publishers, Lulea, pp. 177–186.

Rummel, F., 2002. Crustal stress derived from fluid injection tests in boreholes. In: Sharma, V.M., Saxena, K.R. (Eds.), In-Situ Characterization of Rocks. Balkema, Lisse, pp. 205–244.

Sheorey, P.R., 1994. A theory for in situ stresses in isotropic and transversely isotropic rock. Int. J. Rock Mech. Min. Sci. Geomech. Abstr. 31, 23–34.

Sheorey, P.R., Murali Mohan, G., Sinha, A., 2001. Influence of elastic constants on the horizontal in situ stress. Int. J. Rock Mech. Min. Sci. 38, 1211–1216.

Sjöberg, J., Christiansson, R., Hudson, J.A., 2003. ISRM suggested methods for rock stress estimation—Part 2: overcoring methods. Int. J. Rock Mech. Min. Sci. 40, 999–1010.

Stephansson, O., 1993. Rock stress in the Fennoscandian shield. In: Hudson, J.A. (Ed.), In: Comprehensive Rock Engineering—Principle, Practice & Projects, vol. 3. Pergamon, Oxford, UK, pp. 445–459.

Stephansson, O., Zang, A., 2012. ISRM suggested methods for rock stress estimation—part 5: establishing a model for the in situ stress at a given site. Rock Mech. Rock. Eng. 45, 955–969.

Sugawara, K., Obara, Y., 1993. Measuring rock stress. In: Hudson, J.A. (Ed.), In: Comprehensive Rock Engineering—Principle, Practice & Projects, vol. 3. Pergamon, Oxford, UK, pp. 533–552.

Te Kamp, L., Rummel, F., Zoback, M.D., 1995. Hydrofrac stress profile to 9 km at the German KTP site. In: Proceedings Workshop on Rock Stresses in the North Sea, Trondheim, Norway. NTH and SINTEF Publ, Trondheim, pp. 147–153.

Terzaghi, K., Richard, F.E., 1952. Measurement of stresses in rock. Geotechnique 12, 105–124.

Terzaghi, K., 1962. Measurement of stresses in rock. Geotechnique. 12, 105–124.

Van Heerden, W.L., 1976. Practical application of the CSIR triaxial strain cell for rock stress measurements. In: Proceedings ISRM Symposium on Investigation of Stress in Rock, Advances in Stress Measurement, Sydney. The Institution of Engineers, Australia, pp. 1–6.

Voight, B., 1966. Interpretation of in-situ stress measurements—Panel report on Theme IV. In: Proceedings 1st Congress International Society Rock Mechanics, Lisbon, vol. 3. pp. 332–348.

Worotnicki, G., Denham, D., 1976. The state of stress in the upper part of the earth's crust in Australia according to measurements in mines and tunnels and from seismic observations. In: Proceedings ISRM Symposium on Investigation of Stress in Rock: Advances in Rock Measurement, Sydney. The Institution of Engineers, Australia, pp. 71–82.

Worotnicki, G., Walton, R.J., 1976. Triaxial hollow inclusion gauges for determination of rock stresses in-situ. In: Supplement to Proceedings ISRM Symposium on Investigation of Stress in Rock, Advances in Stress Measurement, Sydney. The Institution of Engineers, Australia, pp. 141–150.

Xu, T., Han, X., Yin, J., 2015. Arrangement of ground stress test and the evaluation of the stress in long highway tunnel. In: Proceedings 2015 Academic Annual Meeting of the National Highway Tunnel. Chongqing University Press, pp. 136–141.

Yokoyama, T., Ogawa, K., Kanagawa, T., Tanak, M., Ishida, T., 2003. Regional in-situ stress in Japan based on measurements. In: Sugawara, K., Obara, Y., Sato, A. (Eds.), Rock Stress: Proceedings 3rd International Symposium on Rock Stress, Kumamoto, Japan. Balkema, Rotterdam, pp. 335–341.

Zang, A., Stephansson, O., 2010. Stress Field of the Earth's Crust. Springer Science and Business Media BV, Dordrecht.

Zobak, M.L., Zobak, M.D., Adams, J., et al., 1989. Global patterns of tectonic stress. Nature 341, 291–298.

Chapter 3

Intact Rock

3.1 INTRODUCTION

Intact rock refers to the unfractured blocks between discontinuities in a typical rock mass. These blocks may range from a few millimeters to several meters in size (Hoek, 1994; Hudson and Harrison, 1997). The properties of intact rock are governed by the physical properties of the materials of which it is composed and the manner in which they are bonded to each other. The parameters which may be used in a description of intact rock include petrological name, color, texture, grain size, minor lithological characteristics, density, porosity, strength, hardness, and deformability.

This chapter describes the classification of intact rocks and presents the typical values of and correlations between different index properties of them.

3.2 CLASSIFICATION OF INTACT ROCKS

Intact rocks may be classified from a geological or an engineering point of view. In the first case the mineral content of the rock is of prime importance, as is its texture and any change which has occurred since its formation. Although geological classifications of intact rocks usually have a genetic basis, they may provide little information relating to the engineering behavior of the rocks concerned since intact rocks of the same geological category may show a large scatter in strength and deformability, say of the order of 10 times. Therefore, engineering classifications of intact rocks are more related to the engineering properties of rocks.

3.2.1 Geological Classification

3.2.1.1 Rock-Forming Minerals

Rocks are composed of minerals, which are formed by the combination of naturally occurring elements. Although there are hundreds of recognized minerals, only a few are common. Table 3.1 summarizes the common rock-forming minerals and their properties. Moh's scale, used in the table, is a standard of 10 minerals by which the hardness of a mineral may be determined. Hardness is defined as the ability of a mineral to scratch another. The scale is one for the softest mineral (talc) and ten for the hardest (diamond).

Engineering Properties of Rocks. http://dx.doi.org/10.1016/B978-0-12-802833-9.00003-1

TABLE 3.1 Common Rock-Forming Minerals and Their Properties

Mineral	Hardness (Moh's Scale, 1–10)	Relative Density	Fracture	Structure
Orthoclase feldspar	6	2.6	Good cleavage at right angles	Monoclinic. Commonly occurs as crystals
Plagioclase feldspar	6	2.7	Cleavage nearly at right angles— very marked	Triclinic. Showing distinct cleavage lamellae
Quartz	7	2.65	No cleavage; choncoidal fracture	Hexagonal
Muscovite	2.5	2.8	Perfect single cleavage into thin easily separated plates	Monoclinic. Exhibiting strong cleavage lamellae
Biotite	3	3	Perfect single cleavage into thin easily separated plates	Monoclinic. Exhibiting strong cleavage lamellae
Hornblende	5–6	3.05	Good cleavage at 120 degrees	Hexagonal— normally in elongated prisms
Augite	5–6	3.05	Cleavage nearly at right angles	Monoclinic
Olivine	6–7	3.5	No cleavage	No distinctive structure
Calcite	3	2.7	Three perfect cleavages. Rhomboids formed	Hexagonal
Dolomite	4	2.8	Three perfect cleavages	Hexagonal
Kaolinite	1	2.6	No cleavage	No distinctive structure (altered feldspar)
Hematite	6	5	No cleavage	Hexagonal

3.2.1.2   Elementary Rock Classification

Intact rocks are classified into three main groups according to the process by which they are formed: igneous, metamorphic and sedimentary.

Igneous rocks are formed by crystallization of molten magma. The mode of crystallization of the magma, at depth in the Earth's crust or by extrusion, and the rate of cooling affect the rock texture or crystal size. The igneous rocks are subdivided into plutonic, hypabyssal, and extrusive (volcanic), according to their texture. They are further subdivided into acid, intermediate, basic, and ultrabasic, according to their silica content. Table 3.2 shows a schematic classification of igneous rocks.

Metamorphic rocks are the result of metamorphism. Metamorphism is the solid-state conversion of pre-existing rocks by temperature, pressure, and/or chemical changes. The great varieties of metamorphic rocks are characterized, classified and named according to their mineral assemblages and textures. Table 3.3 shows a classification of the metamorphic rocks according to their physical structure, that is, massive or foliated.

TABLE 3.2 Geological Classification of Igneous Rocks

	Type			
	Acid	*Intermediate*	*Basic*	*Ultrabasic*
Grain Size	*>65% Silica*	*55–65% Silica*	*45–55% Silica*	*<45% Silica*
Plutonic	Granite	Diorite	Gabbro	Picrite
				Peridotite
	Granodiorite			Serpentinite
				Dunite
Hypabyssal	Quartz	Plagioclase porphyries	Dolerite	Basic dolerites
	Orthoclase porphyries			
Extrusive	Rhyolite	Pichstone	Basalt	Basic olivine basalts
	Dacite	Andesite		
Major mineral constituents	Quartz, orthoclase, sodium-plagioclase, muscovite, biotite, hornblende	Quartz, orthoclase, plagioclase, biotite, hornblende, augite	Calcium-plagioclase, augite, olivine, hornblende	Calcium-plagioclase, olivine, augite

TABLE 3.3 Classification of Metamorphic Rocks

Classification	Rock	Description	Major Mineral Constituents
Massive	Hornfels	Micro-fine grained	Quartz
	Quartzite	Fined grained	Quartz
	Marble	Fine—coarse grained	Calcite or dolomite
Foliated	Slate	Micro-fine grained, laminated	Kaolinite, mica
	Phyllite	Soft, laminated	Mica, kaolinite
	Schist	Altered hypabyssal rocks, coarse grained	Feldspar, quartz, mica
	Gneiss	Altered granite	Hornblende

Sedimentary rocks are formed from the consolidation of sediments. Sedimentary rocks cover three-quarters of the continental areas and most of the sea floor. In the process of erosion, rocks weather and are broken down into small particles or totally dissolved. These detritic particles may be carried away by water, wind or glaciers, and deposited far from their original position. When these sediments start to form thick deposits, they consolidate under their own weight and eventually turn into solid rock through chemical or biochemical precipitation or organic process. As a result of this process, sedimentary rocks almost invariably possess a distinct stratified, or bedded, structure. Table 3.4 shows the classification of sedimentary rocks according to their formation process.

3.2.2   Engineering Classification

The engineering classification of intact rocks is based on strength and/or deformation properties of the rock. Table 3.5 shows the classification system of the International Society of Rock Mechanics (ISRM, 1978c). The ISRM classification is also recommended in the *Canadian Foundation Engineering Manual* (CGS, 1985). In this classification, the rock may range from extremely weak to extremely strong depending on the unconfined compressive strength or the approximate field identification.

Based on laboratory measurements of strength and deformation properties of rocks, Deere and Miller (1966) established a classification system based on the ultimate strength (unconfined compressive strength) and the tangent modulus E_t of elasticity at 50% of the ultimate strength. Fig. 3.1 summarizes the engineering classification of igneous, sedimentary, and metamorphic rocks,

TABLE 3.4 Classification of Sedimentary Rocks

Method of formation	Classification	Rock	Description	Major mineral constituents
Mechanical	Rudaceous	Breccia	Large grains in clay matrix	Various
		Conglomerate		
	Arenaceous	Sandstone	Medium, round grains in calcite matrix	Quartz, calcite (sometimes feldspar, mica)
		Quartzite	Medium, round grains in silica matrix	Quartz
		Gritstone	Medium, angular grains in matrix	Quartz, calcite, various
		Breccia	Coarse, angular grains in matrix	Quartz, calcite, various
	Argillaceous	Claystone	Micro-fine-grained plastic texture	Kaolinite, quartz, mica
		Shale	Harder-laminated compacted clay	Kaolinite, quartz, mica
		Mudstone		
Organic	Calcareous	Limestone	Fossiliferous, coarse or fine grained	Calcite
	Carbonaceous (siliceous, ferruginous, phosphatic)	Coal		
Chemical	Ferruginous	Ironstone	Impregnated limestone or claystone (or precipitated)	Calcite, iron oxide
	Calcareous (siliceous, saline)	Dolomite limestone	Precipitated or replaced limestone, fine grained	Dolomite, calcite

TABLE 3.5 Engineering Classification of Rock by Strength

Grade	Classification	Field Identification	Unconfined Compressive Strength (MPa)	Point Load Index (MPa)	Examples
R0	Extremely weak	Indented by thumbnail	<1	_[a]	Stiff fault gouge
R1	Very weak	Crumbles under firm blows of geological hammer; can be peeled with a pocket knife	1–5	_[a]	Highly weathered or altered rock, shale
R2	Weak	Can be peeled with a pocket knife with difficulty; shallow indentations made by a firm blow with point of geological hammer	5–25	_[a]	Chalk, claystone, potash, marl, siltstone, shale, rock salt
R3	Medium strong	Cannot be scraped or peeled with a pocket knife; specimen can be fractured with a single firm blow of geological hammer	25–50	1–2	Concrete, phyllite, schist, siltstone
R4	Strong	Specimen requires more than one blow of geological hammer to fracture	50–100	2–4	Limestone, marble, sandstone, schist
R5	Very strong	Specimen requires many blows of geological hammer to fracture	100–250	4–10	Amphibolite, sandstone, basalt, gabbro, gneiss, granodiorite, peridotite, rhyolite, tiff
R6	Extremely strong	Specimen can only be chipped with the geological hammer	>250	>10	Fresh basalt, chert, diabase, gneiss, granite, quartzite

[a] Point load tests on rocks with unconfined compressive strength below 25 MPa are likely to yield highly ambiguous results.
Based on ISRM, 1978c. Suggested methods for the quantitative description of discontinuities in rock masses. International Society for Rock Mechanics, Commission on Standardization of Laboratory and Field Tests. Int. J. Rock Mech. Min. Sci. Geomech. Abstr. 15, 319–368; CGS, 1985. Canadian Foundation Engineering Manual. Part 2, second ed. Canadian Geotechnical Society, Vancouver, British Columbia, Canada; Marinos, P., Hoek, E., 2001. Estimating the geotechnical properties of heterogeneous rock masses such as flysch. Bull. Erg. Geol. Environ. 60, 85–92.

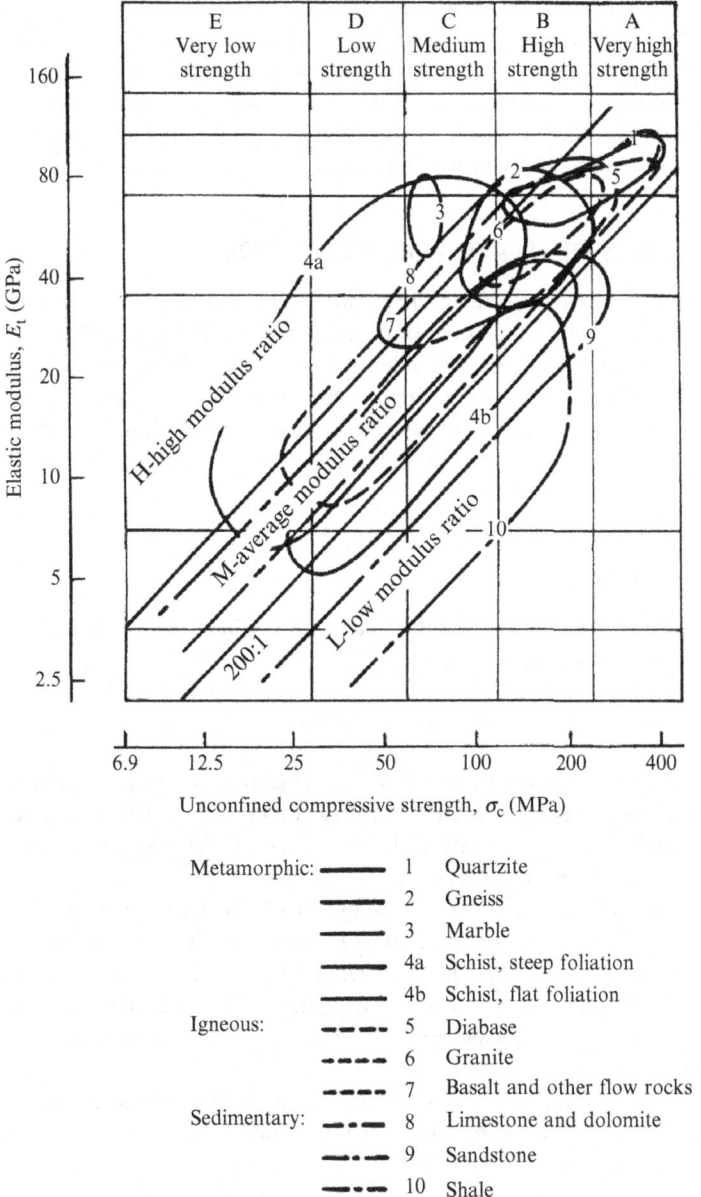

FIG. 3.1 Engineering classification of intact rocks (E_t is the tangent modulus at 50% ultimate strength). *(Modified from Deere, D.U., Miller, R.P., 1966. Engineering classification and index properties for intact rock, Tech. Rep. No. AFWL-TR-65-116. Air Force Weapons Lab, Kirtland Air Force Base, Albuquerque, NM.)*

respectively, as given in Deere and Miller (1966). The modulus ratio (MR) in these figures is that of the elastic modulus to the unconfined compressive strength. A rock may be classified as AM, BH, BL, etc. Voight (1968), however, argued that the elastic properties of intact rock could be omitted from practical classification since the elastic moduli as determined in the laboratory are seldom those required for engineering analysis.

3.3 INDEX PROPERTIES OF INTACT ROCKS

This section describes different index properties of intact rocks, lists their typical values, discusses the effect of different factors on them, and presents the correlations between them. The index properties can help describe rocks quantitatively and can be used for estimating mechanical and hydraulic properties of rocks, as described in later chapters. It needs to be noted, however, that determination of index properties is not a substitute of detailed characterization of rocks.

3.3.1 Porosity

The (total) porosity, n, is defined as the ratio of void or pore volume, V_v, to the total volume, V, of the rock:

$$n = \frac{V_v}{V} = \frac{V - V_s}{V} \tag{3.1}$$

where V_s is the volume of the grains or solid matrix substance. Porosity is usually given as a percentage. The porosity of rocks can be determined by using the method suggested by ISRM (1979). Table 3.6 lists the typical values of porosity of different intact rocks.

Porosity is the result of various geological, physical, and chemical processes and varies significantly for different rock types. Porosity changes significantly even for the same rock type due to different factors such as grain size distribution, grain shape, depth/pressure, and temperature. Fig. 3.2 shows the variation of porosity n with mean grain diameter d_{50} for Bentheim Sandstone (Schön, 1996).

Porosity generally decreases with increasing depth or pressure. Their relationship can be expressed by an exponential function or a logarithmic function (Schön, 1996):

$$n = n_0 e^{-Az} \tag{3.2}$$

$$n = n_0 - B \log z \tag{3.3}$$

where n_0 is the initial porosity at depth $z = 0$ and A and B are empirical factors depending on the compressibility of rocks. Jelic (1984) derived the following relationship for sandstones with an initial porosity of $n_0 = 49.6\%$:

TABLE 3.6 Typical Values of Porosity of Intact Rocks

Rock Type	Age	Depth	Porosity (%)	Reference
Mount Simon sandstone	Cambrian	13,000 ft	0.7	Goodman (1980)
Nugget sandstone (Utah)	Jurassic		1.9	Goodman (1980)
Potsdam sandstone	Cambrian	Surface	11.0	Goodman (1980)
Pottsville sandstone	Pennsylvanian		2.9	Goodman (1980)
Berea sandstone	Mississippian	0–2000 ft	14.0	Goodman (1980)
Keuper sandstone (England)	Triassic	Surface	22.0	Goodman (1980)
Navajo sandstone	Jurassic	Surface	15.5	Goodman (1980)
Sandstone, Montana	Cretaceous	Surface	34.0	Goodman (1980)
Beekmantown dolomite	Ordovician	10,500 ft	0.4	Goodman (1980)
Black River limestone	Ordovician	Surface	0.46	Goodman (1980)
Niagara dolomite	Silurian	Surface	2.9	Goodman (1980)
Limestone, Great Britain	Carboniferous	Surface	5.7	Goodman (1980)
Chalk, Great Britain	Cretaceous	Surface	28.8	Goodman (1980)
Solenhofen limestone		Surface	4.8	Goodman (1980)
Salem limestone	Mississippian	Surface	13.2	Goodman (1980)
Bedford limestone	Mississippian	Surface	12.0	Goodman (1980)
Bermuda limestone	Recent	Surface	43.0	Goodman (1980)
Shale	Pre-Cambrian	Surface	1.6	Goodman (1980)

Continued

TABLE 3.6 Typical Values of Porosity of Intact Rocks—cont'd

Rock Type	Age	Depth	Porosity (%)	Reference
Shale, Oklahoma	Pennsylvanian	1000 ft	17.0	Goodman (1980)
Shale, Oklahoma	Pennsylvanian	3000 ft	7.0	Goodman (1980)
Shale, Oklahoma	Pennsylvanian	5000 ft	4.0	Goodman (1980)
Shale	Cretaceous	600 ft	33.5	Goodman (1980)
Shale	Cretaceous	2500 ft	25.4	Goodman (1980)
Shale	Cretaceous	3500 ft	21.1	Goodman (1980)
Shale	Cretaceous	6100 ft	7.6	Goodman (1980)
Mudstone, Japan	Upper Tertiary	Near surface	22–32	Goodman (1980)
Granite, fresh		Surface	0–1	Goodman (1980)
Granite, weathered			1–5	Goodman (1980)
Decomposed granite (Saprolyte)			20.0	Goodman (1980)
Marble			0.3	Goodman (1980)
Marble			1.1	Goodman (1980)
Bedded tuff			40.0	Goodman (1980)
Welded tuff			14.0	Goodman (1980)
Cedar City tonalite			7.0	Goodman (1980)
Frederick diabase			0.1	Goodman (1980)
San Marcos gabbro			0.2	Goodman (1980)
Basalt, Southeast Greenland		700–1100 ft	2–7	Najibi and Asef (2014)
Basalt, Mid Atlantic Ridge		2700–4400 ft	0.1–10	Najibi and Asef (2014)

Continued

TABLE 3.6 Typical Values of Porosity of Intact Rocks—cont'd

Rock Type	Age	Depth	Porosity (%)	Reference
Shale and sandstone, East Texas		6000–10,200 ft	1–14	Najibi and Asef (2014)
Shale, North America		<9000 ft	1–6.2	Najibi and Asef (2014)
Clastic rocks, Germany		11,000–12,000 ft	1–15	Najibi and Asef (2014)
Clean and shaly sandstones, USA		Subsurface	2–20	Najibi and Asef (2014)
Limestone, Iran		15,100–15,400 ft	0.2–0.4	Najibi and Asef (2014)

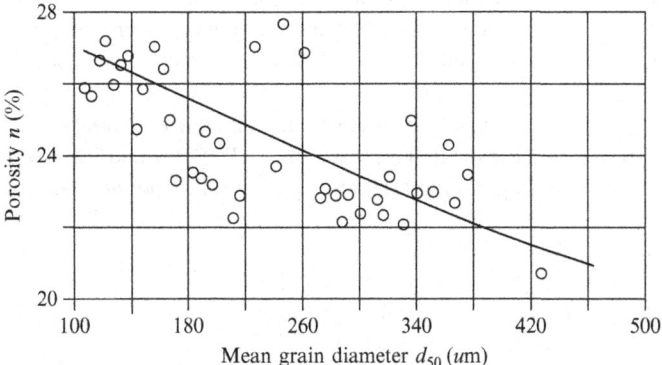

FIG. 3.2   Porosity n versus mean grain diameter d_{50} for Bentheim Sandstone. *(From Schön, J.H., 1996. Physical Properties of Rocks—Fundamentals and Principles of Petrophysics. Pergamon, Oxford.)*

$$n = 49.6 e^{-0.556z} \, (\%) \tag{3.4}$$

in which z is in km.

Porosity also changes with temperature (Chaki et al., 2008; Yavuz et al., 2010). Chaki et al. (2008) heated a granite rock from 105°C at ambient pressure to 200°C, 300°C, 400°C, 500°C, and 600°C and found that the effective porosity (which considers only the interconnected voids) had increased from 0.68% to

0.71%, 0.81%, 0.91%, 1.10%, and 2.85%, respectively. Yavuz et al. (2010) heated two marble and three limestone rocks to 100°C, 200°C, 300°C, 400°C, and 500°C, respectively, and measured the effective porosity of the rocks at different temperatures. The results indicate that the effective porosity either increases with higher temperature or first slightly decreases until 200°C and then increases with higher temperature. Based on best fitting of the test data, they proposed the following relationship between effective porosity and temperature:

$$n = n_i + aT^2 + bT \qquad (3.5)$$

where n_i is the initial effective porosity; T is the temperature in °C; and a and b are the regression coefficients. Table 3.7 shows the values of n_i, a, and b for the five tested rocks.

3.3.2 Density

The density is defined the mass per unit volume of a material. Since a rock contains both grains (solid matrix material) and voids, it is necessary to distinguish between different densities which are related to different parts or components of the rock, as defined in Table 3.8. The density of rocks can be determined using the method suggested by ISRM (1979).

The density of rocks depends on the mineral composition, the porosity and the filling material in the voids. Table 3.9 lists the typical values of density for different intact rocks.

Since, as described earlier, the porosity decreases with increasing depth, the density of rocks increases with depth (see Fig. 3.3). Polak and Rapoport (1961) published the following simple relationship between depth and density:

TABLE 3.7 Values of n_i, a and b for Eq. (3.5)

Rock	n_i (%)	$a(\times 10^{-4})$	$b(\times 10^{-4})$	r^2
Intramicritic limestone	7.86	0.2	−13	0.93
Porous limestone (travertine)	3.43	0.4	−144	0.96
Crystallized limestone	0.66	0.1	−21	0.92
Fine grain (0.1–1 mm) marble	0.30	0.1	−5	0.95
Coarse grain (0.5–2 mm) marble	0.24	0.07	5	0.95

Notes: r^2 is the determination coefficient for the best fitting of the experimental data.
Data from Yavuz, H., Demirdag, S., Caran, S., 2010. Thermal effect on the physical properties of carbonate rocks. Int. J. Rock Mech. Min. Sci. 47, 94–103.

TABLE 3.8 Definition of Various Density Terms

Term	Definition	Remarks
Density (or bulk density)	$\rho = \dfrac{m}{V}$	Mass determined at natural water content
Dry density	$\rho_d = \dfrac{m_s}{V}$	Mass refers to solids only. All moistures dried out of the voids
Saturated density	$\rho_{sat} = \dfrac{m_{sat}}{V}$	Mass refers to solids plus water which fills completely the voids
Grain density (or solid density)	$\rho_s = \dfrac{m_s}{V_s}$	Both mass and volume refer to the grains (solids) only

Notes: $m = m_s + m_w$ and $V = V_s + V_v$ in which m is the bulk sample mass, m_s is the mass of the grains (solids), m_w is the mass of water in the voids, V is the bulk sample volume, V_s is the volume of the grains (solids), and V_v is the volume of the voids.
Modified from Stacey, T.R., van Veerden, W.L., Vogler, U.W., 1987. Properties of intact rock. In: Bell, F.G. (Ed.), Ground Engineer's Reference Book. Butterworths, London.

$$\rho = \rho_{z_0} + A \log \left(\frac{z}{z_0} \right) \tag{3.6}$$

where z is the depth; ρ_{z_0} is the density of the rock at depth z_0; and A is an empirical factor related to the compressibility of the rock.

Stegena (1964) recommended the following relationship between depth and density:

$$\begin{aligned}\rho &= \rho_{z_0} + \left(\rho_{z_m} - \rho_{z_0} \right)\left(1 - e^{-Bz} \right) \\ &= \rho_{z_m} - \left(\rho_{z_m} - \rho_{z_0} \right) e^{-Bz}\end{aligned} \tag{3.7}$$

where z is the depth; ρ_{z_0} is the density of the rock at depth z_0; ρ_{z_m} is the density of the rock at maximum depth z_m; and B is an empirical factor related to the compressibility of the rock. This relation has an asymptotic value ρ_{z_m} when z reaches infinity. Jelic (1984) derived the following relationship for sandstones and siltstones:

$$\rho = 2.72 - 1.244 e^{-0.846z} \tag{3.8}$$

in which ρ is in g/cm^3 and z is in km.

Since, as described earlier, the porosity tends to increase with higher temperature, the density of rocks tends to decrease with temperature (see Fig. 3.4).

3.3.3 Wave Velocity

The velocity of elastic waves in a rock can be determined in laboratory rock testing using one of the three methods: the high frequency ultrasonic pulse

TABLE 3.9 Typical Values of Density of Intact Rocks

Rock Type	Range of Density (kg/m³)	Mean Density (kg/m³)
Igneous rocks		
Granite	2516–2809	2667
Granodiorite	2668–2785	2716
Syenite	2630–2899	2757
Quartz diorite	2680–2960	2806
Diorite	2721–2960	2839
Norite	2720–3020	2984
Gabbro	2850–3120	2976
Diabase	2804–3110	2965
Peridotite	3152–3276	3234
Dunite	3204–3314	3277
Pyroxenite	3100–3318	3231
Anorthosite	2640–2920	2734
Sedimentary rocks		
Sandstone	2170–2700	–
Limestone	2370–2750	–
Dolomite	2750–2800	–
Chalk	2230	–
Marble	2750	–
Shale	2060–2660	–
Sand	1920-1930	–
Metamorphic rocks		
Gneiss	2590–3060	2703
Schist	2700–3030	2790
Slate	2720–2840	2810
Amphibolite	2790–3140	2990
Granulite	2630–3100	2830
Eclogite	3338–3452	3392

Note: The values listed in the table are for the bulk density determined at natural water content. Modified from Lama, R.D., Vutukuri, V.S., 1978. Handbook on Mechanical Properties of Rocks. Trans Tech Publications, Clausthal.

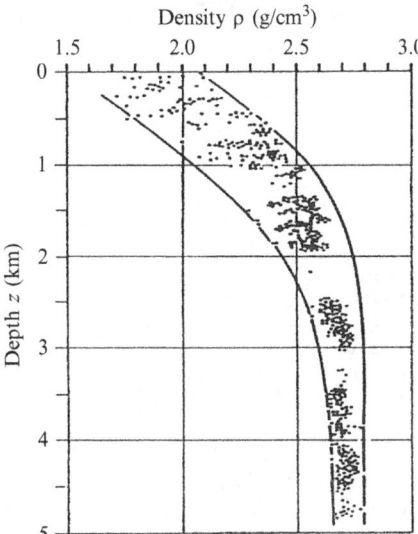

FIG. 3.3 Density ρ versus depth z for sedimentary rocks at the North German-Polish Basin. *(From Schön, J.H., 1996. Physical Properties of Rocks—Fundamentals and Principles of Petrophysics. Pergamon, Oxford.)*

technique, the low frequency ultrasonic pulse technique and the resonant method (ISRM, 1978a). Wave velocity is closely related to rock properties and has been used as one of the most important index properties. Fig. 3.5 shows the range of the P-wave velocity v_p and the S-wave velocity v_s for some of the commonly occurring rocks.

The wave velocity of rocks varies with pressure, temperature and saturation. Increasing pressure leads to closure of voids or microcracks and thus greater wave velocity of rocks. In general, the wave velocity versus pressure curve is composed of two segments, an exponential portion followed by a linear trend (Ji et al., 2007; Sun et al., 2012; Asef and Najibi, 2013; Najibi and Asef, 2014), which can be described by the following single equation proposed by Ji et al. (2007) (Fig. 3.6):

$$v = v_0 + Dp - B_0 e^{-kp} \tag{3.9}$$

where v is the wave velocity (v_p or v_s) at pressure p; v_0 is the projected wave velocity of a nonporous rock at zero pressure, which can be obtained by extrapolation of the linear velocity-pressure relationship at high pressure ($p > p_c$) to zero pressure, in which p_c is a pressure at which all voids and microcracks of the rock have been fully closed and above which the linear velocity-pressure relationship applies; D is a parameter which determines the slope of the linear velocity-pressure relationship; B_0 is the initial velocity drop

2.80
2.70
2.60
2.50
2.40
2.30
2.20
Bulk density (g/cm³)
FL
DT
BB
AW
MW
0
100
200
300
400
500
Temperature (°C)

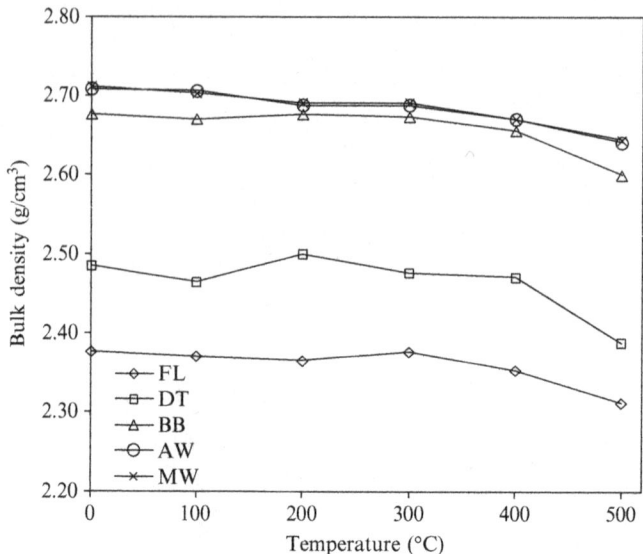

FIG. 3.4 Variation of bulk density with temperature for Finike Lymra limestone (FL), Denizli Travertine (DT), Burdur Beige limestone (BB), Afyon white marble (AW), and Mugla white marble (MW). *(Data from Yavuz, H., Demirdag, S., Caran, S., 2010. Thermal effect on the physical properties of carbonate rocks. Int. J. Rock Mech. Min. Sci. 47, 94–103.)*

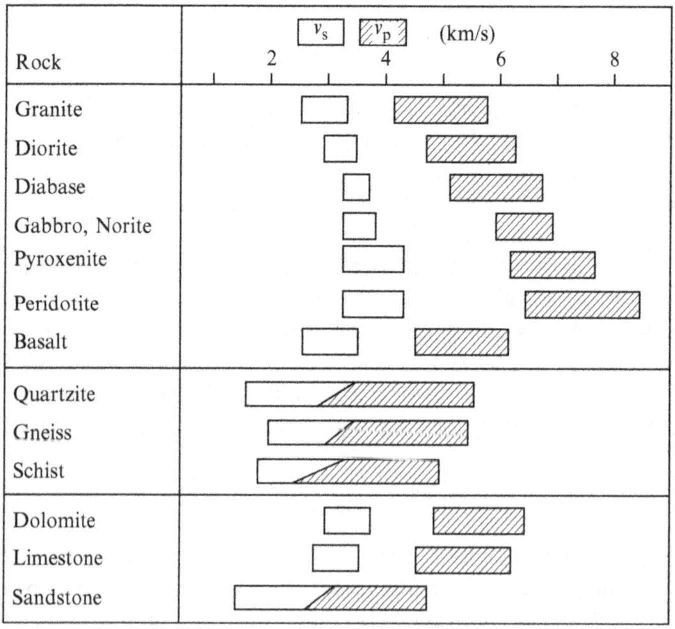

Rock
v_s
v_p
(km/s)
2
4
6
8
Granite
Diorite
Diabase
Gabbro, Norite
Pyroxenite
Peridotite
Basalt
Quartzite
Gneiss
Schist
Dolomite
Limestone
Sandstone

FIG. 3.5 Range of P-wave velocity v_p and S-wave velocity v_s of different rocks. *(Modified from Schön, J.H., 1996. Physical Properties of Rocks—Fundamentals and Principles of Petrophysics. Pergamon, Oxford.)*

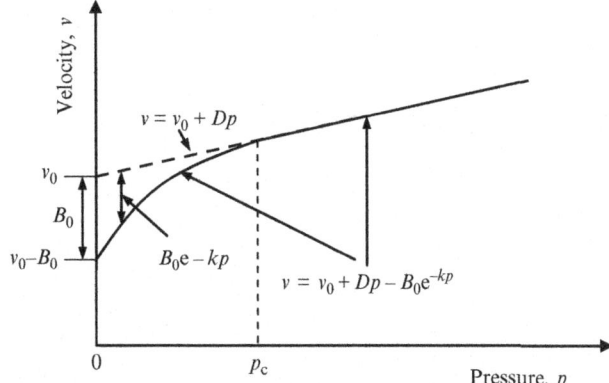

FIG. 3.6 Variation of wave velocity with pressure. *(Modified from Ji, S., Wang, Q., Marcotte, D., Salisbury, M.H., Xu, Z., 2007. P wave velocities, anisotropy and hysteresis in ultrahigh-pressure metamorphic rocks as a function of confining pressure. J. Geophys. Res. 112, B09204.)*

due to the presence of voids and microcracks at zero pressure or the maximum increase of velocity due to the closure of voids and microcracks, and thus $(v_0 - B_0)$ is the velocity of the rock containing voids and microcracks at zero pressure; and k is a decay constant that controls the shape of the nonlinear velocity-pressure relationship. Fig. 3.7 shows a comparison of measured v_p and v_s values with those predicted using Eq. (3.9) for three limestone rocks. The corresponding values of the coefficients in Eq. (3.9) for the three rocks are shown in Table 3.10.

Increasing temperature leads to creation of microcracks and thus the decrease of wave velocity of rocks (Toksoz et al., 1976; Timur, 1977; Kilic, 2006; Chaki et al., 2008; Yavuz et al., 2010). Fig. 3.8 shows the variation of P-wave velocity with temperature for different types of rocks.

The wave velocity of rocks is also affected by the degree of saturation. For many rocks, the P-wave velocity v_p tends to increase with higher degree of saturation (Wyllie et al., 1956; Ramana and Venkatanarayana, 1973; Wang et al., 1975; Lama and Vutukuri, 1978; Kahraman, 2007), although a decreasing tendency in v_p with degree of saturation is also possible for rocks with high clay content and porosity (Karakul and Ulusay, 2013).

3.3.4 Point Load Index

The point load index has often been reported as an indirect measure of the rock strength. The point load test has been widely used in practice because of its

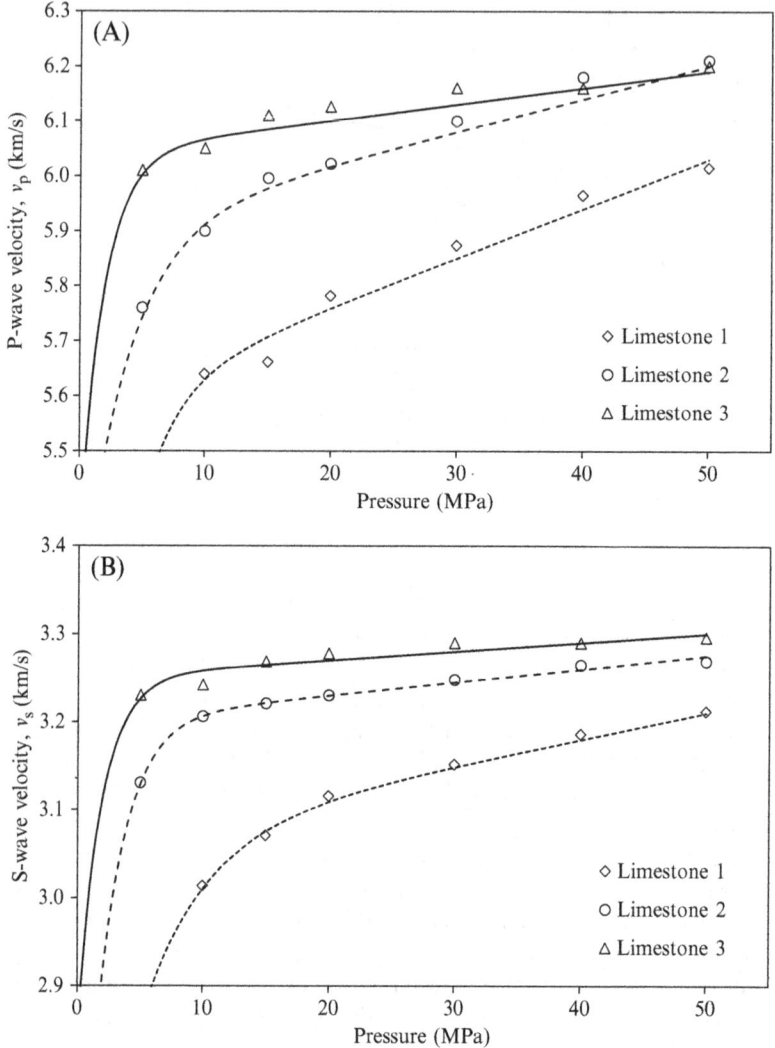

FIG. 3.7 Comparison of measured (points) and predicted (lines) wave velocity values: (A) P-wave velocity v_p; and (B) S-wave velocity v_s. *(Data from Asef, M.R., Najibi, A.R., 2013. The effect of confining pressure on elastic wave velocities and dynamic to static Young's modulus ratio. Geophysics 78 (3), D135–D142.)*

testing ease, simplicity of specimen preparation and field applications (ISRM, 1985; ASTM, 2008a).

For a point load test, a compressive load is applied through two conical platens, which causes the rock to break in tension between these two points. If the breaking load is P, the point load index, I_s, can then be determined by

TABLE 3.10 Values of Coefficients in Eq. (3.9) for Three Limestone Rocks

Rock	v_p				v_s			
	$v_0(km/s)$	D (10^{-4} km/s/MPa)	$B_0(km/s)$	k (10^{-2}/MPa)	$v_0(km/s)$	D (10^{-4} km/s/MPa)	$B_0(km/s)$	k (10^{-2}/MPa)
Limestone 1	5.58	0.009	1.23	0.34	3.06	0.003	0.59	0.20
Limestone 1	5.90	0.006	0.72	0.27	3.20	0.0015	0.70	0.44
Limestone 1	6.04	0.003	0.70	0.52	3.25	0.001	0.40	0.53

Data from Asef, M.R., Najibi, A.R., 2013. The effect of confining pressure on elastic wave velocities and dynamic to static Young's modulus ratio. Geophysics 78 (3), D135–D142.

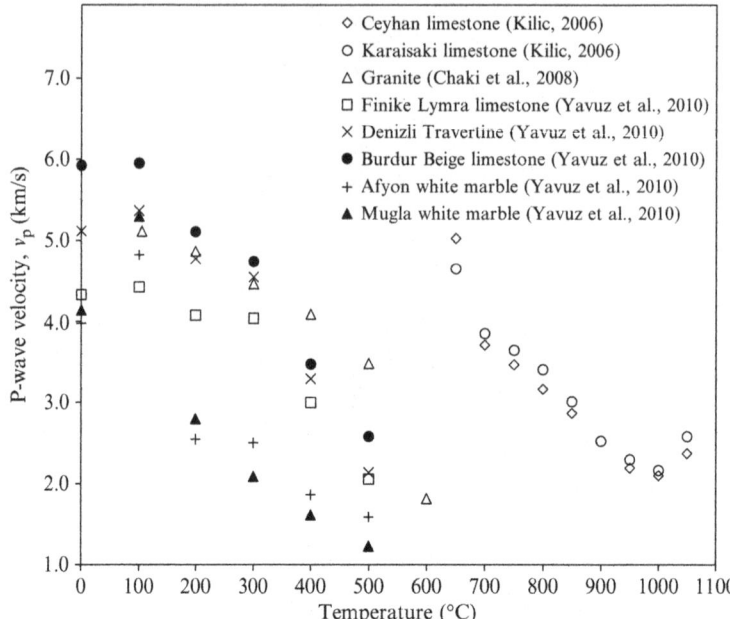

FIG. 3.8 Variation of P-wave velocity v_p with temperature. *(Data from Kilic, Ö., 2006. The influence of high temperatures on limestone P-wave velocity and Schmidt hammer strength. Int. J. Rock Mech. Min. Sci. 43, 980–986; Chaki, S., Takarli, M., Agbodjan, W.P., 2008. Influence of thermal damage on physical properties of a granite rock: porosity, permeability and ultrasonic wave evolutions. Constr. Build. Mater. 22, 1456–1461; Yavuz, H., Demirdag, S., Caran, S., 2010. Thermal effect on the physical properties of carbonate rocks. Int. J. Rock Mech. Min. Sci. 47, 94–103.)*

$$I_s = \frac{P}{D^2} \qquad (3.10)$$

where D is the diameter of the specimen if the load is applied in the diametric direction of a core. In other cases, $D = 2\sqrt{A/\pi}$, where A is the minimum cross-sectional area of the specimen for a plane through the platen contact points.

The size of rock specimen affects the measured value of I_s, greater D leading to higher I_s. To consider the size effect, it has been common to convert the measured I_s to that corresponding to $D = 50$ mm:

$$I_{s(50)} = I_s k_{PLT} \qquad (3.11)$$

where k_{PLT} is the size correction factor, which can be determined by (ISRM, 1985; ASTM, 2008a):

$$k_{PLT} = \left(\frac{D}{50}\right)^{0.45} \qquad (3.12)$$

The shape of the specimen also affects the measured value of I_s. As shown in Fig. 3.9, the measured $I_{s(50)}$ by applying load in the diametric direction of a core

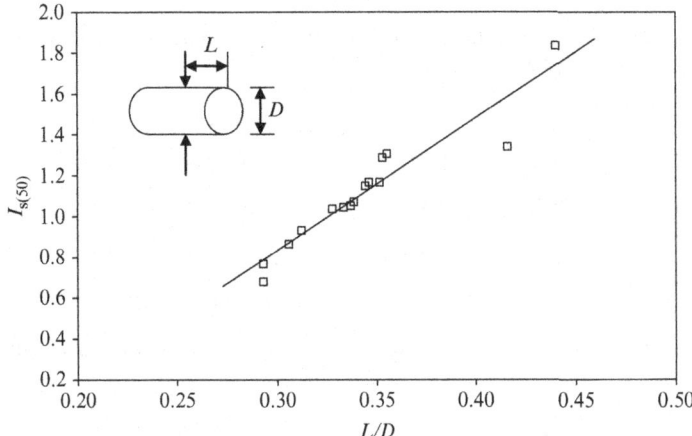

FIG. 3.9 Variation of measured $I_{s(50)}$ using diametric loading with geometric ratio L/D for air-dried gypsum specimens ($D = 56.7$ mm). *(Data from Heidari, M., Khanlari, G.R., Kaveh, M.T., Kargarian, S., 2012. Predicting the uniaxial compressive and tensile strengths of gypsum rock by point load testing. Rock Mech. Rock Eng. 45, 265–273.)*

increases with greater L/D ratio. ASTM (2008a) requires that $L/D > 0.5$ for test specimens.

One of the columns in Table 3.5 lists the typical range of the point load index for different rocks.

3.3.5 Schmidt Hammer Rebound Number

The Schmidt rebound hammer has been used for testing the quality of concretes and rocks. Schmidt hammers are designed in different levels of impact energy, but the types of L and N are commonly adopted for rock property determinations. The L-type has an impact energy of 0.735 N m which is only one third that of the N-type. There are no clear guidelines on the choice of the hammer type. ISRM (1978b) only endorses the L-type hammer while ASTM (2001) does not specify a hammer type. Both types of hammers have been used in practice for estimating the strength of rock. Studies have shown that there are good correlations between the L- and N-type Schmidt hammer rebound numbers, $R_{n(L)}$ and $R_{n(N)}$. Table 3.11 shows some of them.

The size of rock specimen affects the measured Schmidt hammer rebound number. Fig. 3.10 shows the measured $R_{n(L)}$ at various specimen sizes for different rocks by Demirdag et al. (2009) following the ISRM (1978b) test procedure. In general, $R_{n(L)}$ increases with specimen size up to about 110 mm and then stays about the same (Demirdag et al., 2009).

Table 3.12 lists the typical L-type Schmidt hammer rebound numbers $R_{n(L)}$ for some of the commonly occurring rocks.

TABLE 3.11 Correlations Between L- and N-Type Schmidt Hammer Rebound Numbers, $R_{n(L)}$ and $R_{n(N)}$

Relation	Reference
$R_{n(N)} = 1.395 R_{n(L)} - 1.646 \ (r^2 = 0.842)$	Ayday and Goktan (1992)
$R_{n(N)} = 1.300 R_{n(L)} - 1.304 \ (r^2 = 0.879)$	Ayday and Goktan (1992)
$R_{n(N)} = 1.477 R_{n(L)} - 0.894 \ (r^2 = 0.882)$	Ayday and Goktan (1992)
$R_{n(N)} = 1.065 R_{n(L)} + 6.367 \ (r^2 = 0.980)$	Aydin and Basu (2005)
$R_{n(N)} = 1.064 R_{n(L)} + 2.569 \ (r^2 = 0.850)$	del Potro and Hürlimann (2009)

Note: r^2 is the determination coefficient.

3.3.6 Slake Durability Index

The slake durability index is used to describe the resistance of shales and similar weak rocks against breakdown or weathering with time. Franklin and Chandra (1972) developed the procedure for conducting the slake durability test, which was later recommended by ISRM (1979) and standardized by ASTM (2008b). The slake durability test involves (1) taking 10 representative, intact, roughly equidimensional rock fragments weighing 40–60 g each, putting the fragments into a drum of specific dimensions, and determining the oven-dried mass of the drum and the specimen; (2) mounting the drum in a trough, filling the trough with distilled water at room temperature to 20 mm below the drum axis, and rotating the drum at a constant rate of 20 rpm for 10 min; (3) removing the drum from the trough, drying the drum and the specimen, and obtaining the oven-dried mass; and (4) repeating (2) and (3) to obtain a final oven-dried mass for the second cycle. Based on the test, the slake durability index (second cycle) can be determined as follows (ASTM, 2008b):

$$I_d(2) = \frac{W_F - C}{B - C} \times 100 \ (\%) \tag{3.13}$$

where $I_d(2)$ is the slake durability index (second cycle); B is the mass of the drum plus oven-dried specimen before the first cycle; W_F is the mass of the drum plus oven-dried specimen retained after the second cycle; and C is the mass of the drum.

The slake durability index (second cycle) can be classified according to Table 3.13 (Franklin and Chandra, 1972).

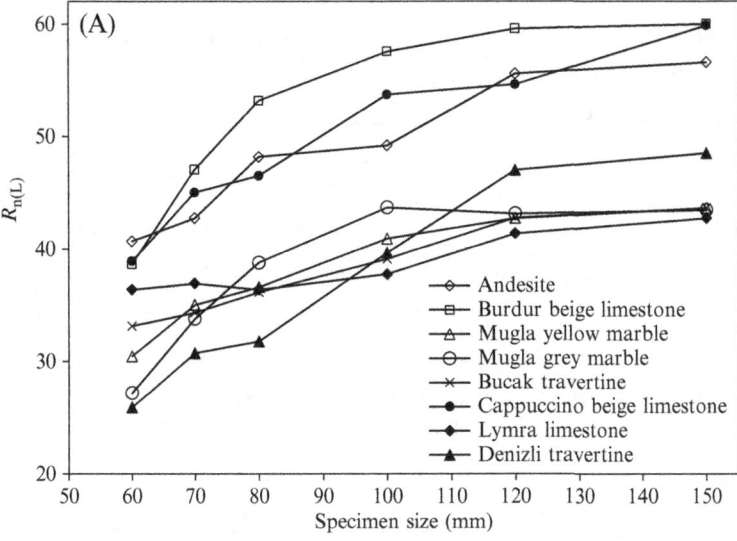

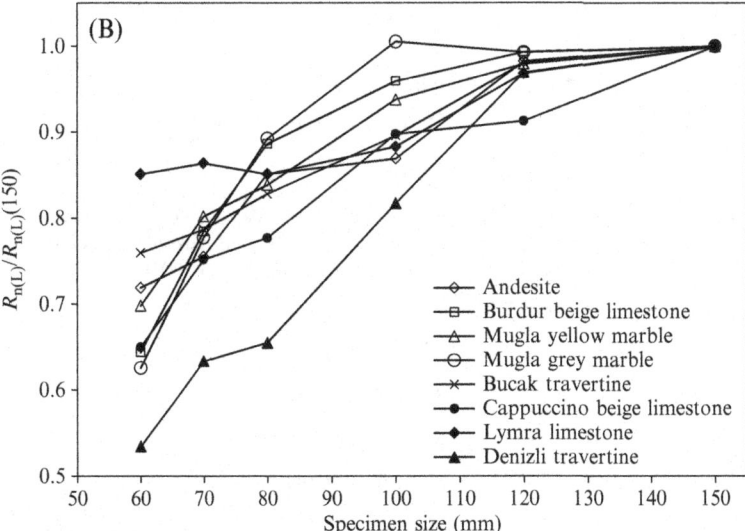

FIG. 3.10 Effect of specimen size on measured Schmidt hammer rebound number: (A) $R_{n(L)}$ versus specimen size; and (B) normalized $R_{n(L)}/R_{n(L)}(150)$ versus specimen size, where $R_{n(L)}(150)$ is the $R_{n(L)}$ at a specimen size of 150 mm. *(Data from Demirdag, S., Yavuz, H., Altindag, R., 2009. The effect of sample size on Schmidt rebound hardness value of rocks. Int. J. Rock Mech. Min. Sci., 46, 725–730.)*

TABLE 3.12 Typical L-Type Schmidt Hammer Rebound Numbers $R_{n(L)}$ for Different Rocks

Rock	$R_{n(L)}$	Reference
Andesite	11–56	Ayday and Goktan (1992); Dincer et al. (2004); Demirdag et al. (2009); Kayabali and Selcuk (2010)
Basalt	35–58	Stacey et al. (1987); Dincer et al. (2004)
Breccia	40	Buyuksagis and Goktan (2007)
Chalk	10–29	Bell et al. (1999)
Diabase	36–59	Stacey et al. (1987); Ayday and Goktan (1992)
Dolomite	40–60	Stacey et al. (1987); Sachpazis (1990)
Gabbro	49–63	Xu et al. (1990); Buyuksagis and Goktan (2007)
Gneiss	48	Stacey et al. (1987)
Granite	45–58	Stacey et al. (1987); Ayday and Goktan (1992); Buyuksagis and Goktan (2007)
Gypsum	13–44	Yilmaz and Sendir (2002); Kayabali and Selcuk (2010)
Ignimbrite	14–48	Kayabali and Selcuk (2010)
Limestone	16–61	Stacey et al. (1987); Demirdag et al. (2009); Buyuksagis and Goktan (2007); Kayabali and Selcuk (2010)
Marble	31–53	Stacey et al. (1987); Ayday and Goktan (1992); Demirdag et al. (2009); Buyuksagis and Goktan (2007); Kayabali and Selcuk (2010)
Marl	18–39	Ayday and Goktan (1992)
Mudstone	15	Xu et al. (1990)
Peridodite	45	Ayday and Goktan (1992)
Prasinite	41	Xu et al. (1990)
Quartzite	39	Stacey et al. (1987)
Rock salt	23	Stacey et al. (1987)
Sandstone	11–47	Stacey et al. (1987); Kayabali and Selcuk (2010)
Schist	29–41	Stacey et al. (1987); Xu et al. (1990)
Serpentinite	45	Xu et al. (1990)
Siltstone	47	Stacey et al. (1987)
Travertine	30–46	Buyuksagis and Goktan (2007)
Tuff	11–40	Stacey et al. (1987); Ayday and Goktan (1992); Dincer et al. (2004); Kayabali and Selcuk (2010)

TABLE 3.13 Slake Durability Index Classification

$I_d(2)$ (%)	Durability Classification
0–25	Very low
25–50	Low
50–75	Medium
75–90	High
90–95	Very high
95–100	Extremely high

Based on Franklin, J.A., Chandra, A., 1972. The slake durability test. Int. J. Rock Mech. Min. Sci. 9, 325–341.

3.3.7 Cerchar Abrasivity Index (CAI)

Although a number of abrasivity tests have been proposed, the most widely accepted is the Cerchar scratch test which is standardized by ASTM (2010) and suggested by ISRM (Alber et al., 2014). In this test, the tip of a steel stylus having a Rockwell Hardness of HRC 55, a diameter at least 6 mm and a conical angle of 90 degrees is slowly drawn 10 mm across the rock surface under a normal, static force of 70 N. The wear surface of the stylus tip is then measured under a microscope to an accuracy of 0.01 mm. The CAI is a dimensionless unit value and is calculated by multiplying the wear surface stated in units of 0.01 mm by 10. In general, a minimum of five test replications must be made and the CAI is taken as the mean value.

The CAI test should be conducted on a rough, freshly broken rock surface. If a smooth, saw-cut rock surface is used, the obtained CAI_s tends to be lower and can be corrected using the following equation (Käsling and Thuro, 2010; Alber et al., 2014):

$$CAI = 1.14CAI_s \qquad (3.14)$$

Based on the measured CAI, the abrasivity of rock can be classified according to Table 3.14.

Fig. 3.11 shows the typical values of CAI for various types of rocks based on data from Plinninger et al. (2003), Maloney (2010), and Deliormanlı (2012).

3.3.8 Needle Penetration Index (NPI)

The needle penetration test was developed by the Japan Society of Civil Engineers (JSCE, 1980, 1991) and later standardized by the Japan Geotechnical Society (JGS, 2012). It is now a suggested method by the ISRM (Ulusay et al., 2014).

TABLE 3.14 Abrasivity Classification Based on CAI

Mean CAI	Abrasivity Classification
0.1–0.4	Extremely low
0.5–0.9	Very low
1.0–1.9	Low
2.0–2.9	Medium
3.0–3.9	High
4.0–4.9	Very high
≥ 0.5	Extremely high

Based on Alber, M., Yaralı, O., Dahl, F., Bruland, A., Käsling, H., Michalakopoulos, T. N., Cardu, M., Hagan, P., Aydın, H., Özarslan, A., 2014. ISRM suggested method for determining the abrasivity of rock by the CERCHAR abrasivity test. Rock Mech. Rock Eng. 47, 261–266.

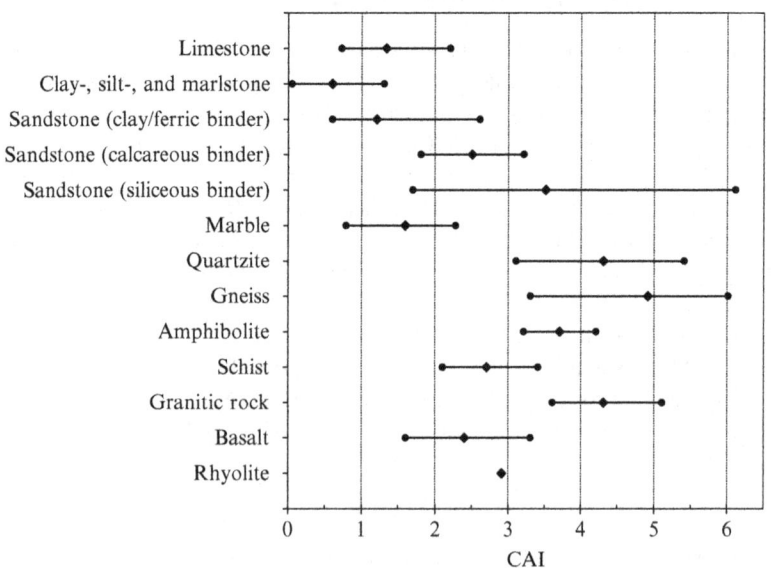

FIG. 3.11 Typical values of Cerchar abrasivity index (CAI) for various types of rocks based on data from Plinninger, R., Käsling, H., Thuro, K., Spaun, G., 2003. Testing conditions and geomechanical properties influencing the CERCHAR abrasiveness index (CAI) value. Int. J. Rock Mech. Min. Sci. 40, 259–263; Maloney, S., 2010. CERCHAR abrasivity testing of argillaceous limestone of the Cobourg formation. Technical Report, MIRARCO/Geomechanics Research Centre, Laurentian University; Deliormanlı, A.H., 2012. Cerchar abrasivity index (CAI) and its relation to strength and abrasion test methods for marble stones. Constr. Build. Mater. 30, 16–21.

The needle penetration test can be performed in the field on rock exposures or in laboratory on rock specimens. To obtain realistic results, it is generally recommended that the needle penetration test be used for rocks with UCS lower than 20 MPa. To conduct a needle penetration test, the needle is pushed into the rock until 100 N is reached and the corresponding penetration depth is measured and then the needle is slowly pulled out. For soft and saturated rocks, it is possible that the maximum penetration depth (10 mm) is reached at a penetration force smaller than 100 N. In this case, the test stops at this penetration depth, the penetration force is read from the load scale and the needle is then slowly pulled out. The test should be repeated between three and five times. Based on the test, the NPI can be calculated from the following equations:

$$\text{NPI} = 100/D \quad (\text{for } F = 100\,\text{N and } D = 10\,\text{mm}) \tag{3.15a}$$

$$\text{NPI} = F/10 \quad (\text{for } D = 10\,\text{mm and } F = 100\,\text{N}) \tag{3.15b}$$

where F is the applied penetration force in N and D is the penetration depth in mm. The unit of NPI is N/mm.

The value of NPI decreases as the degree of saturation increases, as clearly shown in Fig. 3.12 for various tuffs sampled from Cappadocia in Turkey and

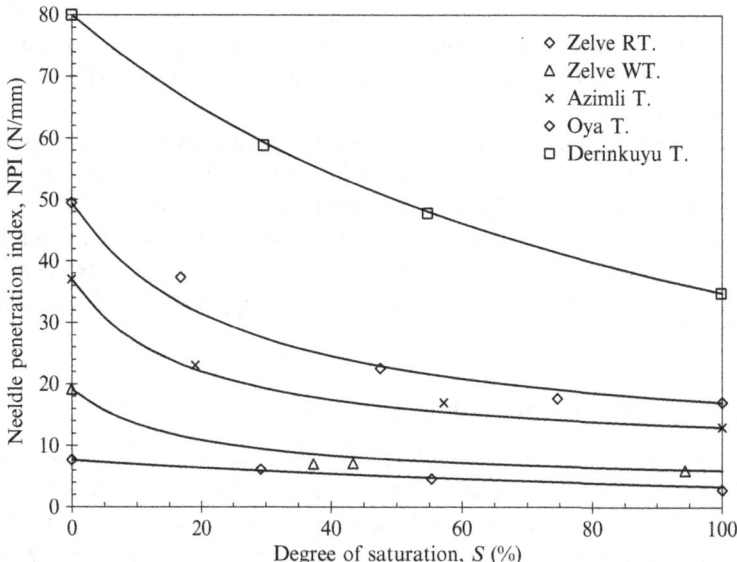

FIG. 3.12 Variation of needle penetration index with degree of saturation—measured data in points and prediction from Eq. (3.16) in *solid lines. (Data from Aydan, Ö., Sato, A., Yagi, M., 2014. The inference of geo-mechanical properties of soft rocks and their degradation from needle penetration tests. Rock Mech. Rock Eng. 47, 1867–1890.)*

Oya in Japan (Aydan, 2012; Aydan et al., 2014). The variation of NPI with the degree of saturation, S, can be described by the following relation:

$$\frac{\text{NPI}_w}{\text{NPI}_d} = \alpha_0 - (\alpha_0 - \alpha_{100})\frac{S}{S + \beta(100 - S)} \qquad (3.16)$$

where NPI_w and NPI_d are the NPI value at degree of saturation S and dry condition, respectively; α_0 and α_{100} are the value of $\text{NPI}_w/\text{NPI}_d$ at degree of saturation S equal to 0% and 100%, respectively; and β is an empirical coefficient to be determined from experiments. The values of β range between 0.15 and 0.50 for Cappadocia and Oya tuffs. Fig. 3.12 shows a good agreement between the prediction from Eq. (3.16) and the measured data.

3.3.9   Shore Sclerscope Hardness

The Shore Sclerscope hardness was originally designed for use on metals. It measures the relative rebound of a diamond-tipped hammer that drops freely from a fixed height onto the surface of a specimen. ISRM (1978b) details the method for Shore Sclerscope hardness testing of rocks using model C-2. The specimen surface should be smooth to within 0.02 mm, and preferably the volume is at least 40 cm^3 and 50 mm thick. Each test should be on a fresh site on the prepared rock surface as the hammer makes a small indentation on impact. 50 readings are recommended, with the highest five and lowest five being discarded before calculating the average rebound height, H, the Shore Sclerscope hardness.

Altındağ (2001) investigated the effects of specimen volume, temperature and water content on H. The results indicate that H increases with specimen volume up to 80 cm^3 and then stays about the same. So Altındağ (2001) suggested a minimum specimen volume of 80 cm^3 in order to obtain reliable H. The results also show that H decreases with higher temperature (Fig. 3.13) and the H of a saturated specimen is lower than that of a dry specimen (Table 3.15).

3.3.10   Cone Indenter Number

The cone indenter test was developed by the National Coal Board (now British Coal Corporation) as a method for estimating the strength of rocks which may be excavated by roadheading equipment (Szlavin, 1974; Brook, 1993). In a cone indenter test, a sharp tungsten carbide conical point is pressed into the rock under a standard force of 40 N, the force being measured by the deflection of a steel strip, and the total travel of the point, enabling the penetration to be calculated, is measured by a micrometer. Small pieces of rock, up to $25 \times 25 \times 6$ mm, are used, either natural chippings in the field, or saw cut thin pieces. This is a modification of the metal hardness tests such as Brinel and Rockwell hardness, such that easily measured penetration occurs and a very flat surface is not essential. The penetration of the cone into the rock specimen, P_s in

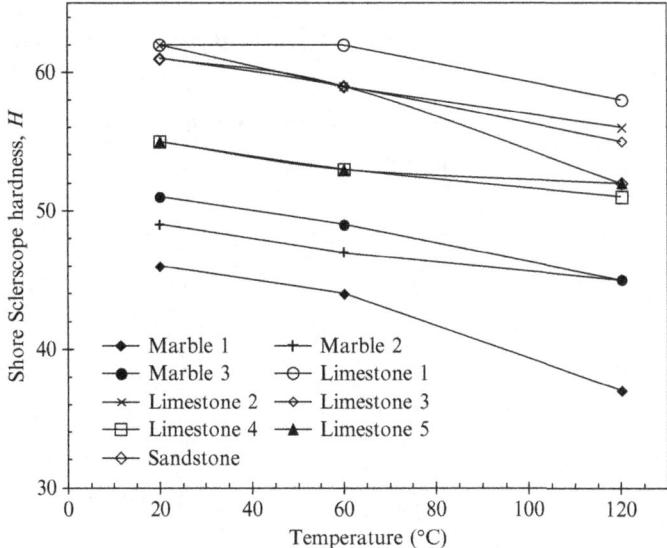

FIG. 3.13 Effect of temperature on Shore Sclerscope hardness. *(Data from Altındağ, R., 2001. The effects of specimen volume, temperature and water content on Shore hardness. 17th International Mining Congress and Exhibition of Turkey-İMCET 2001, pp. 411–415.)*

mm, is compared to the standard spring deflection of 0.635 mm, to give the 40 N force and the standard cone indenter number:

$$I_s = \frac{0.635}{P_s} \tag{3.17}$$

This procedure is repeated on a fresh piece or part of rock and the average of 10 measurements are used to calculate the I_s value.

For some very strong rocks, the penetration of the 40 N force indenter is very small, say less than 0.12 mm. For such rocks, a modified method is suggested in which the applied force is raised to 110 N with spring deflection 1.27 mm. The modified cone indenter number with the penetration P_m is then:

$$I_m = \frac{1.27}{P_m} \tag{3.18}$$

Some very weak rocks may be broken by the point when a 40 N force is applied. For such rocks, a modified method is suggested in which the applied force is reduced to 12 N with spring deflection 0.23 mm. The cone indenter number with the penetration P_w is then:

$$I_w = \frac{0.23}{P_w} \tag{3.19}$$

TABLE 3.15 Comparison of Shore Sclerscope Hardness at Dry and Saturated Conditions

| Rock | Shore Sclerscope Hardness, H | |
	Dry	Saturated
Marble 1	46	42
Marble 2	49	45
Marble 3	51	45
Limestone 1	62	58
Limestone 2	62	59
Limestone 3	61	60
Limestone 4	55	53
Limestone 5	55	53
Sandstone	61	58

Data from Altındağ, R., 2001. The effects of specimen volume, temperature and water content on Shore hardness. 17th International Mining Congress and Exhibition of Turkey-İMCET 2001, pp. 411–415.

Table 3.16 lists the values of cone indenter numbers for several types of rocks.

3.3.11 Correlations Between Different Index Properties

The index properties of rocks are closely related to each other and this section presents the correlations between some of them.

3.3.11.1 Density and porosity

Density is closely related to porosity and their relationship can be expressed by the general function:

$$\rho = (1-n)\rho_s + n[S\rho_f + (1-S)\rho] \tag{3.20}$$

where ρ_s is the grain (or matrix) density; ρ_f and ρ_g are ,respectively, the density of the fluid and gas in the voids; and S is the saturation (or degree of saturation) of the voids which is defined by:

$$S = \frac{V_f}{V_v} \times 100\,(\%) \tag{3.21}$$

where V_f is the volume of the fluid in the voids and V_v is the total volume of the voids.

TABLE 3.16 Values of Cone Indenter Numbers for Several Types of Rocks

Rock	I_s	I_m	Reference
Sandstone (Belmont, field)		3.5–16.9	Cook (2001)
Sandstone (Belmont, block)		4.7–8.3	Cook (2001)
Sandstone (Aviemore, field)		2.3–8.2	Cook (2001)
Sandstone (Aviemore, block)		7.6–8.9	Cook (2001)
Sandstone (Taotaoroa, block)		5.0–8.1	Cook (2001)
Mudstone (Belmont, field)		1.3–4.3	Cook (2001)
Mudstone (Belmont, block)		3.3–4.2	Cook (2001)
Clayey limestone 1	2.79		Keles (2005)
Clayey limestone 2	2.52		Keles (2005)
Calcereous tuff 1	1.94		Keles (2005)
Calcereous tuff 2	2.48		Keles (2005)
Siltstone 1	1.82		Keles (2005)
Siltstone 2	1.94		Keles (2005)
Claystone 1	1.49		Keles (2005)
Claystone 2	1.22		Keles (2005)
Marble	1.31		Keles (2005)
Coal (lower seam)	1.13		Keles (2005)
Coal (upper seam)	0.93		Keles (2005)

For a saturated rock, $S = 1$ and Eq. (3.20) can be rewritten as:

$$\rho = \rho_s - n\left(\rho_s - \rho_f\right) \tag{3.22}$$

Since the density of the fluid in the voids such as water is smaller than that of the grains, the density of rocks decreases with the porosity.

Tuğrul and Zarif (1999) derived the following empirical relation between dry density ρ_d and porosity n for granitic rocks from Turkey:

$$\rho_d = 2695 - 25.48n \quad \left(r^2 = 0.74\right) \tag{3.23}$$

Tuğrul (2004) derived an empirical relation between dry density ρ_d and porosity n for sandstone, limestone, basalt and granodiorite from Turkey:

$$\rho_d = 2765 - 33.64n \quad \left(r^2 = 0.94\right) \tag{3.24}$$

In both Eqs. (3.23), (3.24), ρ_d is in the unit of kg/m^3 and n is in %.

3.3.11.2 P-wave velocity and porosity

The wave velocity of a rock is influenced by the matrix properties, the porosity and the properties of the fluid in the voids. Wyllie et al. (1956) derived the following general relation for saturated porous rock:

$$\frac{1}{v_p} = \frac{1-n}{v_{ps}} + \frac{n}{v_{pf}}$$ (3.25)

where v_p, v_{ps}, and v_{pf} are the P-wave velocity, respectively, of the rock, the grains and the fluid in the voids.

Raymer et al. (1980) derived the following empirical relation for consolidated rocks:

$$v_p = (1-n)^2 v_{ps} + n v_{pf}$$ (3.26)

where v_p, v_{ps}, and v_{pf} are the same as defined earlier.

Table 3.17 lists some of the empirical correlations between P-wave velocity v_p and porosity n based fitting analyses of test data.

Fig. 3.14 shows the variation of P-wave velocity with porosity for water saturated sandstone from Rotliegenes, Northern Germany.

3.3.11.3 P-wave velocity and density

The P-wave velocity v_p increases with the density ρ of rocks. Fig. 3.15 shows the variation of v_p with ρ for igneous and metamorphic rocks. Many researchers have developed closed-form empirical correlations between wave velocity and density. Table 3.18 lists some of them.

TABLE 3.17 Empirical Correlations Between P-Wave Velocity v_p and Porosity n

Correlation	Rock Type	Reference
$v_p = 6.32 n^{-0.016}$ $(r^2 = 0.76)$	Vesicular basalt	Al-Harthi et al. (1999)
$v_p = 6.52 - 0.36n$ $(r^2 = 0.66)$	Granitic rocks	Tuğrul and Zarif (1999)
$v_p = 4.08 n^{-0.42}$ $(r^2 = 0.79)$	Granites	Sousa et al. (2005)
$v_p = 6.21 - 0.21n$ $(r^2 = 0.88)$	Limestone, travertine	Kahraman and Yeken (2008)
$v_p = 4.32 - 0.021n$ $(r^2 = 0.89)$	Calcarenite rocks (dry)	Rahmouni et al. (2013)
$v_p = 4.40 - 0.022n$ $(r^2 = 0.94)$	Calcarenite rocks (saturated)	Rahmouni et al. (2013)

Notes: v_p is in the unit of km/s and n is in %; and r^2 is the determination coefficient.

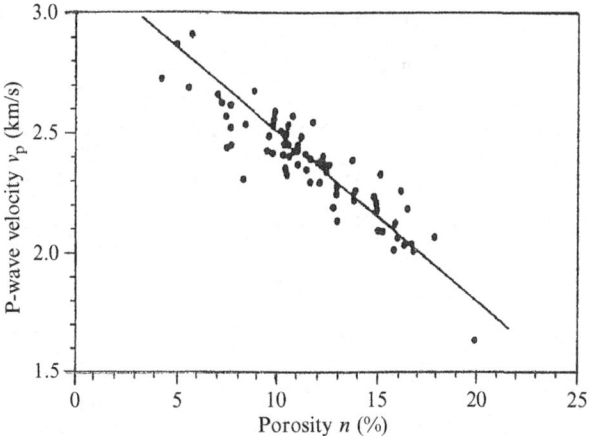

FIG. 3.14 Variation of P-wave velocity with porosity for water saturated sandstone from Rotlie-genes, Northern Germany. *(From Schön, J.H., 1996. Physical Properties of Rocks—Fundamentals and Principles of Petrophysics. Pergamon, Oxford.)*

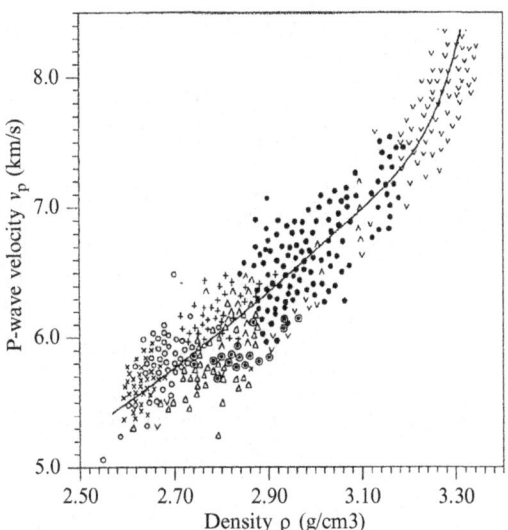

FIG. 3.15 Variation of P-wave velocity v_p with density ρ for igneous and metamorphic rocks. *(From Schön, J.H., 1996. Physical Properties of Rocks—Fundamentals and Principles of Petrophysics. Pergamon, Oxford.)*

TABLE 3.18 Correlations Between P-Wave Velocity v_p and Density ρ

Correlation	Rock Type	Reference
$v_p = 2.76\rho - 0.98$	Igneous rocks	Birch (1961)
$v_p = 2.33 + 0.08\rho^{3.63}$	Basalts	Christensen and Salisbury (1975)
$v_p = 2.67\rho - 1.08$	Igneous rocks	Volarovich and Bajuk (1977)
$v_p = 3.10\rho - 2.98$	Plutonic rocks: granite, diorite, gabbro	Marle (1978); Kopf (1977, 1980)
$v_p = 2.30\rho - 0.91$	Volcanic rocks: porphyrite, keratophyrite, diabase, basalt	Marle (1978); Kopf (1977, 1980)
$v_p = 3.66\rho - 4.46$	Mudstone (Type I)	Gaviglio (1989)
$v_p = 3.66\rho - 4.80$	Mudstone (Type III)	Gaviglio (1989)
$v_p = 3.66\rho - 4.87$	Mudstone (Type IV)	Gaviglio (1989)
$v_p = 3.66\rho - 4.11$	Wackestone (Type V)	Gaviglio (1989)
$v_p = 2.61\rho - 1.0 \pm 0.4$	Mantle rocks	Henkel et al. (1990)
$v_p = 5.00\rho - 8.65 \ (r^2 = 0.55)$	Crystalline rocks	Starzec (1999)
$v_p = 4.32\rho - 7.51 \ (r^2 = 0.81)$	Carbonate rocks	Yasar and Erdoğan (2004b)
$v_p = 4.69\rho - 5.90 \ (r^2 = 0.82)$	Limestone, travertine	Kahraman and Yeken (2008)
$v_p = 1.25\rho + 1.61 \ (r^2 = 0.85)$	Calcarenite rocks (dry)	Rahmouni et al. (2013)
$v_p = 2.99\rho - 2.15 \ (r^2 = 0.87)$	Calcarenite rocks (saturated)	Rahmouni et al. (2013)

Notes: v_p is in the unit of km/s and ρ is in the unit of g/cm^3; and r^2 is the determination coefficient.

3.3.11.4   P-wave velocity and point load index

Fig. 3.16 shows the variation of point load index with the P-wave velocity for fresh and weathered crystalline rocks. In general, the point load index increases as the P-wave velocity increases.

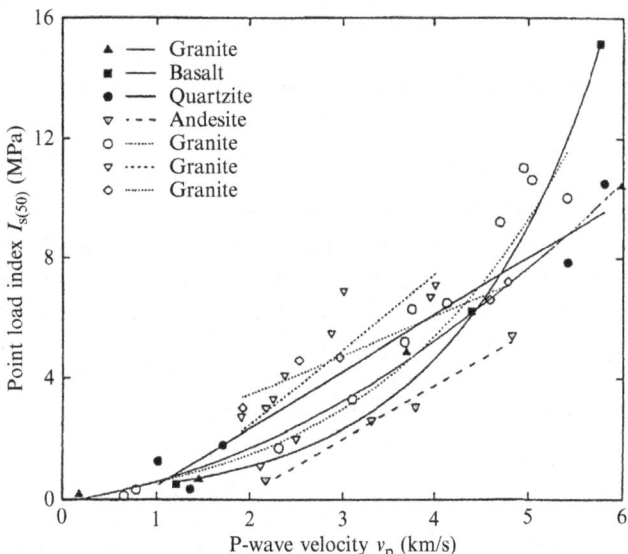

FIG. 3.16 Variation of point load index with P-wave velocity for fresh and weathered crystalline rock. *(From Gupta, A.S., Rao, K.S., 1998. Index properties of weathered rocks: interrelationships and applicability. Bull. Eng. Geol. Environ. 57, 161–172.)*

3.3.11.5 P-wave velocity and Schmidt hammer rebound number

Kahraman (2001) derived the following empirical correlation between P-wave velocity v_p and the N-type Schmidt hammer rebound number $R_{n(N)}$:

$$v_p = 0.11R_{n(N)} - 4.41 \quad \left(r^2 = 0.69\right) \tag{3.27}$$

where v_p is in the unit of km/s and r^2 is the determination coefficient.

3.3.11.6 P-wave velocity and NPI

Based on the test data for tuff, sandstone, soapstone, pumice limestone and lignite measures (lignite, mudstone, siltstone, marl, loam), Aydan et al. (2014) derived the following empirical correlation between P-wave velocity v_p and NPI:

$$v_p = 0.3NPI^{0.5} + 0.33 \quad \left(r^2 = 0.58\right) \tag{3.28}$$

where v_p and NPI are in the unit of km/s and N/mm, respectively and r^2 is the determination coefficient.

3.3.11.7 S-wave velocity and NPI

Using the test data for tuff, sandstone, soapstone, pumice limestone and lignite measures (lignite, mudstone, siltstone, marl, loam), Aydan et al. (2014) also derived the following empirical correlation between S-wave velocity v_p and NPI:

$$v_s = 0.18 \text{NPI}^{0.5} + 0.10 \quad \left(r^2 = 0.56\right) \tag{3.29}$$

where v_s and NPI are in the unit of km/s and N/mm, respectively and r^2 is the determination coefficient.

3.3.11.8 Point load index and porosity

Fig. 3.17 shows the variation of point load index with porosity for fresh and weathered crystalline rocks. As the porosity increases, the point load index decreases. The relationship between point load index and porosity can be generalized as being negatively exponential (Gupta and Rao, 1998). The following is the empirical correlation between point load index $I_{s(50)}$ and porosity n derived by Palchik and Hatzor (2004) for porous chalks:

$$I_{s(50)} = 7.74 e^{-0.039n} \quad \left(r^2 = 0.84\right) \tag{3.30}$$

where $I_{s(50)}$ is in the unit of MPa and n is in % and r^2 is the determination coefficient.

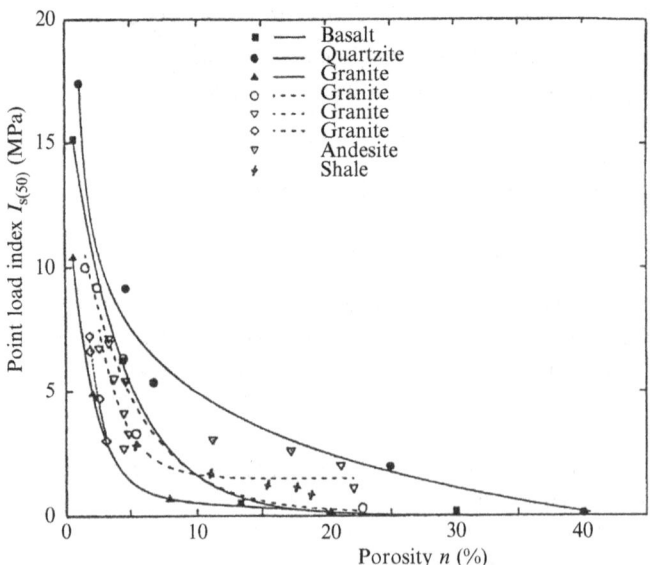

FIG. 3.17 Variation of point load index with porosity for fresh and weathered crystalline rock. *(From Gupta, A.S., Rao, K.S., 1998. Index properties of weathered rocks: inter-relationships and applicability. Bull. Eng. Geol. Environ. 57, 161–172.)*

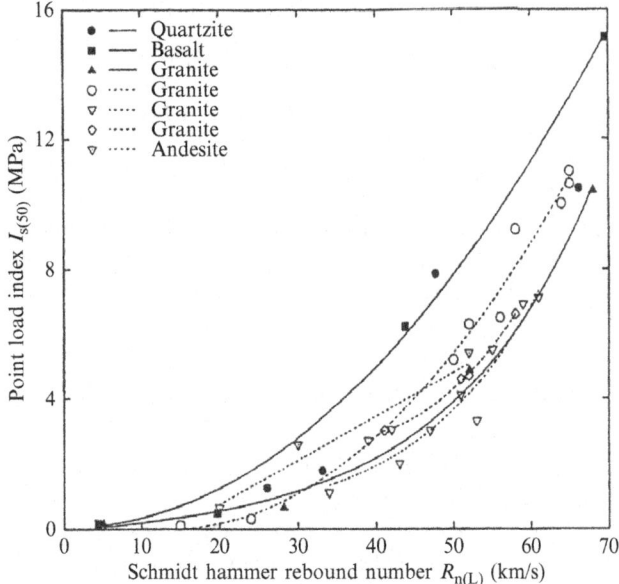

FIG. 3.18 Variation of point load index with Schmidt hammer rebound number for fresh and weathered crystalline rock. *(From Gupta, A.S., Rao, K.S., 1998. Index properties of weathered rocks: inter-relationships and applicability. Bull. Eng. Geol. Environ. 57, 161–172.)*

3.3.11.9 Point load index and Schmidt hammer rebound number

Fig. 3.18 shows the variation of point load index with the L-type Schmidt hammer rebound number for fresh and weathered crystalline rocks. The point load index increases as the Schmidt hammer rebound number increases.

3.3.11.10 Schmidt hammer rebound number and porosity

Schmidt hammer rebound number decreases as porosity increases. Yasar and Erdoğan (2004a) derived the following empirical correlation between them based on the test results of six different rock types: Ceyhan limestone, Barbaros marble, Antique Cream limestone, Osmaniye marble, Toprakkale Basalts, and Handere sandstone:

$$R_{n(L)} = 56.08 - 5.00n \quad (r^2 = 0.80) \tag{3.31}$$

where $R_{n(L)}$ is the L-type Schmidt hammer rebound number; n is the porosity in %; and r^2 is the determination coefficient.

Based on the tests on granitic rocks of various weathering grades, Aydin and Basu (2005) derived the following correlations between Schmidt hammer rebound number and porosity:

$$R_{n(L)} = 70.82 - 2.33n \quad \left(r^2 = 0.79\right) \tag{3.32a}$$

$$R_{n(L)} = 61.12 - 3.13n_e \quad \left(r^2 = 0.81\right) \tag{3.32b}$$

$$R_{n(N)} = 82.55 - 2.56n \quad \left(r^2 = 0.74\right) \tag{3.33a}$$

$$R_{n(L)} = 76.78 - 3.33n_e \quad \left(r^2 = 0.83\right) \tag{3.33b}$$

where $R_{n(L)}$ and $R_{n(N)}$ are, respectively, the L- and N-type Schmidt hammer rebound number; n and n_e are, respectively, the total and effective porosity in %; and r^2 is the determination coefficient.

3.3.11.11   Schmidt hammer rebound number and density or unit weight

Schmidt hammer rebound number is higher as density or unit weight increases. Yasar and Erdoğan (2004a) derived the following empirical correlation between them based on the test results of six different rock types: Ceyhan limestone, Barbaros marble, Antique Cream limestone, Osmaniye marble, Toprakkale Basalts, and Handere sandstone:

$$R_{n(L)} = 3.0e^{1.066\rho} \quad \left(r^2 = 0.84\right) \tag{3.34}$$

where $R_{n(L)}$ is the L-type Schmidt hammer rebound number; ρ is the density in g/cm^3; and r^2 is the determination coefficient.

Using the test data of granitic rocks of various weathering grades, Aydin and Basu (2005) derived the following correlations between Schmidt hammer rebound number and dry density:

$$R_{n(L)} = 98.04\rho_d - 196.5 \quad \left(r^2 = 0.85\right) \tag{3.35}$$

$$R_{n(N)} = 105.3\rho_d - 204.9 \quad \left(r^2 = 0.85\right) \tag{3.36}$$

where $R_{n(L)}$ and $R_{n(N)}$ are, respectively, the L- and N-type Schmidt hammer rebound number; ρ_d is the dry density in g/cm^3; and r^2 is the determination coefficient.

Based on the tests on rock matrices of fresh and altered lavas, welded pyroclastics and autoclastic breccias, del Potro and Hürlimann (2009) derived the following correlation between Schmidt hammer rebound number and unit weight:

$$R_{n(N)} = 2.64\gamma - 16.33 \quad \left(r^2 = 0.98\right) \tag{3.37}$$

where $R_{n(N)}$ is the N-type Schmidt hammer rebound number; γ is the unit weight in kN/m^3; and r^2 is the determination coefficient.

REFERENCES

Alber, M., Yaralı, O., Dahl, F., Bruland, A., Käsling, H., Michalakopoulos, T.N., Cardu, M., Hagan, P., Aydın, H., Özarslan, A., 2014. ISRM suggested method for determining the abrasivity of rock by the CERCHAR abrasivity test. Rock Mech. Rock. Eng. 47, 261–266.

Al-Harthi, A.A., Al-Amri, R.M., Shehata, W.M., 1999. The porosity and engineering properties of vesicular basalt in Saudi Arabia. Eng. Geol. 54, 313–320.

Altındağ, R., 2001. The effects of specimen volume, temperature and water content on Shore hardness. In: 17th International Mining Congress and Exhibition of Turkey-İMCET 2001, pp. 411–415.

Asef, M.R., Najibi, A.R., 2013. The effect of confining pressure on elastic wave velocities and dynamic to static Young's modulus ratio. Geophysics 78 (3), D135–D142.

ASTM, 2001. Standard Test Method for Determination of Rock Hardness by Rebound Hammer Method. D 5873-00, American Society for Testing and Materials, West Conshohocken, PA.

ASTM, 2008a. Standard Test Method for Determination of the Point Load Strength Index of Rock and Application to Rock Strength Classifications. D5731-08. American Society for Testing and Materials, West Conshohocken, PA.

ASTM, 2008b. Standard Test Method for Slake Durability of Shales and Similar Weak Rocks. D4644-08, American Society for Testing and Materials, West Conshohocken, PA.

ASTM, 2010. Standard Test Method for Laboratory Determination of Abrasiveness of Rock Using the CERCHAR Method. D7625-10, American Society for Testing and Materials, West Conshohocken, PA.

Aydan, Ö., 2012. The inference of physico-mechanical properties of soft rocks and the evaluation of the effect of water content and weathering on their mechanical properties from needle penetration tests. In: Proceedings of the Symposium of ARMA, June 2012, Chicago, IL. Paper no. ARMA12-639.

Aydan, Ö., Sato, A., Yagi, M., 2014. The inference of geo-mechanical properties of soft rocks and their degradation from needle penetration tests. Rock Mech. Rock. Eng. 47, 1867–1890.

Ayday, C., Goktan, R.M., 1992. Correlations between L and N-type Schmidt hammer rebound values obtained during field-testing. In: Hudson, J.A. (Ed.), International ISRM Symposium on Rock Characterization, Chester, UK, pp. 47–50.

Aydin, A., Basu, A., 2005. The Schmidt hammer in rock material characterization. Eng. Geol. 81, 1–14.

Bell, F.G., Culshaw, M.G., Cripps, J.C., 1999. A review of selected engineering geological characteristics of English Chalk. Eng. Geol. 54, 237–269.

Birch, F., 1961. The velocity of compressional waves in rocks to 10 kilobars (Part II). J. Geophys. Res. 66, 2199–2224.

Brook, N., 1993. The measurement and estimation of basic rock strength. In: Hudson, J.A. (Ed.), Rock Testing and Site Characterization-Comprehensive Rock Engineering. Pergamon Press, Oxford. vol. 3, pp. 41–66.

Buyuksagis, I.S., Goktan, R.M., 2007. The effect of Schmidt hammer type on uniaxial compressive strength prediction of rock. Int. J. Rock Mech. Min. Sci. 44, 299–307.

CGS, 1985. Canadian Foundation Engineering Manual. Part 2, second ed. Canadian Geotechnical Society, Vancouver, British Columbia, Canada.

Chaki, S., Takarli, M., Agbodjan, W.P., 2008. Influence of thermal damage on physical properties of a granite rock: porosity, permeability and ultrasonic wave evolutions. Constr. Build. Mater. 22, 1456–1461.

Christensen, N.J., Salisbury, U.H., 1975. Structure and constitution of the lower oceanic crust. Rev. Geophys. Space Phys. 13, 57086.

Cook, G.K., 2001. Rock mass structure and intact rock strength of New Zealand greywackes. M.Sc. thesis, University of Canterbury.

Deere, D.U., Miller, R.P., 1966. Engineering classification and index properties for intact rock. Tech. Rep. No. AFWL-TR-65-116, Air Force Weapons Lab, Kirtland Air Force Base, Albuquerque, NM.

del Potro, R., Hürlimann, M., 2009. A comparison of different indirect techniques to evaluate volcanic intact rock strength. Rock Mech. Rock. Eng. 42, 931–938.

Deliormanlı, A.H., 2012. Cerchar abrasivity index (CAI) and its relation to strength and abrasion test methods for marble stones. Constr. Build. Mater. 30, 16–21.

Demirdag, S., Yavuz, H., Altindag, R., 2009. The effect of sample size on Schmidt rebound hardness value of rocks. Int. J. Rock Mech. Min. Sci. 46, 725–730.

Dincer, I., Acar, A., Cobanoglu, I., Uras, Y., 2004. Correlation between Schmidt hardness, uniaxial compressive strength and Young's modulus for andesites, basalts and tuffs. Bull. Eng. Geol. Environ. 63, 141–148.

Franklin, J.A., Chandra, A., 1972. The slake durability test. Int. J. Rock Mech. Min. Sci. 9, 325–341.

Gaviglio, P., 1989. Longitudinal wave propagation in a limestone: the relationship between velocity and density. Rock Mech. Rock. Eng. 22, 299–306.

Goodman, R.E., 1980. Introduction to Rock Mechanics. Wiley, New York.

Gupta, A.S., Rao, K.S., 1998. Index properties of weathered rocks: inter-relationships and applicability. Bull. Eng. Geol. Environ. 57, 161–172.

Henkel, H., Lee, M.K., Lund, C.E., 1990. An integrated geophysical interpretation of the 200 km FEN-NOLORA section of the Baltic Shield. In: Freeman, R., Giese, P., Mueller, St. (Eds.), The European Geotraverse: Integrative Studies. European Science Foundation, Strasbourg, pp. 1–47.

Hoek, E., 1994. Strength of rock and rock masses. News J. ISRM 2, 4–16.

Hudson, J.A., Harrison, J.P., 1997. Rock Engineering Mechanics—An Introduction to the Principles. Elsevier, Oxford.

ISRM, 1978a. Suggested methods for determining sound wave velocity. International Society for Rock Mechanics, Commission on Standardization of Laboratory and Field Tests. Int. J. Rock Mech. Min. Sci. Geomech. Abstr. 15, 53–58.

ISRM, 1978b. Suggested methods for determining hardness and abrasiveness of rocks. International Society for Rock Mechanics, Commission on Standardization of Laboratory and Field Tests. Int. J. Rock Mech. Min. Sci. Geomech. Abstr. 15, 89–97.

ISRM, 1978c. Suggested methods for the quantitative description of discontinuities in rock masses. International Society for Rock Mechanics, Commission on Standardization of Laboratory and Field Tests. Int. J. Rock Mech. Min. Sci. Geomech. Abstr. 15, 319–368.

ISRM, 1979. Suggested methods for determining water content, porosity, density, absorption and related properties and swelling and slake-durability index properties. International Society for Rock Mechanics, Commission on Standardization of Laboratory and Field Tests. Int. J. Rock Mech. Min. Sci. Geomech. Abstr. 16, 145–156.

ISRM, 1985. Suggested methods for determining the point load strength. International Society for Rock Mechanics, Commission on Standardization of Laboratory and Field Tests. Int. J. Rock Mech. Min. Sci. Geomech. Abstr. 22, 51–60.

Jelic, K., 1984. Odnos gustoce I poroznosti s dubinom litostratigrafskih formacija savske I dravske potoline. Nafta (Zagreb) 35, 637–643.

JGS, 2012. Method for Needle Penetration Test (JGS: 3431-2012). Japanese Standards and Explanations of Geotechnical and Geoenvironmental Investigation Methods, vol. 1. Japanese Geotechnical Society (JGS) Publication, Tokyo, pp. 426–432 (in Japanese).

Ji, S., Wang, Q., Marcotte, D., Salisbury, M.H., Xu, Z., 2007. P wave velocities, anisotropy and hysteresis in ultrahigh-pressure metamorphic rocks as a function of confining pressure. J. Geophys. Res. 112.

JSCE, 1980. A Suggested Method for Investigation and Testing of Soft Rocks. Japan Society of Civil Engineers (JSCE), Rock Mechanics Committee, the 4th Sub-committee, Tokyo (in Japanese).

JSCE, 1991. A Suggested Method for Investigation and Testing of Soft Rocks. Committee on Rock Mechanics of Japan Society of Civil Engineers (JSCE), Tokyo (in Japanese).

Kahraman, S., 2001. A correlation between P-wave velocity, number of joints and Schmidt hammer rebound number. Int. J. Rock Mech. Min. Sci. 38, 729–733.

Kahraman, S., 2007. The correlations between the saturated and dry P-wave velocity of rocks. Ultrasonics 46, 341–348.

Kahraman, S., Yeken, T., 2008. Determination of physical properties of carbonate rocks from P-wave velocity. Bull. Eng. Geol. Environ. 67, 277–281.

Karakul, H., Ulusay, R., 2013. Empirical correlations for predicting strength properties of rocks from P-wave velocity under different degrees of saturation. Rock Mech. Rock. Eng. 46, 981–999.

Käsling, H., Thuro, K., 2010. Determining rock abrasivity in the laboratory. In: Rock Mechanics in Civil and Environmental Engineering—Proceedings of EUROCK 2010. Taylor & Francis, London, pp. 425–428.

Kayabali, K., Selcuk, L., 2010. Nail penetration test for determining the uniaxial compressive strength of rock. Int. J. Rock Mech. Min. Sci. 47, 265–271.

Keles, S., 2005. Cutting performance of a medium weight roadheader at Cayirhan coal mine. MS thesis, Middle East Technical University.

Kilic, Ö., 2006. The influence of high temperatures on limestone P-wave velocity and Schmidt hammer strength. Int. J. Rock Mech. Min. Sci. 43, 980–986.

Kopf, M., 1977. Fortschritte der Petrophysik. In: Lauterbach, R. (Ed.), Physik der Erdkruste. Akademite-Verlag, Berlin.

Kopf, M., 1980. Anwendung der Korrelation bei der Ermittlung von Dichte- und Geschwindigkeitswerten für Gravimetrie und Seismik. Zeitschr. Geol. Wiss., Berlin 8, 449–465.

Lama, R.D., Vutukuri, V.S., 1978. Handbook on Mechanical Properties of Rocks. Trans Tech Publications, Clausthal.

Maloney, S., 2010. CERCHAR abrasivity testing of argillaceous limestone of the Cobourg formation. Technical Report, MIRARCO/Geomechanics Research Centre, Laurentian University.

Marle, C., 1978. Geophysik und subsequenter Vulkanismus (dargestellt am Beispiel des NW-sächsischen Vulkanitkomplexes). Thesis, Universität Leipzig (unpublished).

Najibi, A.R., Asef, M.R., 2014. Prediction of seismic-wave velocities in rock at various confining pressures based on unconfined data. Geophysics 79, D235–D242.

Palchik, V., Hatzor, Y.H., 2004. Influence of porosity on tensile and compressive strength of porous chalks. Rock Mech. Rock. Eng. 37, 331–341.

Plinninger, R., Käsling, H., Thuro, K., Spaun, G., 2003. Testing conditions and geomechanical properties influencing the CERCHAR abrasiveness index (CAI) value. Int. J. Rock Mech. Min. Sci. 40, 259–263.

Polak, L.S., Rapoport, M.B., 1961. Osvjasi skorosti uprugich prodolnych voln s nekotorymi fiziceskimi svoistvami osadocnych porod. Prikladnaja geofizika, Moskva 29, 12–19.

Rahmouni, A., Boulanouar, A., Boukalouch, M., Géraud, Y., Samaouali, A., Harnafi, M., Sebbani, J., 2013. Prediction of porosity and density of calcarenite rocks from P-wave velocity measurements. Int. J. Geosci. 4, 1292–1299.

Ramana, Y.V., Venkatanarayana, B., 1973. Laboratory studies on Kolar rocks. Int. J. Rock Mech. Min. Sci. Geomech. Abstr. 10, 465–489.

Raymer, D.S., Hunt, E.R., Gardner, J.S., 1980. An improved sonic transit time-to-porosity transform. In: Proceedings of SPWLA 21st Annual Meeting. paper P.

Sachpazis, C.I., 1990. Correlating Schmidt hardness with compressive strength and Young's modulus of carbonate rocks. Bull. Int. Assoc. Eng. Geol. 42, 75–84.

Schön, J.H., 1996. Physical Properties of Rocks—Fundamentals and Principles of Petrophysics. Pergamon, Oxford.

Sousa, L.M.O., del Rio, L.M.S., Calleja, L., de Argandona, V.G.R., Rey, A.R., 2005. Influence of microfractures and porosity on the physico-mechanical properties and weathering of ornamental granites. Eng. Geol. 77, 153–168.

Stacey, T.R., van Veerden, W.L., Vogler, U.W., 1987. Properties of intact rock. In: Bell, F.G. (Ed.), Ground Engineer's Reference Book. Butterworths, London.

Starzec, P., 1999. Dynamic elastic properties of crystalline rocks from south-west Sweden. Int. J. Rock Mech. Min. Sci. 36, 265–272.

Stegena, L., 1964. The structure of the earth crust in Hungary. Acta Geol., Budapest 8, 413–431.

Sun, S., Ji, S., Wang, Q., Xu, Z., Salisbury, M., Long, C., 2012. Seismic velocities and anisotropy of core samples from the Chinese Continental Scientific Drilling borehole in the Sulu UHP terrane, eastern China. J. Geophys. Res. 117.

Szlavin, J., 1974. Relationships between some physical properties of rock determined by laboratory tests. Int. J. Rock Mech. Min. Sci. Geomech. Abstr. 11, 57–66.

Timur, A., 1977. Temperature dependence of compressional and shear wave velocities in rocks. Geophysics 42, 950–956.

Toksoz, M.N., Cheng, C.H., Timur, A., 1976. Velocities of seismic waves in porous rocks. Geophysics 41, 621–645.

Tuğrul, A., 2004. The effect of weathering on pore geometry and compressive strength of selected rock types from Turkey. Eng. Geol. 75, 215–227.

Tuğrul, A., Zarif, I.H., 1999. Correlation of mineralogical and textural characteristics with engineering properties of selected granitic rocks from Turkey. Eng. Geol. 51, 303–317.

Ulusay, R., Aydan, Ö., Erguler, Z.A., Ngan-Tillard, D.J.M., Seiki, T., Verwaal, W., Sasaki, Y., Sato, A., 2014. ISRM suggested method for the needle penetration test. Rock Mech. Rock. Eng. 47, 1073–1085.

Voight, B., 1968. On the functional classification of rocks for engineering purposes. In: International Symposium on Rock Mechanics, Madrid, pp. 131–135.

Volarovich, M.P., Bajuk, E.I., 1977. Elastic properties of rocks. In: Volarovic, M.P., Stiller, H., Lebedev, T.S. (Eds.), Issledovanie Fiziceskich svoit sv Mineralnogo Vescestva Zemli pri Vysokich Termodinamiceskich Parametrach. Izd. Nakova dumka, Kiev, pp. 43–49.

Wang, C., Lin, W., Wenk, H., 1975. The effects of water and pressure on velocities of elastic waves in a foliated rock. J. Geophys. Res. 80, 1065–1169.

Wyllie, M.R.J., Gregory, A.R., Gardner, L.W., 1956. Elastic wave velocities in heterogeneous and porous media. Geophysics 21, 41–70.

Xu, S., Grasso, P., Mahtab, A., 1990. Use of Schmidt hammer for estimating mechanical properties of weak rock. In: Sixth International IAEG Congress. Balkema, Rotterdam, pp. 511–519.

Yasar, E., Erdoğan, Y., 2004a. Estimation of rock physicomechanical properties using hardness methods. Eng. Geol. 71, 281–288.

Yasar, E., Erdoğan, Y., 2004b. Correlating sound velocity with the density, compressive strength and Young's modulus of carbonate rocks. Int. J. Rock Mech. Min. Sci. 41, 871–875.

Yavuz, H., Demirdag, S., Caran, S., 2010. Thermal effect on the physical properties of carbonate rocks. Int. J. Rock Mech. Min. Sci. 47, 94–103.

Yilmaz, I., Sendir, H., 2002. Correlation of Schmidt hardness with unconfined compressive strength and Young's modulus in gypsum from Sivas (Turkey). Eng. Geol. 66, 211–219.

Chapter 4

Rock Discontinuities

4.1 INTRODUCTION

Discontinuity is a general term that denotes any separation in a rock mass having zero or low tensile strength. It is a collective term for most types of joints, weak bedding planes, weak schistocity planes, weakness zones, and faults (ISRM, 1978). It is mainly the discontinuities which make the rock mass different from other engineering materials. Therefore, to determine the engineering properties of rocks, it is important to characterize discontinuities in detail. As Palmström (2002) noted: "The engineering properties of a rock mass often depend far more on the system of geological defects within the rock mass than on the strength of the rock itself. Thus, from an engineering point of view, a knowledge of the type and frequency of the joints and fissures are often more important than the types of rock involved. The observations and characterization of the joints should therefore be done carefully."

This chapter will first describe the different types of discontinuities and then discuss the geometrical properties of discontinuities. The mechanical and hydrological properties of discontinuities will be discussed in later chapters.

4.2 TYPES OF DISCONTINUITIES

Discontinuities and their origins are well described in several textbooks on general, structural and engineering geology. From an engineering point of view, the discussions by Price (1966), Hills (1972), Blyth and de Freitas (1974), Hobbs (1976), Priest (1993), and Wyllie and Mah (2004) are particularly helpful. The following lists the major types of discontinuities and briefly describes their key engineering properties.

(a) Faults

Faults are discontinuities on which identifiable shear displacement has taken place. They may be recognized by the relative displacement of the rock on the opposite sides of the fault plane. The sense of this displacement is often used to classify faults (Fig. 4.1). Faults may be pervasive features which traverse a large area or they may be of relatively limited local extent on the scale of meters; they often occur in echelon or in groups. Fault thickness may vary from meters in the case of major, regional structures to millimeters in the case of local faults. This fault thickness may contain weak

Engineering Properties of Rocks. http://dx.doi.org/10.1016/B978-0-12-802833-9.00004-3
81

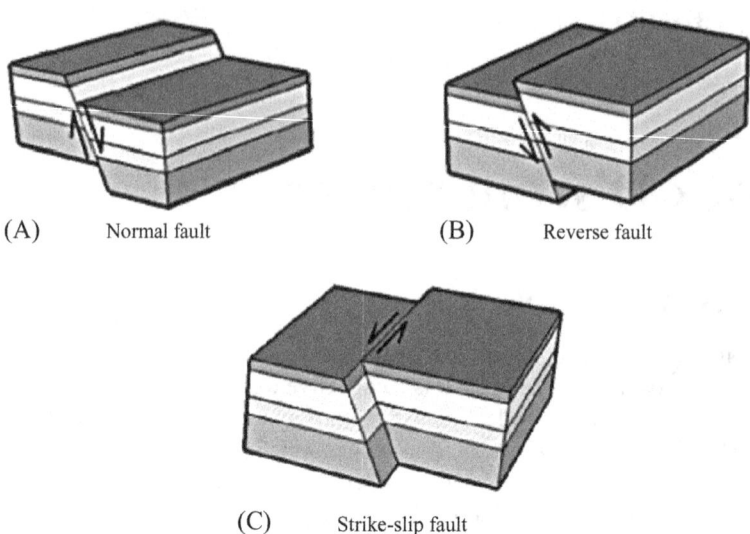

(A)   Normal fault   (B)   Reverse fault

(C)   Strike-slip fault

FIG. 4.1 Classification of faults based on relative displacement of rock on the opposite sides of fault plane: (A) Normal fault; (B) Reverse fault; and (C) Strike-slip fault.

materials such as fault gouge (clay), fault breccia (recemented), rock flour or angular fragments. The wall rock is frequently slickensided and may be coated with minerals such as graphite and chlorite which have low frictional strengths. The ground adjacent to the fault may be disturbed and weakened by associated discontinuities such as drag folds or secondary faults. These factors result in faults being zones of low shear strength on which slip may readily occur.

(b) Bedding planes

Bedding planes divide sedimentary rocks into beds or strata. They represent interruptions in the course of deposition of the rock mass. Bedding planes are generally highly persistent features, although sediments laid down rapidly from heavily laden wind or water currents may contain cross or discordant bedding. Bedding planes may contain parting material of different grain size from the sediments forming the rock mass, or may have been partially healed by low-order metamorphism. In either of these two cases, there would be some "cohesion" between the beds; otherwise, the shear resistance on bedding planes would be purely frictional. Arising from the depositional process, there may be a preferred orientation of particles in the rock, giving rise to planes of weakness parallel to the bedding planes.

(c) Joints

Joints are the most common and generally the most geotechnically significant discontinuities in rocks. Joints are breaks of geological origin along which there has been no visible relative displacement. A group of parallel or sub-parallel joints is called a joint set, and joint sets intersect to form a joint system. Joints may be open, filled or healed. Discontinuities frequently form parallel to bedding planes, foliations or slaty cleavage, and they may be termed bedding joints, foliation joints or cleavage joints. Sedimentary rocks often contain two sets of joints approximately orthogonal to each other and to the bedding planes. These joints sometimes end at bedding planes, but others, called master joints, may cross several bedding planes.

(d) Cleavage

There are two broad types of rock cleavages: fracture cleavage and flow cleavage. Fracture cleavage (also known as false cleavage and strain slip cleavage) is a term describing incipient, cemented or welded parallel discontinuities that are independent of any parallel alignment of minerals. Spencer (1969) lists six possible mechanisms for the formation of fracture cleavage. In each mechanism, lithology and stress conditions are assumed to have produced shearing, extension or compression, giving rise to numerous closely spaced discontinuities separated by thin slivers of intact rock. Fracture cleavage is generally associated with other structural features such as faults, folds and kink bands. Flow cleavage, which can occur as slaty cleavage or schistosity, is dependent upon the recrystallization and parallel alignment of platy minerals such as mica, producing inter-leaving or foliation structure. It is generally accepted that flow cleavage is produced by high temperatures and/or pressures associated with metamorphism in fine-grained rocks.

Although cleavage is usually clearly visible in slates, phyllites and schists, most cleavage planes possess significant tensile strength and do not, therefore, contribute to the discontinuity network. Cleavage can, however, create significant anisotropy in the deformability and strength of such rocks. Geological processes, such as folding and faulting, subsequent to the formation of the cleavage can exploit these planes of weakness and generate discontinuities along a proportion of the better developed cleavage planes. The decision as to whether a particular cleavage plane is a discontinuity presents one of the most challenging problems to those undertaking discontinuity surveys in cleaved rocks.

4.3 DESCRIPTION OF DISCONTINUITIES

The International Society for Rock Mechanics (ISRM) publication *Suggested methods for the quantitative description of discontinuities in rock masses* (ISRM, 1978) defines 10 parameters to describe the characteristics of discontinuities:

1. Orientation: the attitude of a discontinuity in space. It is described by the dip direction (azimuth) and dip of the line of steepest declination in the plane of the discontinuity.

2. Spacing: the perpendicular distance between adjacent discontinuities. It normally refers to the mean or modal spacing of a set of discontinuities.
3. Persistence: the discontinuity trace length as observed in an exposure. It may give a crude measure of the areal extent or penetration length of a discontinuity. Termination in solid rock or against other discontinuities reduces the persistence.
4. Roughness: the inherent surface roughness and waviness relative to the mean plane of a discontinuity. Both roughness and waviness contribute to the shear strength. Large-scale waviness may also alter the dip locally.
5. Wall strength: the equivalent compressive strength of the adjacent rock walls of a discontinuity. It may be lower than rock block strength due to weathering or alteration of the walls. It is an important component of shear strength if rock walls are in contact.
6. Aperture: the perpendicular distance between adjacent rock walls of a discontinuity, in which the intervening space is air or water filled.
7. Filling: the material that separates the adjacent rock walls of a discontinuity and that is usually weaker than the parent rock. Typical filling materials are sand, silt, clay, breccia, gouge, and mylonite. It also includes thin mineral coatings and healed discontinuities such as quartz and calcite veins.
8. Seepage: the water flow and free moisture visible in individual discontinuities or in the rock mass as a whole.
9. Number of Sets: the number of discontinuity sets comprising the intersecting discontinuity system. The rock mass may be further divided by individual discontinuities.
10. Block Size: the rock block dimensions resulting from the mutual orientation of intersecting discontinuity sets, and resulting from the spacing of the individual sets. Individual discontinuities may further influence the block size and shape.

The following sections describe the geometrical properties of discontinuities, including orientation, intensity, spacing, frequency, persistence, shape, size, roughness, aperture, discontinuity sets and block size.

4.4 DISCONTINUITY ORIENTATION

Orientation, or attitude of a discontinuity in space, is described by the dip of the line of maximum declination on the discontinuity surface measured from the horizontal, and the dip direction or azimuth of this line, measured clockwise from true north. Some geologists record the strike of the discontinuity rather than the dip direction. For rock mechanics purposes, it is usual to quote orientation data in the form of dip direction (three digits)/dip (two digits) such as 035°/75° and 290°/30°. The orientation of discontinuities relative to an engineering structure largely controls the possibility of unstable conditions or excessive deformations developing. The importance of orientation increases when other conditions for deformation are present, such as low shear strength and a sufficient number of discontinuities for slip to occur. The mutual orientation of discontinuities will determine the shape of the individual blocks, beds or mosaics comprising the rock mass.

The orientation of discontinuities can be measured from cores or from exposures using one or two dimensional scanlines. The measured orientation data can be plotted on stereonets. Fig. 4.2 shows such a plot on a polar stereonet of the poles of 351 individual discontinuities whose orientations were measured at a particular field site (Hoek and Brown, 1980a). Different symbols have been used for three different types of discontinuities—bedding planes, joints and a fault. The fault has a dip direction of 307° and a dip of 56°. Contours of pole concentrations may be drawn for the bedding planes and joints to give an indication of the preferred orientations of the various discontinuity sets present. Fig. 4.3 shows the contours of pole concentrations determined from the data shown in Fig. 4.2. The central orientations of the two major joint sets are 347°/22° and 352°/83°, and that of the bedding planes is 232°/81°.

Computer programs such as the one by Mahtab et al. (1972) are also available for plotting and contouring discontinuity orientation data. For a large number of discontinuities, it will be more convenient to use computer programs to plot and contour orientation data.

The assignment of poles into discontinuity sets is usually achieved by a combination of contouring, visual examination of the stereonet and knowledge of

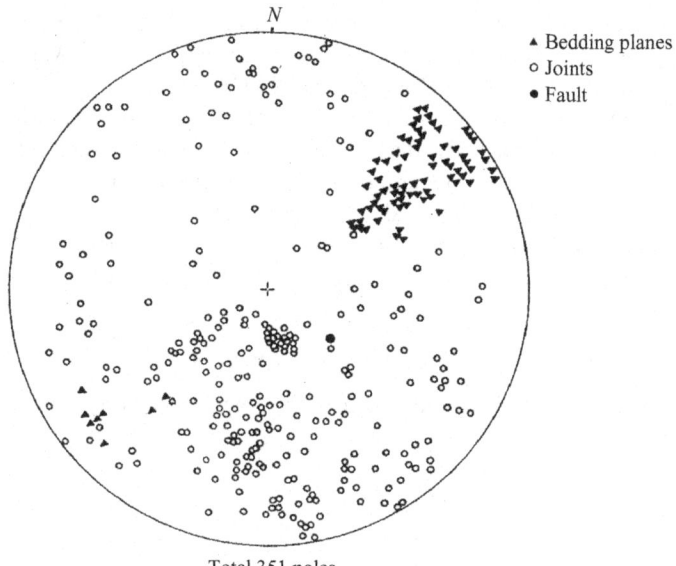

Total 351 poles

FIG. 4.2 Plot of poles of 351 discontinuities. *(From Hoek, E., Brown, E.T., 1980. Underground Excavation in Rock. Institution of Mining and Metallurgy, London, UK.)*

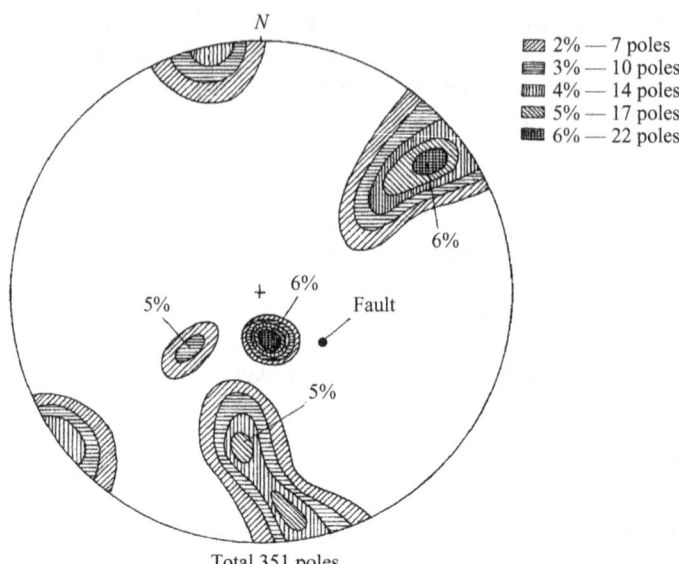

FIG. 4.3 Contours of pole concentrations for the data plotted in Fig. 4.1. *(From Hoek, E., Brown, E.T., 1980. Underground Excavation in Rock. Institution of Mining and Metallurgy, London, UK.)*

geological conditions at the site. However, in many cases visual clustering is very difficult due to the overlap of clusters. A number of algorithms which are based on statistical or fuzzy-set approaches are available for numerically clustering orientation data (Einstein et al., 1979; Miller, 1983; Mahtab and Yegulalp, 1984; Harrison, 1992; Kulatilake, 1993).

As seen in Fig. 4.2, there is scatter of the poles of discontinuities when they are plotted on the stereonet. The mean orientation of a number of discontinuities can be calculated from the direction cosines as described in the following. The sampling bias on orientation can also be considered.

The pole of a discontinuity in three-dimensional (3D) space can be represented by a unit vector (u_x, u_y, u_z) associated with the direction cosines as shown in Fig. 4.4:

$$u_x = \cos\alpha_n \cos\beta_n, \quad u_y = \sin\alpha_n \cos\beta_n, \quad u_z = \sin\beta_n \qquad (4.1)$$

where α_n and β_n are the trend and plunge of the pole, respectively, which can be obtained by:

$$\alpha_n = \arctan\left(\frac{u_y}{u_x}\right) + Q \qquad (4.2a)$$

$$\beta_n = \arctan\left(\frac{u_z}{\sqrt{u_x^2 + u_y^2}}\right) \qquad (4.2b)$$

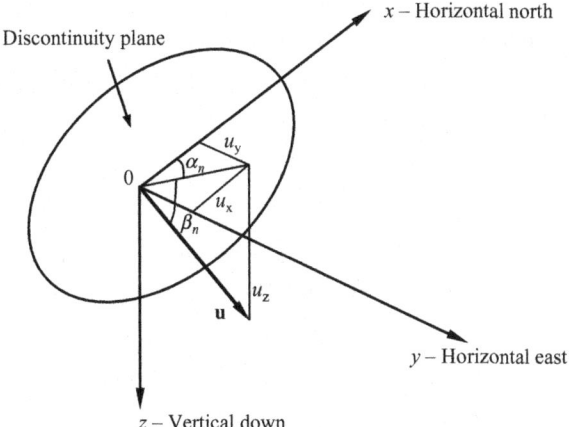

FIG. 4.4 Pole of a discontinuity represented by a unit vector **u**.

TABLE 4.1 The Quadrant Parameter Q in Eq. (4.2a)

u_x	u_y	Q
≥ 0	≥ 0	0
< 0	≥ 0	180°
< 0	< 0	180°
≥ 0	< 0	360°

The parameter Q is an angle, in degrees, that ensures that α_n lies in the correct quadrant and in the range of 0° to 360° (see Table 4.1).

The dip direction and dip angle α/β of a discontinuity are related to the trend and plunge α_n/β_n of its normal by the following expressions:

$$\begin{aligned} \alpha_n &= \alpha + 180° \quad \text{(for } \alpha \leq 180°\text{)} \\ \alpha_n &= \alpha - 180° \quad \text{(for } \alpha \geq 180°\text{)} \end{aligned} \tag{4.3a}$$

$$\beta_n = 90° - \beta \tag{4.3b}$$

The mean orientation of a set of discontinuities intersecting a sampling line of trend/plunge α_s/β_s can be obtained using the procedure outlined below. This procedure corrects for orientation sampling bias through the introduction of weighted direction cosines.

1. For discontinuity i, calculate the angle δ_i between its normal and the sampling line:

$$\cos \delta_i = \left| u_{xi} u_{xs} + u_{yi} u_{ys} + u_{zi} u_{zs} \right| \tag{4.4}$$

where (u_{xi}, u_{yi}, u_{zi}) and (u_{xs}, u_{ys}, u_{zs}) are the direction cosines of the normal to discontinuity i and the sampling line, respectively.

2. For discontinuity i, calculate the weighting factor w_i based on the angle δ_i obtained in step 1:

$$w_i = \frac{1}{\cos \delta_i} \quad (\delta_i < 90°) \qquad (4.5)$$

3. After the weighting factor for each discontinuity is obtained, calculate the total weighted sample size N_w for a sample of size N by:

$$N_w = \sum_{i=1}^{N} w_i \qquad (4.6)$$

4. Calculate the normalized weighting factor w_{ni} for each discontinuity by:

$$w_{ni} = \frac{w_i N}{N_w} \qquad (4.7)$$

5. Calculate the corrected direction cosines (n_{xi}, n_{yi}, n_{zi}) for the normal of each discontinuity by:

$$\left(n_{xi}, n_{yi}, n_{zi}\right) = w_{ni}\left(u_{xi}, u_{yi}, u_{zi}\right) \qquad (4.8)$$

6. Calculate the resultant vector (r_x, r_y, r_z) of the corrected normal vectors (n_{xi}, n_{yi}, n_{zi}), $i = 1–N$:

$$r_x = \sum_{i=1}^{N} n_{xi}, \quad r_y = \sum_{i=1}^{N} n_{yi}, \quad r_z = \sum_{i=1}^{N} n_{zi} \qquad (4.9)$$

7. The mean orientation of the N discontinuities is the orientation of the resultant vector whose trend and plunge can be found by replacing u_x, u_y and u_z by r_x, r_y and r_z in Eq. (4.2).

Several probability distributions have been suggested in the literature to represent the discontinuity orientations, including hemispherical uniform, hemispherical normal or Fisher, bivariate Fisher, Bingham, bivariate normal and bivariate lognormal (Shanley and Mahtab, 1976; Zanbak, 1977; Einstein et al., 1979; Kulatilake, 1985a, 1986). The best means to check if a certain distribution is applicable to represent the orientation of a discontinuity set is to perform goodness-of-fit tests. Shanley and Mahtab (1976) and Kulatilake (1985a, 1986) performed χ^2 goodness-of-fit tests for Bingham, hemispherical normal and bivariate normal distributions, respectively. Einstein et al. (1979) tried all the aforementioned distributions to represent the statistical distributions for 22 data sets. They reported that they could not find a probability distribution which satisfied χ^2 goodness-of-ft test at 5% significance level for 18 of these data sets. This indicates clearly the inadequacy of the currently available analytical distributions in representing the discontinuity orientations. In the case that no analytical distribution can represent the discontinuity orientation data, empirical distributions can be used.

4.5   DISCONTINUITY INTENSITY

Discontinuity intensity is one of the most important parameters for describing discontinuities in a rock mass. Intensity can be expressed in terms of different measures in one, two or three dimensions, including discontinuity spacing, linear, areal and volumetric frequency, Rock Quality Designation (RQD), discontinuity trace length per unit area of rock exposure, and discontinuity area per unit volume of rock mass, as described in detail in the following subsections.

4.5.1   Discontinuity Spacing and Linear Frequency

Discontinuity spacing is the distance between adjacent discontinuities measured along a sampling line (scanline). If the discontinuities are from a particular discontinuity set, the spacing is called the set spacing. When the sampling line is normal to the discontinuity planes, the set spacing is called the normal set spacing (Priest, 1993). Table 4.2 presents the terminology used by ISRM (1978) for describing the magnitude of discontinuity spacing.

Discontinuity frequency is most commonly expressed in terms of the linear frequency λ defined as the number of discontinuities intersected by a unit length of sampling line. Linear frequency is the reciprocal of the mean spacing. Like the spacing, the frequency can be specified as set frequency or normal set frequency.

If the frequency of a discontinuity set along a sampling line that makes an acute angle φ to the discontinuity planes is λ_s (Fig. 4.5), the normal set frequency of the discontinuity set is (Terzaghi, 1965):

TABLE 4.2 Classification of Discontinuity Spacing

Description	Spacing (mm)
Extremely close spacing	< 20
Very close spacing	20–60
Close spacing	60–200
Moderate spacing	200–600
Wide spacing	600–2000
Very wide spacing	2000–6000
Extremely wide spacing	>6000

Based on ISRM, 1978. Suggested methods for the quantitative description of discontinuities in rock masses. International Society for Rock Mechanics, Commission on Standardization of Laboratory and Field Tests. Int. J. Rock Mech. Min. Sci. Geomech. Abstr. 15, 319–368.

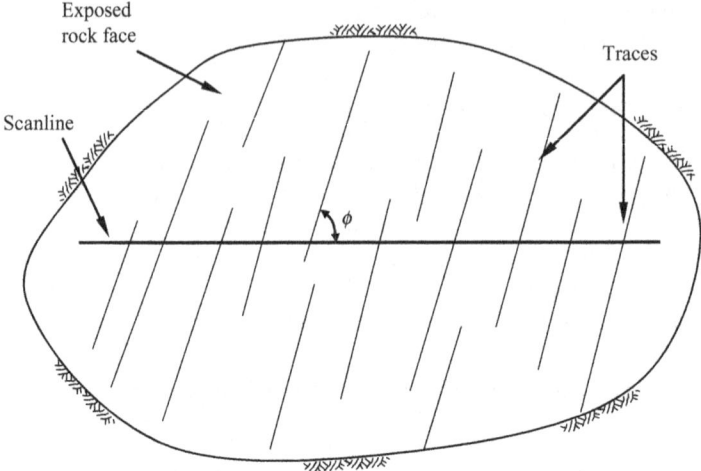

FIG. 4.5   Scanline sampling of discontinuities on an exposed rock face.

$$\lambda_n = \frac{\lambda_s}{\sin \varphi} \tag{4.10}$$

The term $1/\sin \varphi$ in Eq. (4.10) is a correction factor for sampling bias with scanline sampling (Terzaghi, 1965).

For a sampling line intersecting N sets of discontinuities, the total discontinuity frequency λ is determined by:

$$\lambda = \sum_{i=1}^{N} \lambda_{ni} \sin \theta_i \tag{4.11}$$

where λ_{ni} is the normal set frequency of the ith discontinuity set; and θ_i is the acute angle between the sampling line and the ith set of discontinuity planes.

Like all other characteristics of rock masses, discontinuity spacings will not have uniquely defined values but, rather, will take a range of values, possibly according to some form of statistical distribution. The two major discontinuity spacing distribution forms used in the literature are negative exponential and lognormal (Rives et al., 1992). Priest and Hudson (1976) made measurements on a number of sedimentary rock masses in the United Kingdom and found that, in each case, the discontinuity spacing histogram gave a probability density distribution that could be approximated by the negative exponential distribution. The same conclusion has been reached by others, notably Wallis and King (1980) working on a Precambrian porphyritic granite, and Einstein and Baecher (1983) working on a variety of igneous, sedimentary and metamorphic rocks. Thus the frequency $f(s)$ of a given discontinuity spacing value s is given by the function:

$$f(s) = \lambda e^{-\lambda s} \tag{4.12}$$

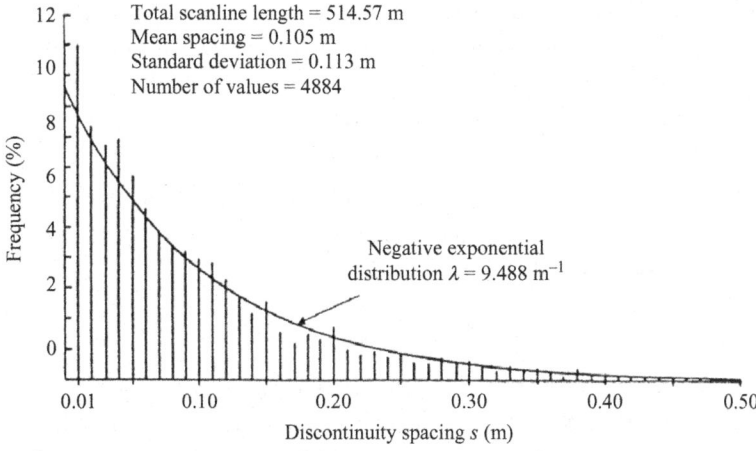

FIG. 4.6 Discontinuity spacing histogram, Lower Chalk, Chinnor, Oxfordshire, UK. *(From Priest, S.D., Hudson, J., 1976. Discontinuity spacing in rock. Int. J. Rock Mech. Min. Sci. Geomech. Abstr. 13, 135–148.)*

where $\lambda = 1/\bar{s}$ is the mean discontinuity frequency of a large discontinuity population and $\bar{s}$ is the mean spacing.

Fig. 4.6 shows the discontinuity spacing histogram and the corresponding negative exponential distribution calculated from Eq. (4.12) for the Lower Chalk, Chinnor, Oxfordshire, UK (Priest and Hudson, 1976).

Seismic velocity measurements have been used for assessing the discontinuity frequency by different researchers (Savic et al., 1969; Sjögren et al., 1979; Idziak, 1981; Jamscikov et al., 1985; Palmström, 1995). Palmström (1995) presented the following two relationships between the linear discontinuity frequency λ and the P-wave velocity:

$$\lambda = \frac{v_{p0}^{3.4}}{v_{pF}^{2.8}} \tag{4.13}$$

$$\lambda = 3 \left(\frac{v_{p0}}{v_{pF}} \right)^{v_{p0}/2} \tag{4.14}$$

where v_{p0} is the P-wave velocity of intact rock under the same conditions as in the field; and v_{pF} is the P-wave velocity of in situ rock mass. Both v_{p0} and v_{pF} are in the unit of km/s. Where v_{p0} is not known, it can be estimated from the value ranges shown in Fig. 3.5.

The following general equation was also used to fit the experimental data of linear discontinuity frequency λ and P-wave velocities (Schön, 1996):

$$v_{pF} = \frac{v_{p0}}{1 + a\lambda^m} \tag{4.15}$$

where a and m are empirical constants. For sedimentary rocks (limestone, dolomite) of the Uppersilesian Coal Basin, Poland, the following empirical relation was obtained (Idziak, 1981):

$$v_{pF} = \frac{7.67}{1 + 0.252\lambda^{3/2}} \qquad (4.16)$$

Sjögren et al. (1979) and Palmström (1995) proposed the following hyperbolic expression for calculating the linear discontinuity frequency λ from measured P-wave velocities:

$$\lambda = \frac{v_{pN} - v_{pF}}{v_{pN} \times v_{pF} \times k_s} \qquad (4.17)$$

where v_{pN} is the natural or maximum P-wave velocity of crack- and discontinuity-free rock; and k_s is a parameter taking into account the actual conditions of the in situ rock mass. v_{pN} and k_s can be determined using the procedure described in the following.

Since the rocks near the surface are seldom free from discontinuities, cracks and pores, it is seldom possible to find v_{pN} of rocks near the surface directly from seismic measurements. The best way to determine v_{pN} is conducting calculations when two sets of measured λ and v_{pF} data are available. From Sjögren et al. (1979) and Palmström (1995),

$$v_{pN} = \frac{v_{pF1} \times v_{pF2}(\lambda_2 - \lambda_1)}{\lambda_2 \times v_{pF2} - \lambda_1 \times v_{pF1}} \qquad (4.18)$$

$$k_s = \frac{1}{\lambda_1}\left(\frac{1}{v_{pF1}} - \frac{1}{v_{pN}}\right) \qquad (4.19)$$

where λ_1, v_{pF1} and λ_2, v_{pF2} are corresponding values of measured linear discontinuity frequency and in situ rock P-wave velocity, respectively, from two pairs of measurements.

Based on regression analysis of the data obtained for heavily fractured calcareous rock masses out-cropping in southern Italy (Fig. 4.7), Budetta et al. (2001) obtained v_{pN} and k_s as $v_{pN} = 6.33$ km/s and $k_s = 0.025$, respectively, ie,

$$\lambda = \frac{6.33 - v_{pF}}{6.33 \times v_{pF} \times 0.025}. \qquad (4.20)$$

It is noted that the discontinuity frequency λ varies with the direction of the sampling line (Harrison and Hudson, 2000; Choi and Park, 2004). As an example, Fig. 4.8 shows the variation of estimated discontinuity frequency by Choi and Park (2004) for a site in the west-southern part of Korea on the lower hemisphere equal-angle stereo projection net. The variation of discontinuity frequency with direction can be clearly seen. Therefore, it is important to specify the corresponding direction when stating a discontinuity frequency value. The other way is to use the volumetric frequency as discussed in Section 4.5.3.

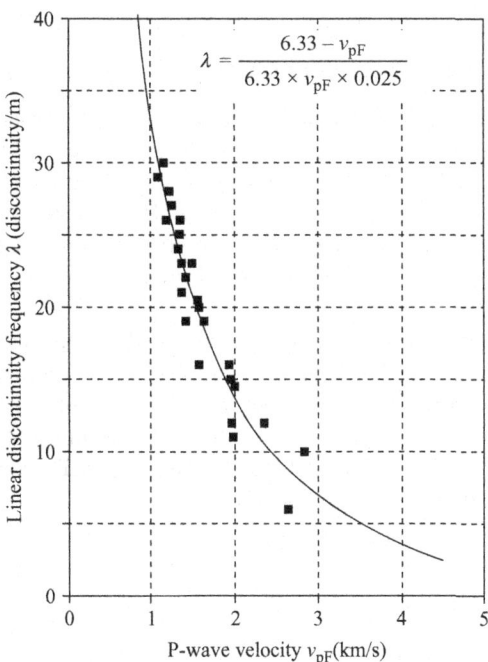

$$\lambda = \frac{6.33 - v_{pF}}{6.33 \times v_{pF} \times 0.025}$$

FIG. 4.7 Correlation between linear discontinuity frequency λ and P-wave velocity v_{pF} for heavily fractured calcareous rock in southern Italy. *(From Budetta, P., de Riso, R., Luca, C. de, 2001. Correlations between jointing and seismic velocities in highly fractured rock masses. Bull. Eng. Geol. Env. 60, 185–192.)*

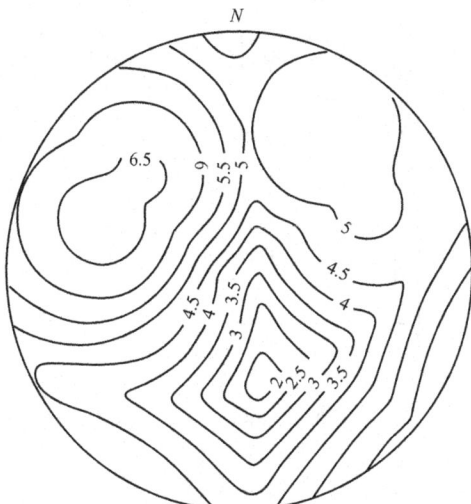

FIG. 4.8 Variation of discontinuity frequency with scanline direction. *(From Choi, S.Y., Park, H.D., 2004. Variation of rock quality designation (RQD) with scanline orientation and length: a case study in Korea. Int. J. Rock Mech. Min. Sci. 41, 207–221.)*

4.5.2 Rock Quality Designation (RQD)

RQD was proposed by Deere (1964) as a measure of the quality of borehole core. The RQD is defined as the ratio (in percent) of the total length of sound core pieces that are 0.1 m (4 in.) or longer to the length of the core run. The value 0.1 m is referred to as the threshold value. RQD is perhaps the most commonly used method for characterizing the jointing in borehole cores, although this parameter may also implicitly include other rock mass features such as weathering and core loss.

For RQD determination, the ISRM recommends a core size of at least NX (size 54.7 mm) drilled with double-tube core barrel using a diamond bit. Artificial fractures can be identified by close fitting of cores and unstained surfaces. All the artificial fractures should be ignored while counting the core length for RQD. A slow rate of drilling will also give better RQD. The correct procedure for measuring RQD is shown in Fig. 4.9.

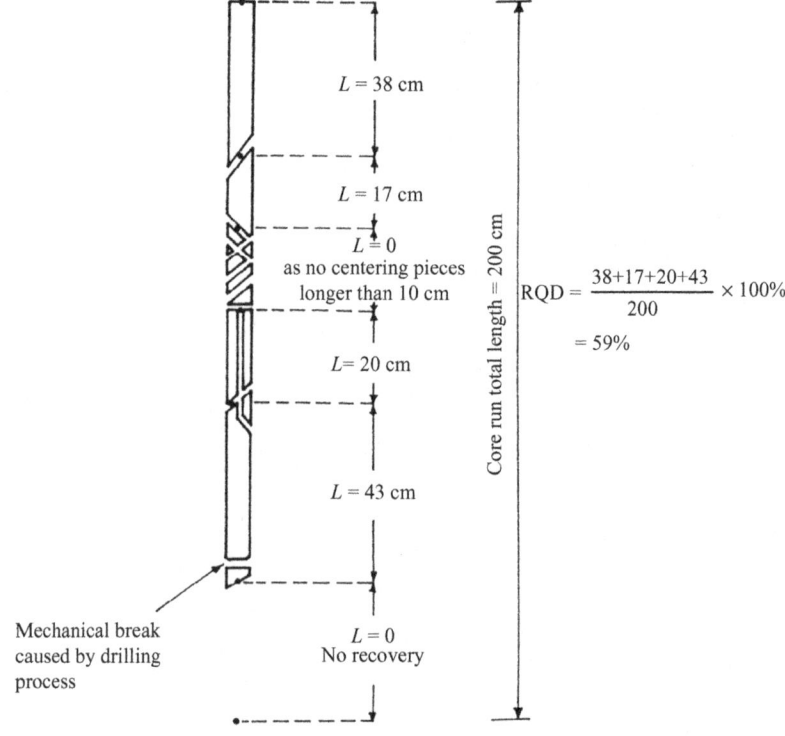

FIG. 4.9 Procedure for measurement and calculation of rock quality designation RQD. *(Based on Deere, D.U., 1989. Rock quality designation (RQD) after twenty years. U.S. Army Corps of Engineers Contract Report GL-89-1, Waterways Experiment Station, Viksburg, MS.)*

Correlations between RQD and linear discontinuity frequency λ have been derived for different discontinuity spacing distribution forms (Priest and Hudson, 1976; Sen and Kazi, 1984; Sen, 1993). For a negative exponential distribution of discontinuity spacings, Priest and Hudson (1976) derived the following relationship between RQD and linear discontinuity frequency λ

$$RQD = 100e^{-\lambda t}(\lambda t + 1) \tag{4.21}$$

where t is the length threshold. For $t = 0.1$ m as for the conventional RQD defined earlier, Eq. (4.21) can be expressed as:

$$RQD = 100e^{-0.1\lambda}(0.1\lambda + 1) \tag{4.22}$$

Fig. 4.10 shows the relations obtained by Priest and Hudson (1976) between measured values of RQD and λ, and the values calculated using Eq. (4.22).

For values of λ in the range $6\text{--}16$ m^{-1}, a good approximation to measured RQD values was found to be given by the linear relation:

$$RQD = 110.4 - 3.68\lambda \tag{4.23}$$

Fig. 4.11 plots the relationship between RQD and mean discontinuity spacing proposed by Bieniawski (1989) for determining the combined RQD and spacing ratings in the evaluation of Rock Mass Rating (RMR) (see Chapter 5).

It is noted that Eq. (4.21) is derived with the assumption that the length of the sampling line L is large so that the term $e^{-\lambda L}$ is negligible. For a short sampling line of length L, Sen and Kazi (1984) derived the following expression for RQD with a length threshold t:

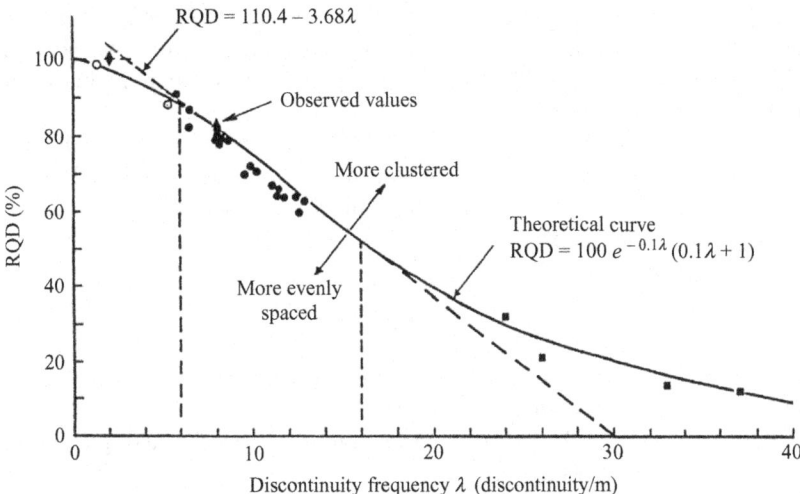

FIG. 4.10 Relationship between RQD and discontinuity frequency λ. *(From Priest, S.D., Hudson, J., 1976. Discontinuity spacing in rock. Int. J. Rock Mech. Min. Sci. Geomech. Abstr. 13, 135–148.)*

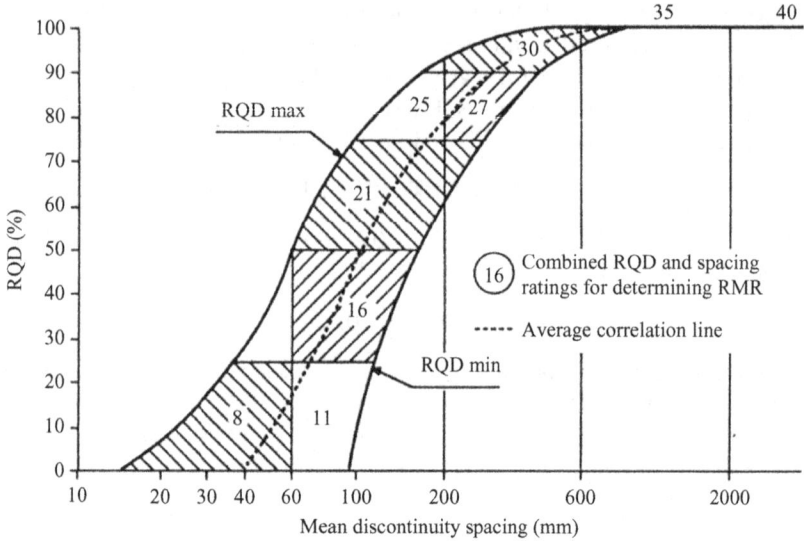

FIG. 4.11 Relationship between RQD and mean discontinuity frequency. *(From Bieniawski, Z.T., 1989. Engineering Rock Mass Classifications. John Wiley, Rotterdam.)*

$$RQD = \frac{100}{1 - e^{-\lambda L} - \lambda L e^{-\lambda L}} \left[e^{-\lambda t}(\lambda t + 1) - e^{-\lambda L}(\lambda L + 1) \right] \qquad (4.24)$$

Fig. 4.12 shows the variation of RQD with the length of the sampling line L for discontinuity frequency $\lambda = 10 \text{ m}^{-1}$ and length threshold $t = 0.1$ m. It can be seen that when L is smaller than about 0.5 m or when $\lambda L < 5$, RQD increases significantly when L increases. When L is larger than 0.5 or when $\lambda L > 5$, RQD changes little with L. Therefore, it is important to use sampling lines that are long so that $\lambda L > 5$ and $e^{-\lambda L}$ is negligible.

Seismic velocity measurements have also been used to estimate RQD. By comparing the P-wave velocity of in situ rock mass with laboratory P-wave velocity of intact drill core obtained from the same rock mass, the RQD can be estimated by (Deere et al., 1967):

$$RQD(\%) = \left(\frac{v_{pF}}{v_{p0}} \right)^2 \times 100 \qquad (4.25)$$

where v_{pF} is the P-wave velocity of in situ rock mass; and v_{p0} is the P-wave velocity of the corresponding intact rock.

Bery and Saad (2012) derived a correlation based on the data of granites in Penang and Sarawak, Malaysia:

$$RQD(\%) = 0.97 \left(\frac{v_{pF}}{v_{p0}} \right)^2 \times 100 \quad (r = 0.99) \qquad (4.26)$$

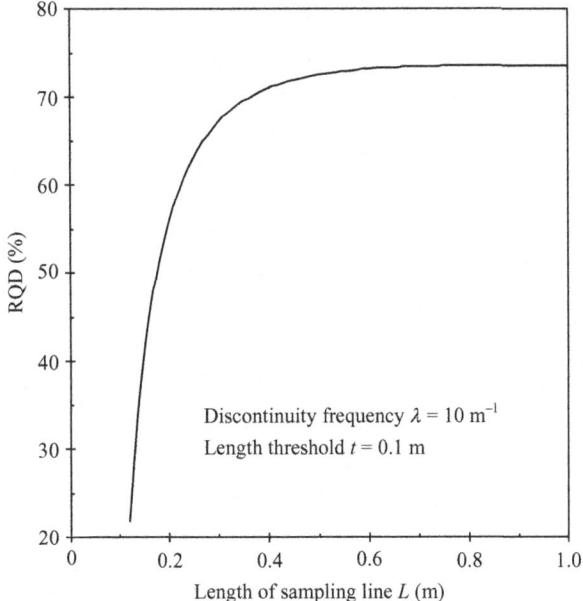

FIG. 4.12 Variation of RQD with the length of sample line L. *(Based on Sen, Z., Kazi, A., 1984. Discontinuity spacing and RQD estimates from finite length scanlines. Int. J. Rock Mech. Min. Sci. Geomech. Abstr. 21, 203–212; Priest, S.D., 1993. Discontinuity Analysis for Rock Engineering. Chapman & Hall.)*

where r is the correlation coefficient. It can be seen that Eq. (4.26) is very similar to Eq. (4.25)

Based on the data of limestones, mudstones, marls and shales at a dam site in Wadi Mujib, Jordan, El-Naqa (1996) obtained the following empirical correlation between RQD and P-wave velocities:

$$\mathrm{RQD}(\%) = 0.77 \left(\frac{v_{pF}}{v_{p0}}\right)^{1.05} \times 100 \quad (r = 0.89) \tag{4.27}$$

where v_{pF}, v_{p0}, and r are as defined earlier.

Sjögren et al. (1979) and Palmström (1995) proposed the following hyperbolic correlation between RQD and P-wave velocities:

$$\mathrm{RQD}(\%) = \frac{v_{pq} - v_{pF}}{v_{pq} \times v_{pF} \times k_q} \times 100 \tag{4.28}$$

where v_{pF} is the P-wave velocity of in situ rock mass; v_{pq} is the P-wave velocity of a rock mass with RQD $= 0$; and k_q is a parameter taking into account the actual conditions of the in situ rock mass. Based on regression analysis of the data obtained for heavily fractured calcareous rock masses out-cropping

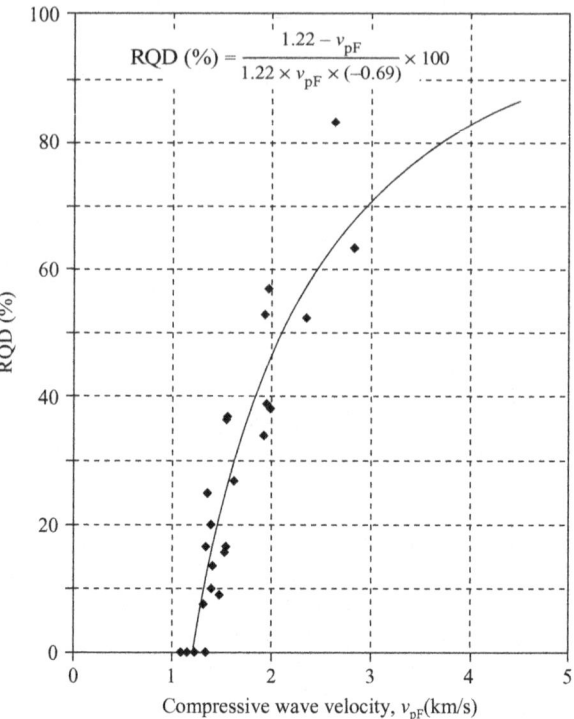

FIG. 4.13 Correlation between RQD and P-wave velocity v_{pF} for heavily fractured calcareous rock in southern Italy. *(From Budetta, P., de Riso, R., de Luca, C., 2001. Correlations between jointing and seismic velocities in highly fractured rock masses. Bull. Eng. Geol. Env. 60, 185–192.)*

in southern Italy (Fig. 4.13), Budetta et al. (2001) obtained v_{pq} and k_q as $v_q = 1.22$ km/s and $k_q = -0.69$, respectively, ie,

$$RQD(\%) = \frac{1.22 - v_{pF}}{1.22 \times v_{pF} \times (-0.69)} \times 100. \qquad (4.29)$$

Like the discontinuity frequency, RQD varies with the direction of the borehole or sampling line. As an example, Fig. 4.14 shows the variation of estimated RQD by Choi and Park (2004) for a site in the west-southern part of Korea on the lower hemisphere equal-angle stereo projection net. The variation of RQD with direction can be clearly seen. Therefore, it is important to specify the corresponding direction when stating a RQD value.

4.5.3 Areal and Volumetric Frequency

Discontinuity intensity is also often expressed in terms of areal and/or volumetric frequency. Table 4.3 lists different discontinuity intensity measures defined

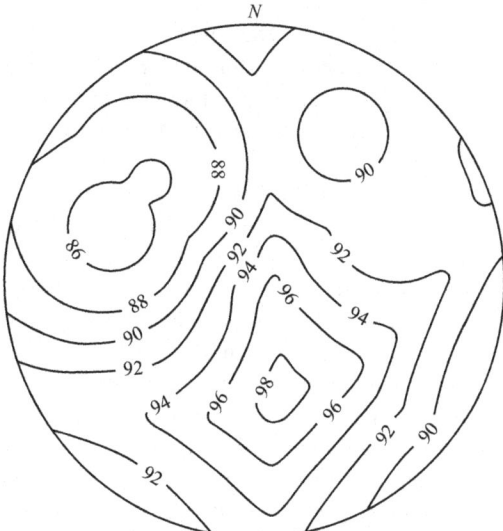

FIG. 4.14 Variation of estimated RQD with scanline direction. *(From Choi, S.Y., Park, H.D., 2004. Variation of rock quality designation (RQD) with scanline orientation and length: a case study in Korea. Int. J. Rock Mech. Min. Sci. 41, 207–221.)*

TABLE 4.3 Different Measures for Discontinuity Intensity

Measured Parameter	Dimension of Sampling Region		
	1. Line (Scanline or Borehole)	*2. Area (Rock Exposure)*	*3. Volume (Rock Mass)*
Number of discontinuities	P_{10} or λ Number of discontinuities per unit length of sampling line $[L^{-1}]$	P_{20} or λ_a Number of discontinuities per unit area of rock exposure $[L^{-2}]$	P_{30} or λ_v Number of discontinuities per unit volume of rock mass $[L^{-3}]$
Dimension one less than that of sampling region		P_{21} Length of discontinuity traces per unit area of rock exposure $[L^{-1}]$	P_{32} Area of discontinuities per unit volume of rock mass $[L^{-1}]$
Dimension equal to that of sampling region			P_{33} Volume of discontinuities per unit volume of rock mass $[-]$

Based on Dershowitz, W. S., Herda, H. H., 1992. Interpretation of fracture spacing and intensity. Proc. 33rd U.S. Symp. on Rock Mech., Santa Fe, NM, pp. 757–766.

by Dershowitz and Herda (1992). The areal frequency P_{20} (or λ_a) is the number of discontinuity traces per unit sampling area. If the number of discontinuity trace midpoints within a sampling window is known, λ_a can be simply obtained by dividing the number of discontinuity trace midpoints by the window area.

The difficulty is that, in general, the midpoints of discontinuity traces within a sampling window cannot be identified. To address this issue, Mauldon (1998) developed an end-point estimator of λ_a based on the principle of associated points (Parker and Cowan, 1976; Laslett, 1982). Since one discontinuity trace midpoint corresponds to two discontinuity trace ends, the spatially averaged density of midpoints is simply half of that of ends, and an unbiased estimate of λ_a is given by one half the total number of ends of discontinuity traces in a sampling window divided by the area of the window. Consider an irregular convex sampling window of area A (Fig. 4.15). The discontinuity traces intersecting the sampling window can be divided into three classes (Pahl, 1981; Priest, 1993; Mauldon, 1998): (1) discontinuity traces with both ends censored (transecting traces), (2) discontinuity traces with one end censored and one end observable (dissecting traces), and (3) discontinuity traces with both ends observable (contained traces). If the numbers of discontinuity traces in each of the above three categories are N_0, N_1 and N_2, respectively, the total number of discontinuity traces, N, will be:

$$N = N_0 + N_1 + N_2 \tag{4.30}$$

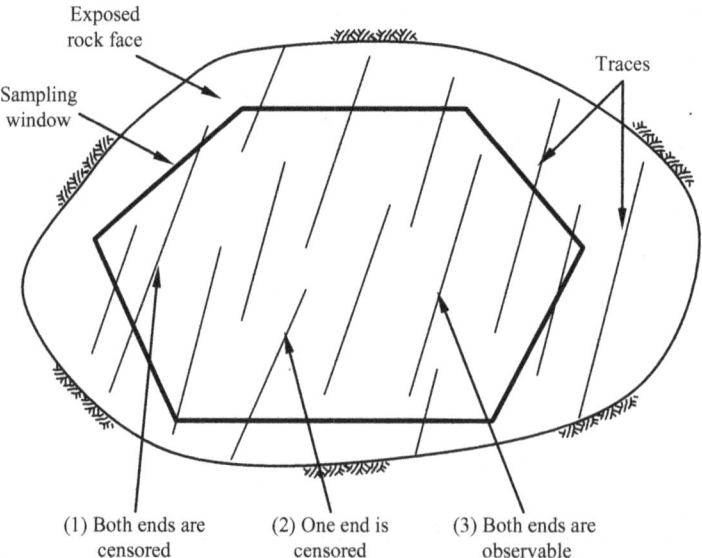

FIG. 4.15   Discontinuity traces intersecting an irregular convex sampling window of area A.

And the total number of discontinuity trace ends within the window is $N - N_0 + N_2$, and λ_a can be determined by:

$$\lambda_a = \frac{1}{2A}(N - N_0 + N_2) \qquad (4.31)$$

It is noted that the areal frequency is dependent on the relative orientation between the discontinuities and the sampling plane. If the area frequency of a set of discontinuities on a sampling plane that makes an acute angle θ to the discontinuity planes is λ_a, the normal areal frequency λ_{an} (ie, the areal frequency when the sampling plane is normal to the discontinuity planes) is:

$$\lambda_{an} = \frac{\lambda_a}{\sin \theta} \qquad (4.32)$$

The term $1/\sin \theta$ in Eq. (4.32) can be regarded as a correction factor for sampling bias in window sampling, similar to the Terzaghi's correction factor (Terzaghi, 1965) for a scanline sampling (see Section 4.5.1).

Considering that in many cases the exposed rock faces such as a tunnel surface are non-planar, Song (2006) presented a method for estimating areal frequency using a non-planar sampling window. The method consists of four steps as described below:

(1) Divide the non-planar sampling window into several sub planar windows. For example, an exposed tunnel surface can be divided into several, say M, rectangular windows (Fig. 4.16).

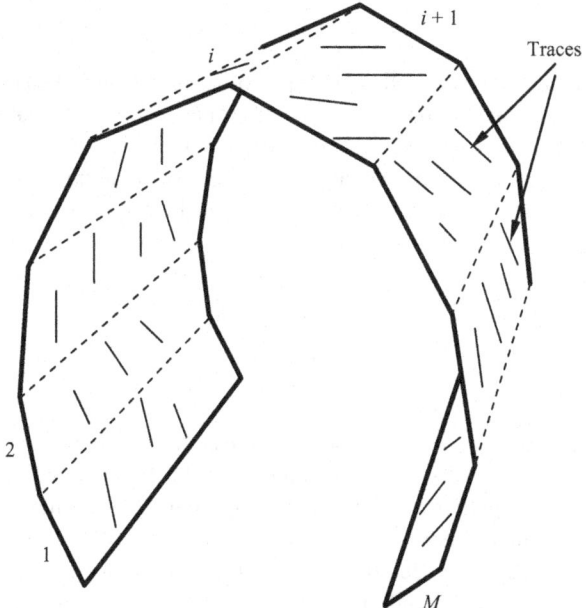

FIG. 4.16 Dividing a non-planar sampling window into M non-parallel rectangular sampling windows.

(2) Calculation the areal frequency in each sub planar window using Eq. (4.31). For example, for the ith sub planar window of area A_i:

$$\lambda_{ai} = \frac{1}{2A_i}(N_i - N_{0i} + N_{2i}) \tag{4.33}$$

where N_i, N_{0i} and N_{2i} are the number of all, transecting and contained discontinuity traces in the ith sub planar window, respectively.

(3) Convert the obtained frequencies into the normal frequencies using Eq. (4.32). For example, for the ith sub planar window which makes an acute angle φ_i to the discontinuity planes:

$$\lambda_{ani} = \frac{1}{2A_i \sin\varphi_i}(N_i - N_{0i} + N_{2i}) \tag{4.34}$$

(4) Determine the overall normal area frequency λ_{an} as a weighted sum of the obtained normal area frequencies in the M sub planar windows:

$$\lambda_{an} = \sum_{i=1}^{M} w_i \lambda_{ani} \tag{4.35}$$

where w_i is a weight factor for the ith sub planar window defined as

$$w_i = \frac{A_i \sin\varphi_i}{\sum_{j=1}^{M} A_j \sin\varphi_j} \tag{4.36}$$

The weight factor w_i as defined in Eq. (4.36) is based on the fact that the areal frequency estimated from a larger window or from a window making a larger angle with the discontinuity planes is more reliable due to the larger sample size which reduces the estimation error. By substituting Eqs. (4.34) and (4.36) into Eq. (4.35), the overall normal areal frequency is as follows:

$$\lambda_{an} = \frac{\sum_{i=1}^{M}(N_i - N_{0i} + N_{2i})}{2\sum_{j=1}^{M} A_j \sin\varphi_j} \tag{4.37}$$

If there are several sets of discontinuities, the above procedure can be applied to each set of discontinuities and then add the obtained normal areal frequency values together to get the total normal areal frequency.

The discontinuity intensity measure P_{21} is the length of discontinuity traces per unit sampling area and can be related to the areal frequency λ_a through the mean discontinuity trace length μ_1:

$$P_{21} = \lambda_a \mu_1 \tag{4.38}$$

The mean discontinuity trace length μ_l can be determined as in Section 4.6.2.

Mauldon et al. (1999) derived a simple expression for estimating discontinuity intensity P_{21} from circular scanline sampling:

$$P_{21} = \frac{N}{4c} \tag{4.39}$$

where N is the number of discontinuity traces intersecting the circular scanline; and c is the radius of the scanline circle (Fig. 4.17). Circular scanline sampling measures only the traces intersecting the line of the circle. One advantage of circular scanlines over straight scanlines is the elimination of directional bias. Circular scanlines have been used for discontinuity sampling at exposed rock faces by different researchers (Einstein et al., 1979; Titley et al., 1986; Davis and Reynolds, 1996; Mauldon et al., 1999; Watkins et al., 2015).

The volume frequency P_{30} (or λ_v) is the number of discontinuities per unit volume of rock mass. Like λ_a, λ_v is scale-dependent and changes with the size of the sampling region for regions at scales smaller than the maximum discontinuity size. Therefore, the intensity measure P_{32}, the area of discontinuities per unit volume of rock mass, can be used. P_{32} is related to λ_v through the mean area $\bar{A}$ of the discontinuities:

$$P_{32} = \lambda_v \bar{A} \tag{4.40}$$

The mean area of the discontinuities can be determined as in Section 4.6.4.

The volumetric frequency λ_v can be determined from the mean discontinuity set spacings within a volume of rock mass as (Palmström, 1982):

$$\lambda_v = \sum_{i=1}^{N} \left(\frac{1}{s_i} \right) \tag{4.41}$$

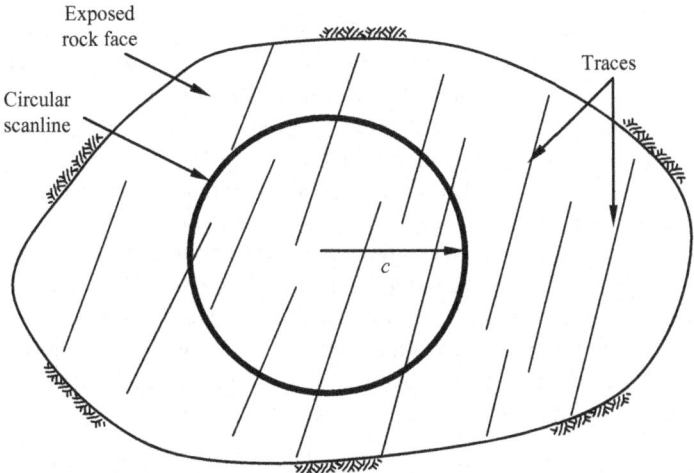

FIG. 4.17   Circular scanline sampling.

where s_i is the mean discontinuity set spacing in meters of the ith discontinuity set; and N is the total number of discontinuity sets.

Random discontinuities in the rock mass can be considered by assuming a random spacing s_r for each of them. According to Palmström (2002), $s_r = 5$ m can be assumed. So the volumetric frequency λ_v can be in general expressed as:

$$\lambda_v = \sum_{i=1}^{N} \left(\frac{1}{s_i}\right) + \frac{N_r}{5} \tag{4.42}$$

where N_r is the number of random discontinuities.

The volumetric frequency λ_v can also be estimated from the areal frequency λ_a using the following empirical expression (Palmström, 2002):

$$\lambda_v = k_a \lambda_a \tag{4.43}$$

where k_a is a correlation factor, which varies mainly between 1 and 2.5 with an average value of 1.5. The highest value is where the sampling plane is parallel to the main discontinuity set.

The ISRM (1978) presented the following approximate correlation between volumetric frequency λ_v and RQD based on the work by Palmström (1974):

$$RQD = 115 - 3.3\lambda_v \tag{4.44}$$

Here RQD $= 0$ for $\lambda_v > 35$, and RQD $= 100$ for $\lambda_v < 4.5$. Palmström (2005) modified Eq. (4.44) as follows, with the new equation giving somewhat better results:

$$RQD = 110 - 2.5\lambda_v \quad (4 \leq \lambda_v \leq 44) \tag{4.45}$$

4.5.4 Block Size

Block size is another important parameter for describing discontinuity intensity and rock mass behavior. Block dimensions are determined by discontinuity spacings, by the number of discontinuity sets and by the persistence of the discontinuities delineating potential blocks. Block shapes are determined by the number of sets and the orientations of the discontinuities delineating potential blocks. Where relatively regular jointing exists such as the jointing in sedimentary rocks, it may be possible to give adequate description of the block shapes. Fig. 4.18 shows the examples of block shapes presented by Dearman (1991). In most cases, however, the block shapes are irregular and can only be roughly described.

Where individual blocks can be observed in a surface, their volumes can be directly measured from relevant dimensions by selecting several representative blocks and measuring their average dimensions. Where three persistent discontinuity sets occur, the block volume can be calculated as:

$$V_b = \frac{s_1 s_2 s_3}{\sin\gamma_1 \sin\gamma_3 \sin\gamma_3} \tag{4.46}$$

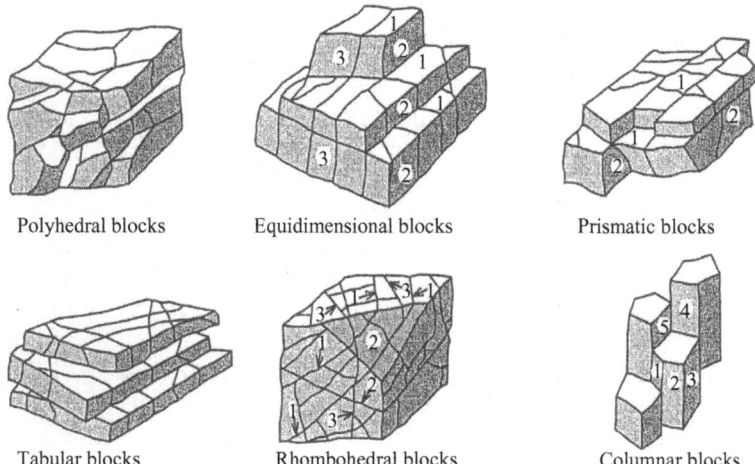

Polyhedral blocks   Equidimensional blocks   Prismatic blocks

Tabular blocks   Rhombohedral blocks   Columnar blocks

FIG. 4.18 Examples of block shapes. *(Based on Dearman, W.R., 1991. Engineering Geological Mapping. Butterworth-Heineman, Oxford.)*

where s_1, s_2, s_3 are the normal set spacing of the three discontinuity sets, respectively; and γ_1, γ_2, γ_3 are the angles between the discontinuity sets. If the discontinuity sets intersect at right angles, the block volume can be simply calculated as:

$$V_b = s_1 s_2 s_3 \qquad (4.47)$$

Eqs. (4.46) and (4.47) are applicable to three persistent discontinuity sets. For non-persistent discontinuities, the equivalent block volume can be determined by (Cai et al., 2004; Kim et al., 2007):

$$V_b = \frac{s_1 s_2 s_3}{\sin\gamma_1 \sin\gamma_3 \sin\gamma_3 \sqrt{PR_1 PR_2 PR_3}} \qquad (4.48)$$

where PR_1, PR_2, PR_3 are the persistence ratio of the three discontinuity sets, respectively, which can determined as in Section 4.6.1.

The block size can also be described based on the volumetric discontinuity frequency λ_v using the terms in Table 4.4 or Table 4.5, which provide finer divisions and the related discontinuity spacing (s) values. To describe both the block size and shape, the adjectives listed in Table 4.6 can be used.

Fig. 4.19 shows the comparison of the possible ranges of RQD, volumetric frequency λ_v and block volume V_b from Palmström (2002). It can be seen that RQD covers only a limited part of the range of jointing.

Palmström (2000) proposed the following general expression relating the block volume V_b to the volumetric frequency λ_v:

$$V_b = \beta \times \lambda_v^{-3} \qquad (4.49)$$

TABLE 4.4 Terms for Describing Block Size Based on Volumetric Discontinuity Frequency λ_v

Volumetric Frequency λ_v (Discontinuity/m³)	Description
<1	Very large blocks
1–3	Large blocks
3–10	Medium-sized blocks
10–30	Small blocks
>30	Very small block

Based on ISRM, 1978. Suggested methods for the quantitative description of discontinuities in rock masses. International Society for Rock Mechanics, Commission on Standardization of Laboratory and Field Tests. Int. J. Rock Mech. Min. Sci. Geomech. Abstr. 15, 319–368.

TABLE 4.5 Terms for Describing Block Size Based on Volumetric Discontinuity Frequency λ_v and Related Discontinuity Spacing s

Volumetric Frequency λ_v (Discontinuity/m³)	Block Volume V_b	Discontinuity Spacing s
Extremely low <0.3	Extremely large size >1000 m³	>10 m
Very low 0.3–1	Very large size 30–1000 m³	3–10 m
Low 1–3	Large size 1–30 m³	1–3 m
Moderately high 3–10	Moderate size 0.03–1 m³	0.3–1 m
High 10–30	Small size 1–30 dm³	10–30 cm
Very high 30–100	Very small size 0.03–1 dm³	3–10 cm
Extremely high >100	Extremely small size <30 cm³	<3 cm

Based on Palmström, A., 2000. Recent developments in rock support estimates by the RMi. J. Rock Mech. Tunneling Technol. 6(1), 1–19.

where β is the block shape factor representing the jointing pattern and can be estimated by:

$$\beta = 20 + 7(S_{max}/S_{min}) \tag{4.50}$$

in which S_{max} and S_{min} are the longest and shortest dimension of the block, respectively.

TABLE 4.6 Adjectives for Describing Block Size and Shape

Adjective	Description
Massive	Few discontinuities or very wide spacing
Blocky	Approximately equidimensional
Tabular	One dimension considerably smaller than the other two
Columnar	One dimension considerably larger than the other two
Irregular	Wide variations of block size and dimensions
Crushed	Heavily jointed to "sugar cube"

Based on ISRM, 1978. Suggested methods for the quantitative description of discontinuities in rock masses. International Society for Rock Mechanics, Commission on Standardization of Laboratory and Field Tests. Int. J. Rock Mech. Min. Sci. Geomech. Abstr. 15, 319–368.

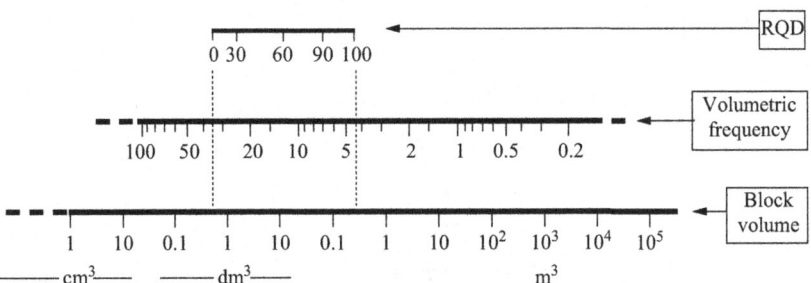

FIG. 4.19 Ranges of RQD, volumetric frequency λ_v and block volume V_b. *(Based on Palmström, A. (2002). Measurement and characterization of rock mass jointing. In V. M. Sharma & K. R. Saxena (Eds.). In-Situ Characterization of Rocks. Balkema, Lisse. pp. 49–98.)*

4.6 DISCONTINUITY PERSISTENCE, TRACE LENGTH AND SIZE

4.6.1 Discontinuity Persistence

Persistence is a term used to describe the areal extent or size of a discontinuity within a plane. It can be crudely quantified by observing discontinuity trace lengths on exposed rock faces. ISRM (1978) uses the most common or modal trace lengths of each set of discontinuities measured on exposed rock faces to classify persistence according to Table 4.7.

Discontinuity persistence is one of the most important rock mass parameters, but one of the most difficult to determine. The discontinuities of one particular set are often more continuous than those of the other sets. The minor sets tend to terminate against the primary features, or they may terminate in solid rock. The sets

TABLE 4.7 Classification of Discontinuity Persistence

Description	Modal Trace Length (m)
Very low persistence	<1
Low persistence	1–3
Medium persistence	3–10
High persistence	10–20
Very high persistence	>20

Based on ISRM, 1978. Suggested methods for the quantitative description of discontinuities in rock masses. International Society for Rock Mechanics, Commission on Standardization of Laboratory and Field Tests. Int. J. Rock Mech. Min. Sci. Geomech. Abstr. 15, 319–368.

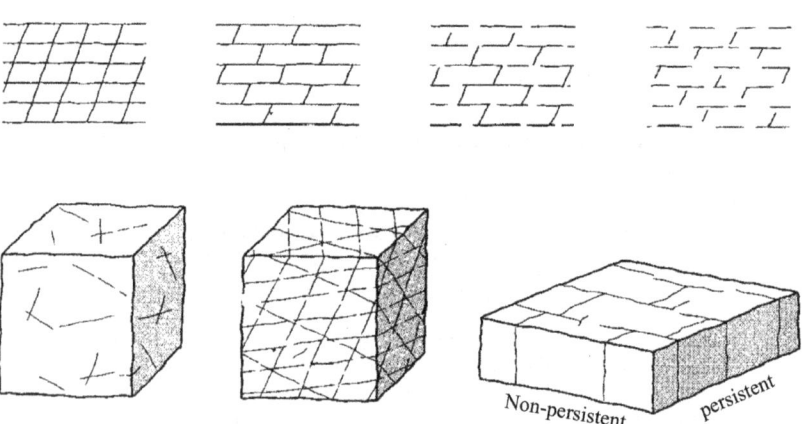

FIG. 4.20 Simple sketches and block diagrams indicating the persistence of various sets of discontinuities. *(Based on ISRM, 1978. Suggested methods for the quantitative description of discontinuities in rock masses. International Society for Rock Mechanics, Commission on Standardization of Laboratory and Field Tests. Int. J. Rock Mech. Min. Sci. Geomech. Abstr. 15, 319–368.)*

of discontinuities can be distinguished by terms of persistent, sub-persistent and non-persistent, respectively. Fig. 4.20 shows a set of simple plane sketches and block diagrams used to help indicate the persistence of various sets of discontinuities in a rock mass. Clearly, the persistence of discontinuities has a major influence on the shear strength developed in the plane of the discontinuity.

Persistence ratio PR is often used to describe the persistence of discontinuities. In the literature, discontinuity persistence ratio PR is usually defined as:

$$PR = \lim_{A_S \to \infty} \frac{\sum_i a_{S_i}}{A_S} \tag{4.51}$$

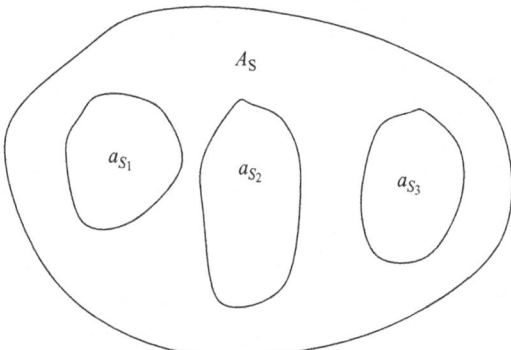

FIG. 4.21 Definition of PR as area ratio. *(Based on Einstein, H. H., Veneziano, D., Baecher, G. B., O'Reilly, K. J., 1983. The effect of discontinuity persistence on rock slope stability. Int. J. Rock Mech. Min. Sci. Geomech. Abstr. 20, 227–236.)*

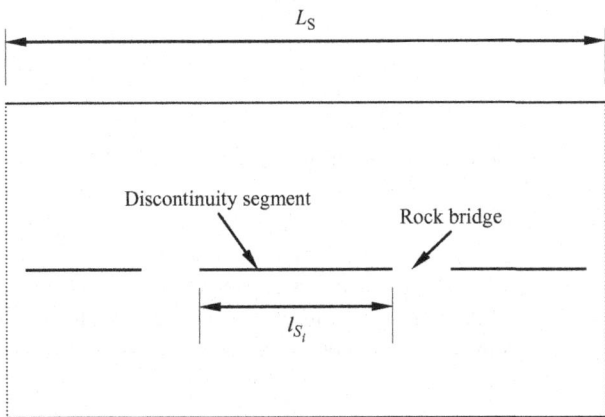

FIG. 4.22 Definition of PR as length ratio.

in which S is a region on the discontinuity plane with area A_S; and a_{S_i} is the area of the ith discontinuity in S (Fig. 4.21). The summation in Eq. (4.51) is over all discontinuities in S. Equivalently, discontinuity persistence ratio PR can be expressed as a limit length ratio along a given line on a discontinuity plane. In this case,

$$PR = \lim_{L_S \to \infty} \frac{\sum_i l_{S_i}}{L_S} \qquad (4.52)$$

in which L_S is the length of a straight line segment S and l_{S_i} is the length of the ith discontinuity segment in S (Fig. 4.22). For a finite sampling length L_S, PR can be simply estimated by (Fig. 4.23):

$$PR = \frac{\sum DL}{\sum DL + \sum RBL} \qquad (4.53)$$

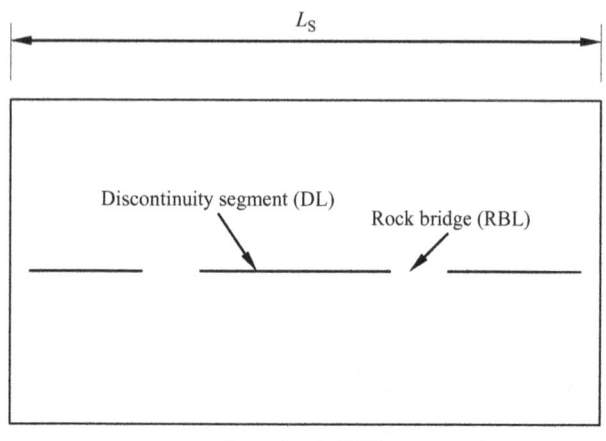

$$L_S = \Sigma DL + \Sigma RBL$$

FIG. 4.23   Estimation of PR for a finite sampling length.

where $\sum DL$ is the sum of the length of all discontinuities; and $\sum RBL$ is the sum of the length of all rock bridges.

The above definition of discontinuity persistence ratio PR considers only the discontinuities in the same plane. However, according to Einstein et al. (1983), when two discontinuities are at a low-angle transition ($\beta < \theta_t$, see Fig. 4.24), the rock bridge may fail by the same mechanism as the in-plane rock bridge (Fig. 4.25), where θ_t is the angle of the tension cracks which can be obtained from Mohr's circle (Fig. 4.25A). For both the in-plane (Fig. 4.25) and the low-angle out-of-plane (Fig. 4.24) transitions, the intact-rock resistance R can be calculated by:

$$R = \tau_a d \tag{4.54}$$

where d is the "in-plane length" of the rock bridge; and τ_a is the peak shear stress mobilized in the direction of discontinuities which can be obtained by (Einstein et al., 1983):

$$\tau_a^2 = \sigma_t(\sigma_t - \sigma_a) \tag{4.55}$$

where σ_t is the tensile strength of the intact rock; and σ_a is the effective normal stress on the discontinuity plane.

Zhang (1999) proposed a definition of discontinuity persistence ratio PR that considers both in-plane and low-angle-transition discontinuities:

$$PR = \lim_{L_S \to \infty} \frac{\sum_i DL_i + \sum_l DL_l}{L_S} \tag{4.56}$$

in which L_S is the total sampling length along the direction of the discontinuity traces; DL_i is the length of the ith in-plane discontinuities; and DL_l is the length

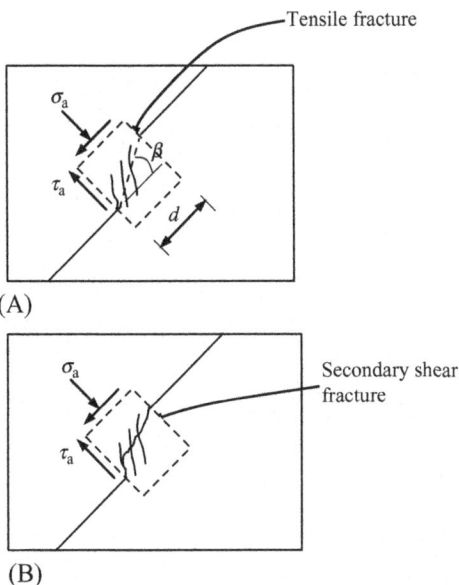

(A)

(B)

FIG. 4.24 Failure of "low-angle" transitions through intact rock: (A) Tensile fracture; and (B) Secondary shear fracture. *(Based on Einstein, H. H., Veneziano, D., Baecher, G. B., O'Reilly, K. J., 1983. The effect of discontinuity persistence on rock slope stability. Int. J. Rock Mech. Min. Sci. Geomech. Abstr. 20, 227–236; Zhang, L., 1999. Analysis and Design of Drilled Shafts in Rock. PhD thesis, Massachusetts Institute of Technology, Cambridge, MA.)*

of the *l*th low-angle-transition discontinuities (Fig. 4.26). For a finite sampling length, PR can be simply approximated by

$$PR = \frac{\displaystyle\sum_{i=1}^{m} DL_i + \sum_{l=1}^{n} DL_l}{L_S} \qquad (4.57)$$

where *m* and *n* are the numbers of the in-plane and low-angle-transition ($\beta < \theta_t$) discontinuities within the sampling length L_S (Fig. 4.26), respectively.

4.6.2 Discontinuity Trace Length

Discontinuity trace length is an important parameter for describing discontinuity size and persistence. When sampling trace lengths on exposed rock surfaces, errors can occur due to the following biases (Baecher and Lanney, 1978; Einstein et al., 1979; Kulatilake and Wu, 1984c; Mauldon, 1998; Priest and Hudson, 1981; Zhang and Einstein, 1998, 2000; Zhang and Ding 2010):

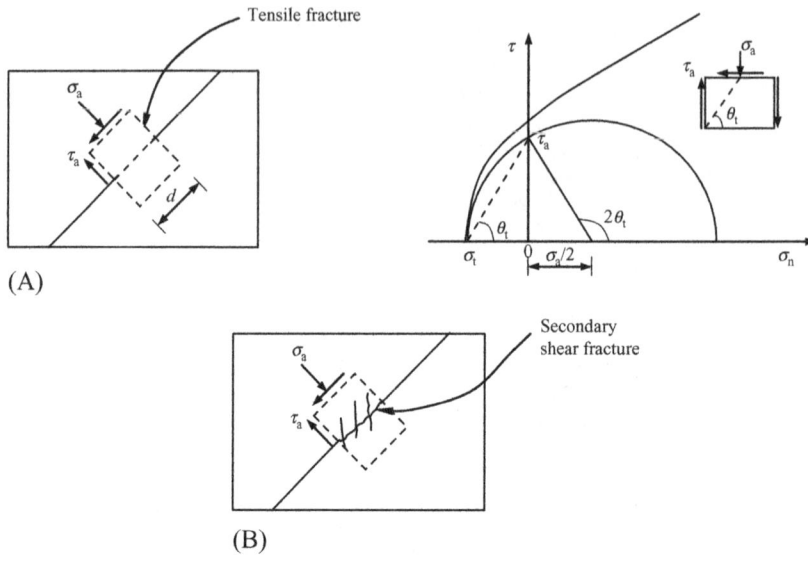

FIG. 4.25 In-plane failure of intact rock: (A) Tensile fracture and corresponding Mohr's circle; and (B) Secondary shear fracture. *(Based on Einstein, H. H., Veneziano, D., Baecher, G. B., O'Reilly, K. J., 1983. The effect of discontinuity persistence on rock slope stability. Int. J. Rock Mech. Min. Sci. Geomech. Abstr. 20, 227–236; Zhang, L., 1999. Analysis and Design of Drilled Shafts in Rock. PhD thesis, Massachusetts Institute of Technology, Cambridge, MA.)*

(1) Orientation bias: the probability of a discontinuity appearing at an exposed rock surface depends on the relative orientation between the rock face and the discontinuity.
(2) Size bias: large discontinuities are more likely to be sampled than small discontinuities. This bias affects the results in two ways: (a) a larger discontinuity is more likely to appear at an exposed rock face than a smaller one; and (b) a longer trace is more likely to appear in a sampling area than a shorter one.
(3) Truncation bias: very small trace lengths are difficult or sometimes impossible to measure. Therefore, trace lengths below some known cutoff length are not recorded.
(4) Censoring bias: long discontinuity traces may extend beyond the visible exposure so that one end or both ends of the discontinuity traces cannot be seen.

In inferring the trace length distribution on an infinite surface from the measured trace lengths on a finite size area on this surface, biases (2b), (3) and (4) should be considered. Biases (1) and (2a) are of interest only in 3D simulations of discontinuities, ie, when inferring discontinuity size distributions as discussed in Section 4.6.4.

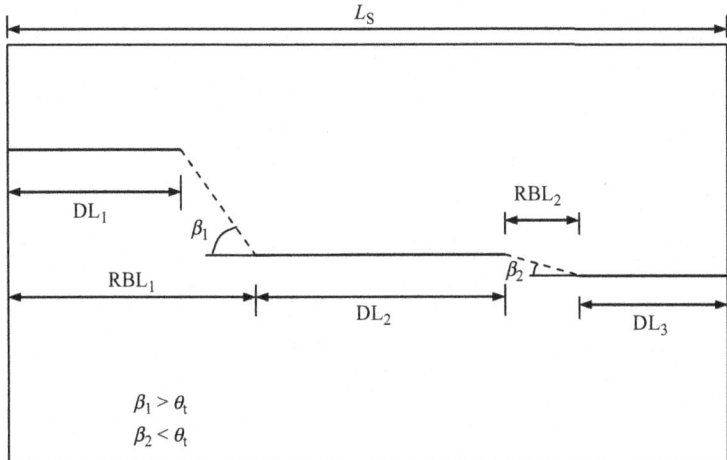

For definition of PR considering only in-plane discontinuities, PR = DL_2/L_S;

For definition of PR considering both in-plane and low-angle-transition discontinuities,
PR = $(DL_2 + DL_3)/L_S$

FIG. 4.26 Definition of PR considering both in-plane and low-angle-transition discontinuities. *(From Zhang, L., 1999. Analysis and Design of Drilled Shafts in Rock. PhD thesis, Massachusetts Institute of Technology, Cambridge, MA.)*

(a) Probability distribution of measured trace lengths

Many investigators have looked into the distribution of trace lengths (Table 4.8). Apart from Baecher et al. (1977), Cruden (1977), Einstein et al. (1979), and Kulatilake (1993), others have based their argument on inspection rather than on goodness-of-fit tests. It seems that only Baecher et al. (1977), Einstein et al. (1979), and Kulatilake (1993) have tried more than one distribution to find the best distribution to represent trace length data.

To find the suitable distribution for the measured trace lengths of each discontinuity set, the distribution forms in Table 4.8 can be checked by using χ^2 and Kolmogorov-Smirnov goodness-of-fit tests.

(b) Corrected mean trace length

In inferring the corrected mean trace length (ie, the mean trace length on an infinite surface), from the measured trace lengths on a finite exposure, biases (2b), (3) and (4) should be considered. Truncation bias (3) can be corrected using the method of Warburton (1980a). Decreasing the truncation level in discontinuity surveys can reduce the effects of truncation bias on trace length estimates. It is practically feasible to observe and measure trace lengths as low as 10 mm both

TABLE 4.8 Distribution Forms of Trace Lengths

Investigator	Distribution
Robertson (1970)	Exponential
McMahon (Mostyn and Li, 1993)	Lognormal
Bridges (1975)	Lognormal
Call et al. (1976)	Exponential
Barton (1977)	Lognormal
Cruden (1977)	Exponential
Baecher et al. (1977)	Lognormal
Einstein et al. (1979)	Lognormal
Priest and Hudson (1981)	Exponential
Kulatilake (1993)	Exponential and Gamma (Gamma better)

in the field and from photographs (Priest and Hudson, 1981). Truncation at this level will have only a small effect on the data, particularly if the mean trace length is in the order of meters (Priest and Hudson, 1981; Einstein and Baecher, 1983). Therefore, the effect of truncation bias on trace length estimates can be ignored. However, biases (2b) and (4) are important (Kulatilake and Wu, 1984c) and need to be considered.

Pahl (1981) suggested a technique to estimate the mean trace length on an infinite surface produced by a discontinuity set whose orientation has a single value, ie, all discontinuities in the set have the same orientation. This technique is based on the categorization of randomly located discontinuities that intersect a rectangular sampling window of width a and height b, and the traces make an angle φ with the height side, as shown in Fig. 4.27. If the numbers of the traces with both ends censored, with one end censored and one end observable, and with both ends observable are N_0, N_1 and N_2, respectively, the mean trace length μ_1 can be determined by (Pahl, 1981):

$$\mu_1 = \frac{ab(N + N_0 - N_2)}{(a\cos\varphi + b\sin\varphi)(N - N_0 + N_2)} \quad (4.58)$$

where N is the total number of traces, which is simply equal to $N_0 + N_1 + N_2$.

Although the approach in Eq. (4.58) is both rigorous and easy to implement, it relies on the discontinuities being grouped into a parallel or nearly parallel set. Kulatilake and Wu (1984c) and Mauldon (1998) extended Pahl's technique to

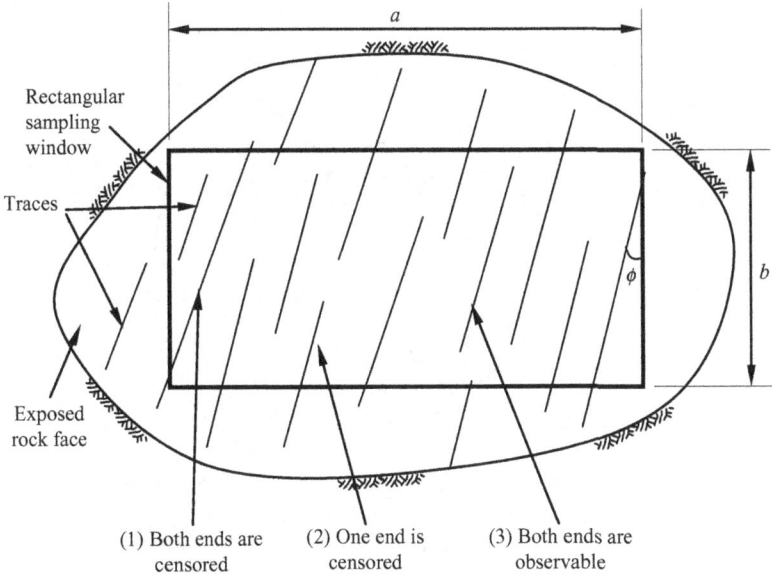

FIG. 4.27   Discontinuities intersecting a rectangular sampling window.

discontinuities whose orientation is described by a probabilistic distribution. A major difficulty in applying the extended technique is to determine the probabilistic distribution function of the orientation of discontinuities.

Using different methods, Mauldon (1998) and Zhang and Einstein (1998) independently derived the following expression for estimating the mean trace length from the observed trace data in a circular sampling window (Fig. 4.28):

$$\mu_1 = \frac{\pi(N + N_0 - N_2)}{2(N - N_0 + N_2)}c \qquad (4.59)$$

where c is the radius of the circular sampling window.

Trace length measurements are not needed when using Eqs. (4.58) and (4.59). In the derivation of Eqs. (4.58) and (4.59), discontinuity trace length l can be anywhere between zero and infinity. Hence, μ_1 obtained by Eqs. (4.58) and (4.59) does not contain errors due to biases (2b) and (4) as described before. It can also be seen that no sampling data about the orientation of discontinuities is needed when using Eq. (4.59) to estimate the mean trace length, ie, Eq. (4.59) is applicable to traces with arbitrary orientation distributions. This is a major advantage of the method using a circular sampling window over the methods using a non-circular sampling window. Eq. (4.59) can be used to estimate the mean trace length of more than one set of discontinuities. The orientation distribution-free nature of this method comes from the symmetric properties of the circular sampling windows.

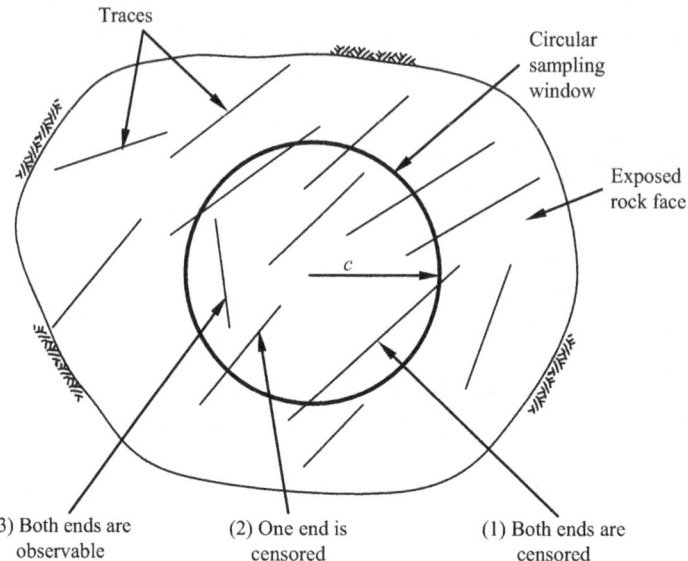

FIG. 4.28   Discontinuities intersecting a circular sampling window.

The parameter μ_1 in Eq. (4.59) is the population (thus correct or true) mean trace length, with N, N_0 and N_2 being the expected total number of traces intersecting the window, the expected number of traces with both ends censored and the expected number of traces with both ends observable, respectively. In practice, the exact values of N, N_0 and N_2 are not known and thus μ_1 has to be estimated using sampled data. From sampling in one circular window, what we get is only one sample of N, N_0 and N_2 and from this sample only a point estimate of μ_1 can be obtained. For example, for a sample of $\hat{N}$ traces intersecting the sampling window, if $\hat{N}_o$ and $\hat{N}_2$ are the numbers of discontinuities that appear on the window with both ends censored and both ends observable, respectively, the mean trace length of the sample, $\hat{\mu}_1$, can be obtained by:

$$\hat{\mu}_1 = \frac{\pi\left(\hat{N} + \hat{N}_0 - \hat{N}_2\right)}{2\left(\hat{N} - \hat{N}_0 + \hat{N}_2\right)} c \tag{4.60}$$

In other words, $\hat{\mu}_1$ of several samples can be used to evaluate μ_1.

When applying Eq. (4.60), the following two special cases may occur:

(1) If $\hat{N}_o = \hat{N}$, then $\hat{\mu}_1 \to \infty$. In this case, all the discontinuities intersecting the sampling window have both ends censored. This implies that the area of the window used for the discontinuity survey may be too small.

(2) If $\hat{N}_2 = \hat{N}$, then $\hat{\mu}_1 = 0$. In this case, all the discontinuities intersecting the sampling window have both ends observable. According to Pahl (1981),

this results is due to violation of the assumption that the midpoints of traces are uniformly distributed in the 2D space.

These two special cases can be addressed by increasing the sampling window size and/or changing the sampling window location (Zhang, 1999). Another method to address these two special cases is to use multiple windows of the same size at different locations and then use the total numbers from those windows to estimate $\hat{\mu}_1$ (Zhang and Einstein, 1998).

For the application of estimator $\hat{\mu}_1$, Eq. (4.60), it is important to know its variance or standard deviation. According to Zhang and Ding (2010), the standard deviation (SD) of $\hat{\mu}_1$ can be estimated by:

$$\mathrm{SD}(\hat{\mu}_1) \cong \frac{\pi c \sqrt{\hat{N}_0/\hat{N} + \hat{N}_2/\hat{N} - \left(\hat{N}_o/\hat{N} - \hat{N}_2/\hat{N}\right)^2}}{\sqrt{\hat{N}}\left[1 - \left(\hat{N}_0/\hat{N} - \hat{N}_2/\hat{N}\right)\right]^2} \tag{4.61}$$

If the exposed rock face such as a tunnel surface is non-planar, it can be divided into several sub planar sampling windows and the overall mean trace length can then be determined based on the obtained values from each window. For example, if a tunnel surface is divided into M rectangular windows as in Fig. 4.16, the mean trace length from window i, $\hat{\mu}_{li}$, can be determined by:

$$\hat{\mu}_{li} = \frac{a_i b_i \left(\hat{N} + \hat{N}_0 - \hat{N}_2\right)}{\left(a_i \cos\varphi_i + b_i \sin\varphi_i\right)\left(\hat{N} - \hat{N}_0 + \hat{N}_2\right)} \tag{4.62}$$

where a_i, b_i, and φ_i are the width, height, and angle between the traces and the height side of window i, respectively, as shown in Fig. 4.27. By applying the weight factor, w_i, from Eq. (4.36), the overall mean trace length, $\hat{\mu}_1$, is calculated as (Song, 2006):

$$\hat{\mu}_1 = \sum_{i=1}^{M} w_i \hat{\mu}_{li} \tag{4.63}$$

By substituting Eqs. (4.36) and (4.62) into Eq. (4.63),

$$\hat{\mu}_1 = \sum_{i=1}^{M}\left[\frac{(a_i b_i)^2 \sin\varphi_i\left(\hat{N} + \hat{N}_0 - \hat{N}_2\right)}{\left(a_i \cos\varphi_i + b_i \sin\varphi_i\right)\left(\hat{N} - \hat{N}_0 + \hat{N}_2\right)}\right] \Big/ \sum_{j=1}^{M} a_i b_i \sin\varphi_j \tag{4.64}$$

(c) Trace length distribution on an infinite surface

Two probability density functions (*pdf*) can be defined for trace lengths as follows:

(1) $f(l) = pdf$ of trace lengths on an infinite surface.
(2) $g(l) = pdf$ of measured trace lengths on a finite exposure subjected to sampling biases.

It is necessary to obtain $f(l)$ from $g(l)$, because 3D size distribution of discontinuities is inferred from $f(l)$. Zhang and Einstein (2000) proposed the following procedure for obtaining $f(l)$:

(1) Use the corrected mean trace length μ as the mean value of $f(l)$
(2) Use the coefficient of variation (COV) value of $g(l)$ as the COV of $f(l)$
(3) Find the distribution of $g(l)$ as discussed earlier and assume that $f(l)$ and $g(l)$ have the same distribution form.

4.6.3   Discontinuity Shape

The planar shape of discontinuities has a profound effect on the connectivity of discontinuities and on the deformability, strength and permeability of rock masses (Dershowitz et al., 1993; Petit et al., 1994). However, since a rock mass is usually inaccessible in three dimensions, the real discontinuity shape is rarely known. Information on discontinuity shape is limited and often open to more than one interpretation (Warburton, 1980a; Wathugala, 1991). As Dershowitz and Einstein (1988) stated: "Shape of joint boundaries can be polygonal, circular, elliptical or irregular. Joints can be planar or non-planar in space and shape should consider this fact also. Since joints are often planar, it is simpler to associate 'shape' with the two-dimensional appearance and treat non-planarity separately."

Discontinuities can be classified into two categories: unrestricted and restricted. Unrestricted discontinuities are blind and effectively isolated discontinuities whose growth has not been perturbed by adjacent geological structures such as other discontinuities and bedding boundaries. In general, the edge of unrestricted discontinuities is or can be approximated by a closed convex curve. In many cases, the growth of discontinuities is limited by adjacent preexisting discontinuities or bedding planes and free surfaces. Such discontinuities are called restricted discontinuities. One way to represent restricted discontinuities is to use polygons, where some of the polygon sides are formed by intersections with the adjacent preexisting discontinuities and/ or bedding boundaries.

Due to the mathematical convenience, many investigators assume that discontinuities are thin circular disks randomly located in space (Baecher et al., 1977; Warburton, 1980a; Chan, 1986; Villaescusa and Brown, 1992; Kulatilake, 1993; Song and Lee, 2001; Song, 2006). For circular discontinuities, the trace patterns in differently oriented sampling planes will be the same. In practice, however, the trace patterns may vary with the orientation of sampling planes (Warburton, 1980b). Therefore, Warburton (1980b) assumed that discontinuities in a set are parallelograms of various sizes. Kulatilake et al. (1990) not only considered joints as circular disks but also as rectangles, squares, right triangles, parallelograms, rhombuses and oblique triangles in their study of the effect of joint orientation, joint size and joint shape on the statistical distribution of the orientation. Dershowitz et al. (1993) used polygons to represent discontinuities in the FracMan discrete fracture code. The polygons are formed by inscribing a polygon in an ellipse. Ivanova (1998) and Meyer (1999) also used polygons to represent discontinuities in their discrete fracture code GeoFrac. It is noted that polygons can be used to effectively represent elliptical discontinuities when the number of polygon sides is large (say > 10) (Dershowitz et al., 1993). Zhang et al. (2002) assumed that discontinuities are elliptical and derived a general stereological relationship between trace length distributions and discontinuity size (expressed by the major axis length of the ellipse) distributions.

Many researchers infer the discontinuity shape from the study of trace lengths in both the strike and dip directions. Based on the fact that the average strike length of a discontinuity set is approximately equal to its average dip length, Robertson (1970) and Barton (1977) assumed that discontinuities are equidimensional (circular). However, the average strike length of a discontinuity set being the same as its average dip length does not necessarily mean that the discontinuities of such a set are equidimensional; instead, there exist the following three possibilities (Zhang et al., 2002):

1. The discontinuities are indeed equidimensional (Fig. 4.29A).
2. The discontinuities are non-equidimensional such as elliptical or rectangular with long axes in a single (or deterministic) orientation. However, the discontinuities are oriented such that the strike length is approximately equal to the dip length (Fig. 4.29B).
3. The discontinuities are non-equidimensional such as elliptical or rectangular with long axes randomly oriented. The random discontinuity orientation distribution makes the average strike length approximately equal to the average dip length (Fig. 4.29C).

Therefore, the conclusion that discontinuities are equidimensional (circular) drawn from the fact that the average strike length of a discontinuity set is about equal to its average dip length is questionable. Investigators assume circular discontinuity shape possibly because of mathematical convenience.

Einstein et al. (1979) measured trace lengths of two sets of discontinuities on both the horizontal and vertical surfaces of excavations and found that discontinuities are non-equidimensional. Petit et al. (1994) presented results of a field study to determine the shape of discontinuities in sedimentary rocks. Pelites with isolated sandstone layers in the red Permian sandstones of the Lodeve Basin were studied. The exposed discontinuities (ie, one of the discontinuity walls had been removed by erosion) appear as rough ellipses with a shape ratio L/H of about 2.0, where L and H are the largest horizontal and vertical dimensions, respectively. For non-exposed discontinuities, the distributions of the dimensions of the horizontal and vertical traces were measured. The ratio of the mean L to the mean H of such traces is 1.9, which is very close to the L/H ratio of the observed individual discontinuity planes.

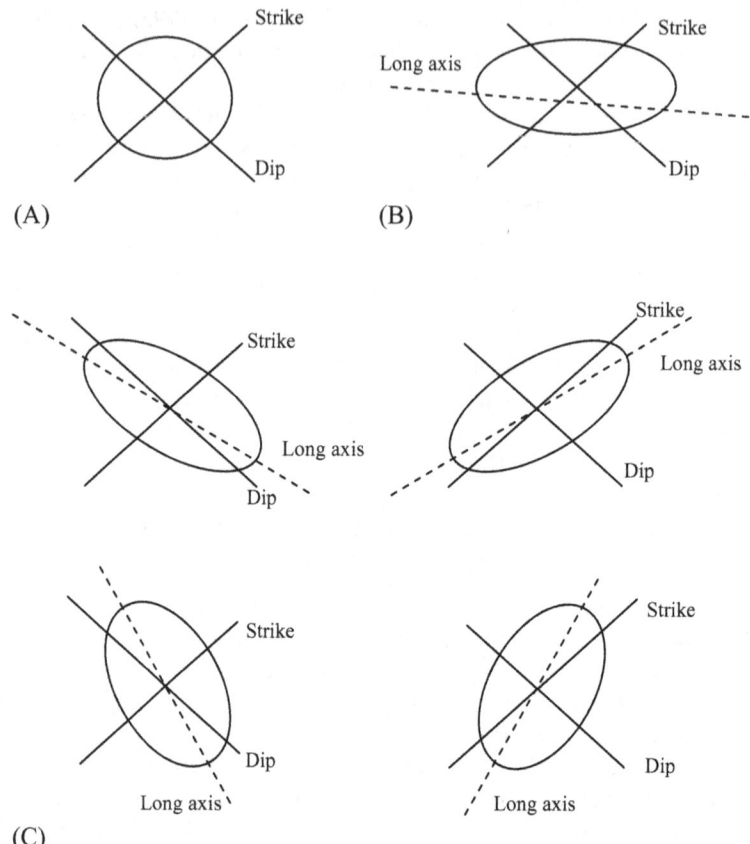

FIG. 4.29 Three possible cases for which the average strike length is about equal to the average dip length: (A) Discontinuities are equidimensional (circular); (B) Discontinuities are non-equidimensional (elliptical), with long axes in a single orientation. The discontinuities are oriented so that the strike length is about equal to the dip length; and (C) Discontinuities are non-equidimensional (elliptical), with long axes randomly orientated so that the average strike length is about equal to the average dip length. *(From Zhang, L., Einstein, H. H., Dershowitz, W. S., 2002. Stereological relationship between trace length distribution and size distribution of elliptical discontinuities. Geotechnique 52, 419–433.)*

Zhang and Einstein (2010) studied the planar shape of discontinuities in detail and concluded that discontinuities not affected by adjacent geological structures such as bedding boundaries tend to be elliptical (or approximately circular but rarely), while discontinuities affected by or intersecting geological structures such as bed boundaries tend to be most likely rectangles or similarly shaped polygons.

4.6.4   Discontinuity Size

Zhang and Einstein (2000) presented a method for inferring the discontinuity size distribution from the corrected trace length distribution obtained from circular window sampling as described in Section 4.6.2, based on the stereological relationship between trace lengths and discontinuity diameter distributions for area sampling of discontinuities (Warburton, 1980a):

$$f(l) = \frac{l}{\mu_D} \int_l^\infty \frac{g(D)}{\sqrt{D^2 - l^2}} dD \tag{4.65}$$

where D is the diameter of discontinuities; l is the trace length of discontinuities; $g(D)$ is the probability density function of the diameter of discontinuities; $f(l)$ is the probability density function of the trace length of discontinuities; and μ_D is the mean of the diameter of discontinuities. Villaescusa and Brown (1992) presented a similar method for inferring the discontinuity size distribution from the corrected trace length distribution obtained from straight scanline sampling. They used the following stereological relationship between trace length and discontinuity diameter distributions for straight scanline sampling of discontinuities (Warburton, 1980a):

$$f(l) = \frac{4l^2}{\pi E(D^2)} \int_l^\infty \frac{g(D)}{\sqrt{D^2 - l^2}} dD \tag{4.66}$$

where $E(D^2)$ is the mean of D^2.

Zhang et al. (2002) derived a general stereological relationship between trace length distributions and discontinuity size (expressed by the major axis length a of the ellipse) distributions for area (or window) sampling, following the methodology of Warburton (1980a, b):

$$f(l) = \frac{l}{M\mu_a} \int_{l/M}^\infty \frac{g(a)}{\sqrt{(Ma)^2 - l^2}} da \quad (l \le aM) \tag{4.67}$$

where

$$M = \frac{\sqrt{\tan^2\beta + 1}}{\sqrt{k^2 \tan^2\beta + 1}} \tag{4.68}$$

in which k is the aspect ratio of the discontinuity, ie, the length of the discontinuity minor axis is a/k (Fig. 4.30); β is the angle between the discontinuity major axis and the trace line (note that β is measured in the discontinuity plane). Obviously, β will change for different sampling planes. For a specific sampling plane, however, there will be only one β value for a discontinuity set with a deterministic orientation.

When $k = 1$ (ie, the discontinuities are circular), $M = 1$ and Eq. (4.67) reduces to Eq. (4.65).

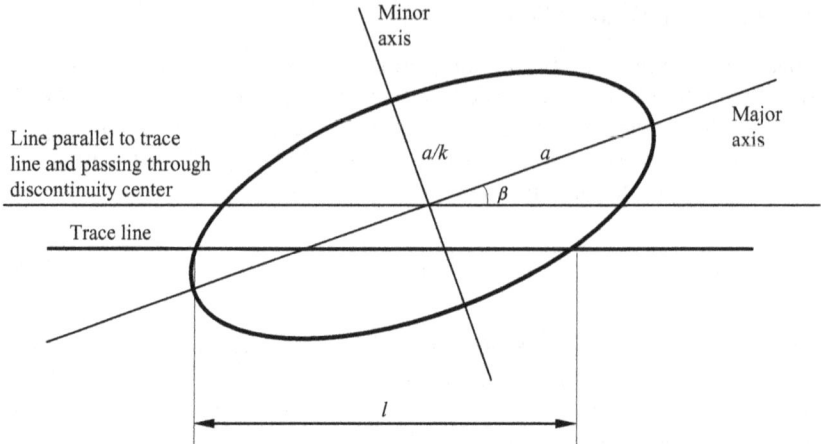

FIG. 4.30 Parameters used in the definition of an elliptical discontinuity. *(From Zhang, L., Einstein, H. H., Dershowitz, W. S., 2002. Stereological relationship between trace length distribution and size distribution of elliptical discontinuities. Geotechnique 52, 419–433.)*

TABLE 4.9 Expressions for Determining μ_a and σ_a From μ_l and σ_l

Distribution Form of g (a)	μ_a	$(\sigma_a)^2$
Lognormal	$\dfrac{128(\mu_l)^3}{3\pi^3 M\left[(\mu_l)^2+(\sigma_l)^2\right]}$	$\dfrac{1536\pi^2\left[(\mu_l)^2+(\sigma_l)^2\right](\mu_l)^4-128^2(\mu_l)^6}{9\pi^6 M^2\left[(\mu_l)^2+(\sigma_l)^2\right]^2}$
Negative exponential	$\dfrac{2}{\pi M}\mu_l$	$\left[\dfrac{2}{\pi M}\mu_l\right]^2$
Gamma	$\dfrac{64(\mu_l)^2-3\pi^2\left[(\mu_l)^2+(\sigma_l)^2\right]}{8\pi M\mu_l}$	$\dfrac{\left\{64(\mu_l)^2-3\pi^2\left[(\mu_l)^2+(\sigma_l)^2\right]\right\}\times\left\{3\pi^2\left[(\mu_l)^2+(\sigma_l)^2\right]-32(\mu_l)^2\right\}}{64\pi^2 M^2(\mu_l)^2}$

From Zhang, L., Einstein, H. H., Dershowitz, W. S., 2002. Stereological relationship between trace length distribution and size distribution of elliptical discontinuities. Geotechnique 52, 419–433.

Based on Eq. (4.67), Zhang et al. (2002) extended the method of Zhang and Einstein (2000) to elliptical discontinuities. Table 4.9 summarizes the expressions for determining the mean μ_a and standard deviation σ_a of discontinuity size a from the mean μ_l and standard deviation σ_l of trace length l, respectively

TABLE 4.10 Expressions for Determining μ_l and σ_l From μ_a and σ_a

Distribution Form of g(a)	μ_l	$(\sigma_l)^2$
Lognormal	$\dfrac{\pi M \left[(\mu_a)^2 + (\sigma_a)^2\right]}{4\mu_a}$	$\dfrac{32 M^2 \left[(\mu_a)^2 + (\sigma_a)^2\right]^3 - 3\pi^2 M^2 (\mu_a)^2 \left[(\mu_a)^2 + (\sigma_a)^2\right]^2}{48(\mu_a)^4}$
Negative exponential	$\dfrac{\pi M}{2}\mu_a$	$\dfrac{(16-\pi^2)M^2}{4}(\mu_a)^2$
Gamma	$\dfrac{\pi M \left[(\mu_a)^2 + (\sigma_a)^2\right]}{4\mu_a}$	$\dfrac{32 M^2 \left[(\mu_a)^2 + (\sigma_a)^2\right]\left[(\mu_a)^2 + 2(\sigma_a)^2\right] - 3\pi^2 M^2 \left[(\mu_a)^2 + (\sigma_a)^2\right]^2}{48(\mu_a)^2}$

From Zhang, L., Einstein, H. H., Dershowitz, W. S., 2002. Stereological relationship between trace length distribution and size distribution of elliptical discontinuities. Geotechnique 52, 419–433.

for the lognormal, negative exponential and Gamma distribution of discontinuity size a. Conversely, with known μ_a and σ_a, and the distribution form of $g(a)$, μ_l and σ_l can also be obtained (Table 4.10).

Consider a discontinuity set having a lognormal size distribution with $\mu_a = 8.0$ m and $\sigma_a = 4.0$ m (For other distribution forms, similar conclusions can be obtained). Fig. 4.31 shows the variation of the mean trace length and the standard deviation of trace lengths with β. Since β is the angle between the trace line and the discontinuity major axis, it is related to the sampling plane orientation relative to the discontinuity. It can be seen that, despite the considerable difference between the maximum and the minimum, respectively, of the mean trace length and the standard deviation of trace lengths, there are extensive ranges of sampling plane orientations, reflected by β, over which both the mean trace length and the standard deviation of trace lengths show little variation, especially for large k values. The results in Fig. 4.31 could well explain why Bridges (1976), Einstein et al. (1979) and McMahon (Mostyn and Li, 1993) found different mean trace lengths on differently oriented sampling planes, whereas Robertson (1970) and Barton (1977) observed them to be approximately equal. In each of these papers or reports, the number of differently oriented sampling planes was very limited and, depending on the relative orientations of the sampling planes, the authors could observe either

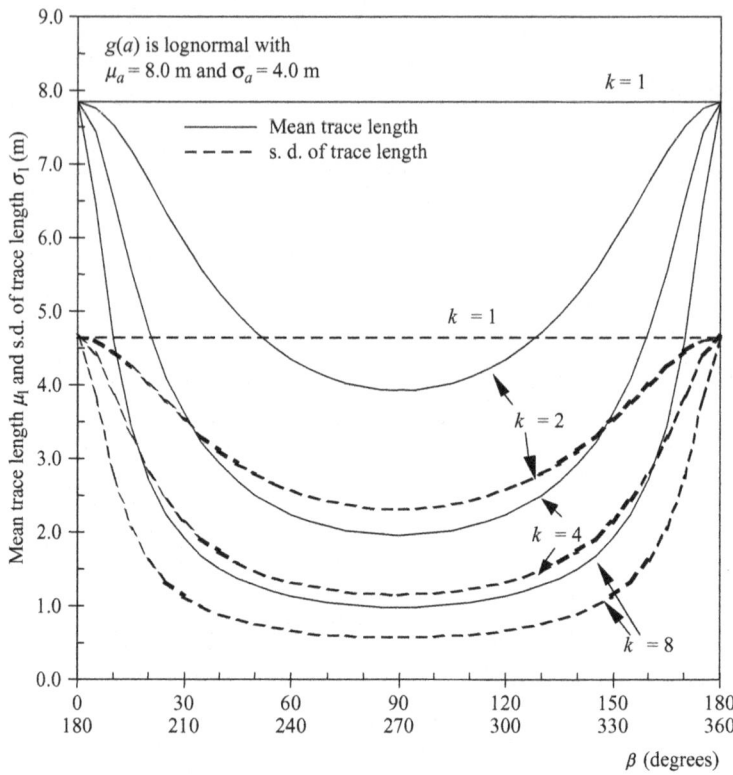

FIG. 4.31 Variation of mean trace length and standard deviation (s.d.) of trace length with β. *(From Zhang, L., Einstein, H. H., Dershowitz, W. S., 2002. Stereological relationship between trace length distribution and size distribution of elliptical discontinuities. Geotechnique 52, 419–433.)*

approximately equal mean trace lengths or significantly different mean trace lengths. For example, in Bridges (1976), Einstein et al. (1979) and McMahon (Mostyn and Li, 1993), the strike and dip sampling planes might be in the $\beta = 0°$–$20°$ (or $160°$–$180°$) range and the $\beta = 40°$–$140°$ range, respectively, or vice versa. From Fig. 4.31, this would result in very different mean trace lengths. On the other hand, in Robertson (1970) and Barton (1977), the strike and dip sampling planes might be both in the $\beta = 40°$–$140°$ range (ie, in the "flat" trace length part of Fig. 4.31) or, respectively, in some β ranges approximately symmetrical about $\beta = 90°$. It should be noted that the comments above are assumptions because no information about the β values can be found in the original papers or reports.

The implications of Fig. 4.31 about field sampling are as follows:

If different sampling planes are used to collect trace (length) data, the sampling planes should be oriented such that significantly different mean trace lengths can *be obtained from different planes. For example, if two sampling planes are used, one should be oriented in the $\beta=0°-20°$ (or $160°-180°$) range and the other in the $\beta=60°-120°$ range.*

It is noted that, with the same μ_1 and σ_1, one can have different μ_a and σ_a if the assumed distribution form of $g(a)$ is different. This means that the estimation of discontinuity size distributions from the equations in Table 4.9 may not be robust. To overcome the problem of uniqueness, a relationship between the ratio of the 4th and 1st moments of the discontinuity size distribution and the third moment of the trace length distribution is used to check the suitability of the assumed discontinuity size distribution form:

$$\frac{E(a^4)}{E(a)}=\frac{16E(l^3)}{3\pi M^3} \tag{4.69}$$

For the three distribution forms of $g(a)$ discussed above, Eq. (4.69) can be rewritten as:

(a) If $g(a)$ is lognormally distributed with mean μ_a and standard deviation σ_a,

$$\frac{\left[(\mu_a)^2+(\sigma_a)^2\right]^6}{(\mu_a)^9}=\frac{16E(l^3)}{3\pi M^3} \tag{4.70}$$

(b) If $g(a)$ has a negative exponential distribution with mean μ_a,

$$24(\mu_a)^2=\frac{16E(l^3)}{3\pi M^3} \tag{4.71}$$

(c) If $g(a)$ has a Gamma distribution with mean μ_a and standard deviation σ_a,

$$\frac{\left[(\mu_a)^2+(\sigma_a)^2\right]\left[(\mu_a)^2+2(\sigma_a)^2\right]\left[(\mu_a)^2+3(\sigma_a)^2\right]}{(\mu_a)^3}=\frac{16E(l^3)}{3\pi M^3} \tag{4.72}$$

The procedure for inferring the major axis orientation, aspect ratio k and size distribution $g(a)$ (probability density function of the major axis length a) of elliptical discontinuities from trace length sampling on different sampling windows is summarized as follows (The reader can refer to Zhang et al., 2002 for details):

1. Sampling
   (a) Trace length: Use two or more sampling windows at different orientations to conduct trace (length) sampling. The sampling windows (planes) should be oriented such that significantly different mean trace lengths can be obtained from different windows.
   (b) Orientation: Use exposed rock surface or borehole sampling so that the normal orientation of each discontinuity set can be obtained.

2. Conduct trace length analysis to estimate the true trace length distribution $f(l)$ on different sampling windows: μ_1, σ_1 and form of $f(l)$.
3. Infer the major axis orientation, aspect ratio k and size distribution $g(a)$ of discontinuities from trace length sampling on different sampling windows:
   (a) Assume a major axis orientation and compute the β (the angle between discontinuity major axis and trace line) value for each sampling window.
   (b) For the assumed major axis orientation, compute μ_a and σ_a from μ_1 and σ_1 of each sampling window, by assuming aspect ratios $k = 1, 2, 4, 6, 8$ and lognormal, negative exponential and Gamma distribution forms of $g(a)$. The results are then used to draw the curves relating μ_a (and σ_a) to k, respectively, for the lognormal, negative exponential and Gamma distribution forms of $g(a)$.
   (c) Repeat steps (a) and (b) until the curves relating μ_a (and σ_a) to k for different sampling windows intersect at one point. The major axis orientation for this case is the inferred actual major axis orientation. The k, μ_a and σ_a values at the intersection point are the corresponding possible characteristics of the discontinuities.
4. Find the best distribution form of $g(a)$ by checking the equality of Eq. (4.69). The k, μ_a and σ_a values found in Step (c) and corresponding to the best distribution form of $g(a)$ are the inferred characteristics of the discontinuity size.

4.7 FRACTURE TENSOR

Tensors have been used by several researchers to describe discontinuity geometry including intensity and orientation. Kachanov (1980) introduced a tensor α_{ij} to quantify the geometry of microcracks in rocks:

$$\alpha_{ij} = \frac{1}{V} \sum_{k=1}^{m^{(V)}} \left[S^{(k)} \right]^{3/2} u_i^{(k)} u_j^{(k)} \tag{4.73}$$

where V is the volume of the rock mass considered; $S^{(k)}$ is the area of the kth discontinuity; $m^{(V)}$ is the number of discontinuities in volume V; $u_i^{(k)}$ and $u_j^{(k)}$ $(i, j = x, y, z)$ are components of the unit normal vector of the kth discontinuity with respect to orthogonal reference axes i and j $(i, j = x, y, z)$ respectively (see Fig. 4.4 about the definition of the normal or pole direction of a discontinuity).

Oda (1982) also proposed a tensor F_{ij} (called the crack tensor) for describing discontinuity geometry:

$$F_{ij} = \frac{1}{V} \sum_{k=1}^{m^{(V)}} S^{(k)} r^{(k)} u_i^{(k)} u_j^{(k)} \tag{4.74}$$

where $r^{(k)}$ is the radius of the kth discontinuity.

Kawamoto et al. (1988) regarded discontinuities as damages, and defined a tensor Ω_{ij} called the damage tensor:

$$\Omega_{ij} = \frac{\bar{l}}{V} \sum_{k=1}^{m^{(V)}} S^{(k)} u_i^{(k)} u_j^{(k)} \tag{4.75}$$

where $\bar{l}$ is a characteristic length for a given discontinuity system.

The tensors described above are non-dimensional due to some arbitrary operation included in their definitions: in Eq. (4.73) the area $S^{(k)}$ of a discontinuity is multiplied by the square root of $S^{(k)}$; in Eq. (4.74) the area $S^{(k)}$ of a discontinuity is multiplied by its radius $r^{(k)}$; and in Eq. (4.75) a characteristic length $\bar{l}$ for a given discontinuity system is included. Because of the arbitrary operation, the physical meaning of the discontinuity intensity expressed by those definitions is not clear and thus a little confusing (eg, what is the physical meaning of $[S^{(k)}]^{3/2}$?).

P_{32}, the mean area of discontinuities per unit volume of rock mass, as defined earlier, is the most useful measure of discontinuity intensity (Dershowitz and Herda, 1992; Mauldon, 1994). However, P_{32} does not include the effect of discontinuity orientations. Zhang (1999) and Zhang et al. (2002) introduced the fracture tensor F_{ij}, which is a combined measure of discontinuity intensity and orientation, defined as follows:

$$F_{ij} = \frac{1}{V} \sum_{k=1}^{m^{(V)}} S^{(k)} u_i^{(k)} u_j^{(k)} \tag{4.76}$$

F_{ij} can be determined with the data obtained in the previous sections. Fracture tensor F_{ij} can also be written in a matrix form as follows:

$$F(F_{ij}) = \begin{bmatrix} F_{xx} & F_{xy} & F_{xz} \\ & F_{yy} & F_{yz} \\ \text{Sym.} & & F_{zz} \end{bmatrix} \tag{4.77}$$

F_{ij} has three principal values F_1, F_2 and F_3, which can be obtained by finding the eigenvalues of F_{ij}. The principal orientation of F_{ij} can be obtained by finding the eigenvectors corresponding to F_1, F_2 and F_3.

The first invariant of F_{ij} is just P_{32}, ie,

$$P_{32} = I_1^{(F)} = F_1 + F_2 + F_3 = F_{xx} + F_{yy} + F_{zz}. \tag{4.78}$$

In contrast to the tensors proposed by Kachanov (1980), Oda (1982), and Kawamoto et al. (1988), the fracture tensor defined in Eq. (4.76) has a clear physical meaning. It represents the ratio of the total area of discontinuities and the volume of the rock mass considered. The fracture tensor defined in Eq. (4.76) keeps the advantage of P_{32}, ie, P_{32} does not depend on the size of the sampled region as long as it is representative of the discontinuity network.

4.8   DISCONTINUITY ROUGHNESS

Roughness is a measure of the inherent surface unevenness and waviness of the discontinuity relative to its mean plane. The wall roughness of a discontinuity has an important influence on its shear strength, especially in the case of undisplaced and interlocked features such as unfilled joints. The importance of roughness declines with increasing aperture, filling thickness or previous shear displacement.

When the properties of discontinuities are being recorded from observations made on either boring cores or exposed faces, it is usual to distinguish between small-scale surface irregularity or unevenness and large-scale undulations or waviness of the surface (Fig. 4.32). Each of these types of roughness may be quantified on an arbitrary scale of, say, one to five. Descriptive terms may also be used particularly in the preliminary stages of mapping. For example, ISRM (1978) suggests that the terms listed in Table 4.11 and illustrated in Fig. 4.33 may be used to describe roughness on two scales: the small scale (several centimeters) and the intermediate scale (several meters). Large-scale waviness may be superimposed on such small- and intermediate-scale roughness.

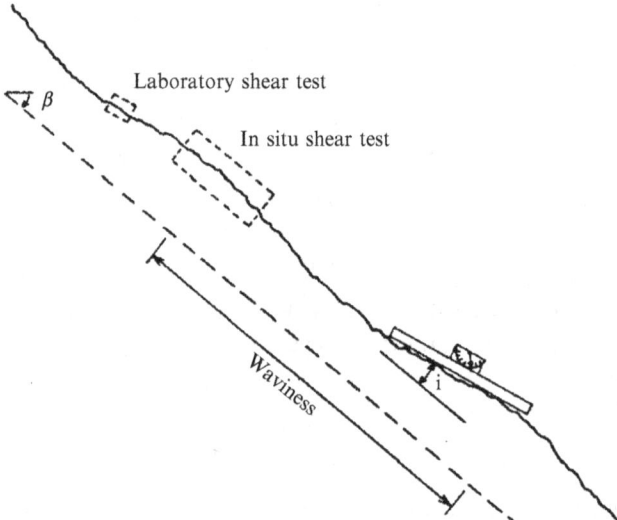

FIG. 4.32   Different scales of discontinuity roughness are sampled by different scales of test. Waviness can be characterized by the angle i. *(Based on ISRM, 1978. Suggested methods for the quantitative description of discontinuities in rock masses. International Society for Rock Mechanics, Commission on Standardization of Laboratory and Field Tests. Int. J. Rock Mech. Min. Sci. Geomech. Abstr. 15, 319–368.)*

TABLE 4.11 Classification of Discontinuity Roughness

Class	Description
I	Rough or irregular, stepped
II	Smooth, stepped
III	Slickensided, stepped
IV	Rough or irregular, undulating
V	Smooth, undulating
VI	Slickensided, undulating
VII	Rough or irregular, planar
VIII	Smooth, planar
IX	Slickensided, planar

Based on ISRM, 1978. Suggested methods for the quantitative description of discontinuities in rock masses. International Society for Rock Mechanics, Commission on Standardization of Laboratory and Field Tests. Int. J. Rock Mech. Min. Sci. Geomech. Abstr. 15, 319–368.

More detailed description of the methods for determining discontinuity roughness will be presented in Chapter 6.

4.9 DISCONTINUITY APERTURE

Aperture is the perpendicular distance separating the adjacent rock walls of an open discontinuity in which the intervening space is filled with air or water. Aperture is thereby distinguished from the width of a filled discontinuity (Fig. 4.34). Large apertures can result from shear displacement of discontinuities having appreciable roughness, from outwash of filling materials (eg, clay), from tensile opening, and/or from solution. In most subsurface rock masses, apertures are small, probably less than half a millimeter. Table 4.12 lists terms describing aperture dimensions suggested by ISRM (1978). Clearly, aperture and its areal variation will have an influence on the deformability, shear strength and hydraulic conductivity of discontinuities (see Chapters 6–8 for details).

4.10 DISCONTINUITY FILLING

Filling is a term used to describe the material separating the adjacent rock walls of discontinuities, such as calcite, chlorite, clay, silt, fault gouge, breccia, quartz and pyrite. The perpendicular distance between the adjacent rock walls is

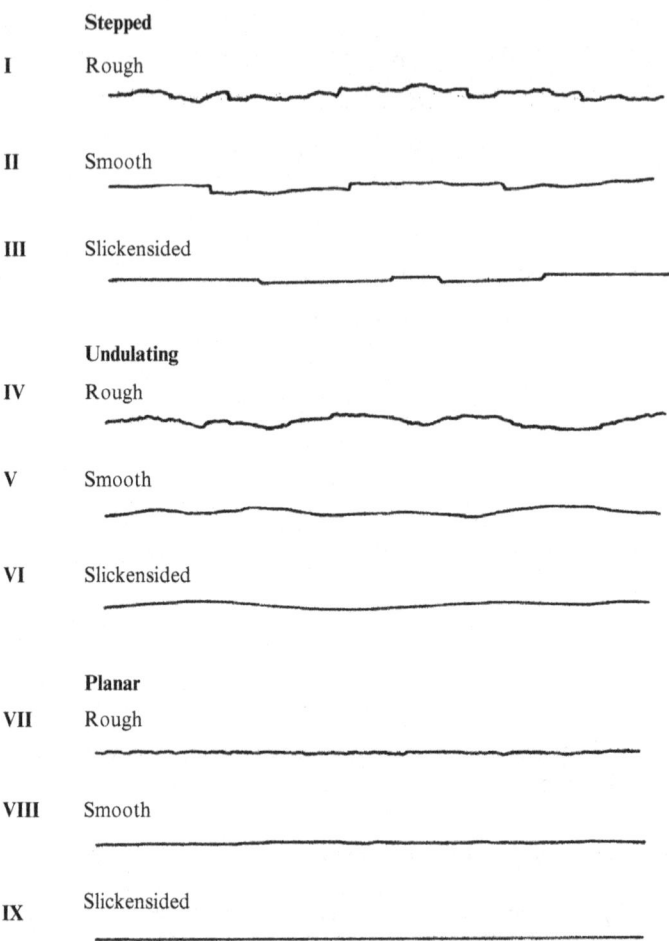

FIG. 4.33   Typical roughness profiles and suggested nomenclature. The length of each profile is in the range of 1–10 m. The vertical and horizontal scales are equal. *(Based on ISRM, 1978. Suggested methods for the quantitative description of discontinuities in rock masses. International Society for Rock Mechanics, Commission on Standardization of Laboratory and Field Tests. Int. J. Rock Mech. Min. Sci. Geomech. Abstr. 15, 319–368.)*

termed the width of the filled discontinuity, as opposed to the aperture of a gapped or open discontinuity.

Filling materials have a major influence on the shear strength of discontinuities. With the exception of discontinuities filled with strong vein materials (calcite, quartz, pyrite), filled discontinuities generally have lower shear strengths than

Closed discontinuity

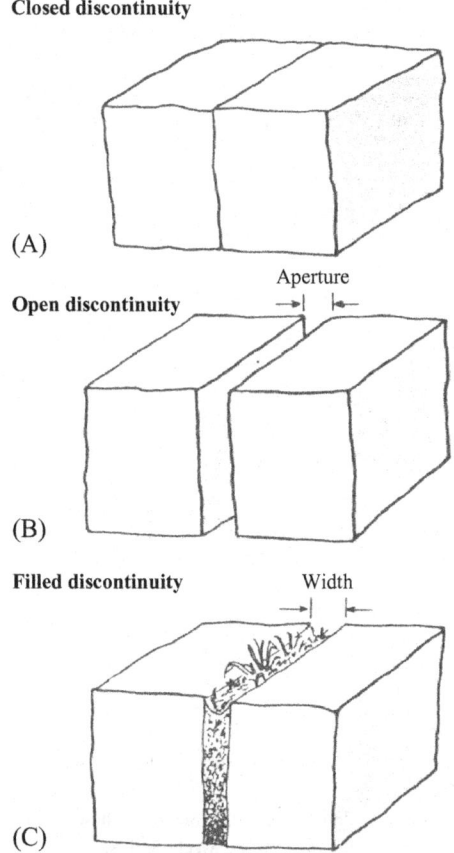

(A)

Open discontinuity

Aperture

(B)

Filled discontinuity   Width

(C)

FIG. 4.34 Suggested definitions of the aperture of open discontinuities and the width of filled discontinuities. *(Based on ISRM, 1978. Suggested methods for the quantitative description of discontinuities in rock masses. International Society for Rock Mechanics, Commission on Standardization of Laboratory and Field Tests. Int. J. Rock Mech. Min. Sci. Geomech. Abstr. 15, 319–368.)*

comparable clean, closed discontinuities. The behavior of filled discontinuities depends on many factors of which the following are probably the most important:

(1) mineralogy of filling material
(2) grading or particle size
(3) over-consolidation ratio
(4) water content and permeability
(5) previous shear displacement
(6) wall roughness.

TABLE 4.12 Classification of Discontinuity Aperture

Description		Aperture (mm)
"Closed" features	Very tight	<0.1
	Tight	0.1–0.25
	Partly open	0.25–0.5
"Gapped" features	Open	0.5–2.5
	Moderately wide	2.5–10
	Wide	>10
"Open" features	Very wide	10–100
	Extremely wide	100–1000
	Cavernous	>1000

Based on ISRM, 1978. Suggested methods for the quantitative description of discontinuities in rock masses. International Society for Rock Mechanics, Commission on Standardization of Laboratory and Field Tests. Int. J. Rock Mech. Min. Sci. Geomech. Abstr. 15, 319–368.

REFERENCES

Baecher, G.N., Lanney, N.A., 1978. Trace length biases in joint surveys. In: Proc. 19th U.S. Symp. on Rock Mech. vol. 1, pp. 56–65.

Baecher, G.N., Lanney, N.A., Einstein, H.H., 1977. Statistical description of rock properties and sampling. In: Proc. 18th U.S. Symp. on Rock Mech. 5C1-8.

Barton, C.M., 1977. Geotechnical analysis of rock structure and fabric in CSA Mine NSW. Applied Geomechanics Technical Paper 24, Commonwealth Scientific and Industrial Research Organization, Australia.

Bery, A.A., Saad, R., 2012. Correlation of seismic P-wave velocities with engineering parameters (N value and rock quality) for tropical environmental study. Int. J. Geosci. 3, 749–757.

Bieniawski, Z.T., 1989. Engineering Rock Mass Classifications. John Wiley, Rotterdam.

Blyth, G.G.H., de Freitas, M.H., 1974. A Geology for Engineers, sixth ed. Edward Arnold, London.

Bridges, M.C., 1975. Presentation of fracture data for rock mechanics. In: Proc. 2nd Australia-New Zealand Conf. on Geomech., Brisbane, Australia, pp. 144–148.

Budetta, P., de Riso, R., de Luca, C., 2001. Correlations between jointing and seismic velocities in highly fractured rock masses. Bull. Eng. Geol. Env. 60, 185–192.

Cai, M., Kaiser, P.K., Uno, H., Tasaka, Y., Minami, M., 2004. Estimation of rock mass deformation modulus and strength of jointed hard rock masses using the GSI System. Int. J. Rock Mech. Min. Sci. 41, 3–19.

Call, R.D., Savely, J.P., Nicholas, D.E., 1976. Estimation of joint set characteristics from surface mapping data. In: Proc. 17th U. S. Symp. on Rock Mech, vol. 2B2. pp. 1–9.

Chan, L.P., 1986. Application of Block Theory and Simulation Techniques to Optimum Design of Rock Excavations. Ph.D. thesis, University of California, Berkeley, CA.

Choi, S.Y., Park, H.D., 2004. Variation of rock quality designation (RQD) with scanline orientation and length: a case study in Korea. Int. J. Rock Mech. Min. Sci. 41, 207–221.

Cruden, D.M., 1977. Describing the size of discontinuities. Int. J. Rock Mech. Min. Sci. Geomech. Abstr. 14, 133–137.

Davis, G.H., Reynolds, S.J., 1996. Structural Geology of Rocks and Regions. John Wiley and Sons, New York.

Dearman, W.R., 1991. Engineering Geological Mapping. Butterworth-Heineman, Oxford.

Deere, D.U., 1964. Technical description of rock cores for engineering purposes. Rock Mech. Rock. Eng. 1, 107–116.

Deere, D.U., Hendron, A.J., Patton, F.D., Cording, E.J., 1967. Design of surface and near surface construction in rock. In: Fairhurst, C. (Ed.), Failure and Breakage of Rock.Proc. 8th U.S. Symp. Rock Mech, pp. 237–302.

Dershowitz, W.S., Einstein, H.H., 1988. Characterizing rock joint geometry with joint system models. Rock Mech. Rock. Eng. 21, 21–51.

Dershowitz, W.S., Herda, H.H., 1992. Interpretation of fracture spacing and intensity. In: Proc. 33rd U.S. Symp. on Rock Mech., Santa Fe, NM, pp. 757–766.

Dershowitz, W. S, Lee., G., Geier, J., Hitchcock, S., LaPointe, P., 1993. FracMan Version 2.306, Interactive Discrete Feature Data Analysis, Geometric Modeling, and Exploration Simulation. User Documentation. Golder Associates Inc., Seattle, Washington.

Einstein, H.H., Baecher, G.B., 1983. Probabilistic and statistical methods in engineering geology (part I). Rock Mech. Rock. Eng. 16, 39–72.

Einstein, H.H., Baecher, G.B., Veneziano, D., et al., 1979. Risk analysis for rock slopes in open pit mines. Parts I–V, USBM Technical Report J0275015.

Einstein, H.H., Veneziano, D., Baecher, G.B., O'Reilly, K.J., 1983. The effect of discontinuity persistence on rock slope stability. Int. J. Rock Mech. Min. Sci. Geomech. Abstr. 20, 227–236.

El-Naqa, A., 1996. Assessment of geotechnical characterization of a rock mass using a seismic geophysical technique. Geotech. Geol. Eng. 14, 291–305.

Harrison, J.P., 1992. Fuzzy objective functions applied to the analysis of discontinuity orientation data. In: Hudson, J.A. (Ed.), ISRM Symp., Eurock '92, Rock Characterization, Chester, UK, pp. 25–30.

Harrison, J.P., Hudson, J.A., 2000. Engineering Rock Mechanics: Illustrative Worked Examples. Pergamon, Elsevier, Oxford.

Hills, E.S., 1972. Elements of Structural Geology, second ed. Chapman and Hall, London.

Hobbs, B.E., 1976. An Outline of Structural Geology. Wiley, New York.

Hoek, E., Brown, E.T., 1980a. Underground Excavation in Rock. Institution of Mining and Metallurgy, London, UK.

Idziak, A., 1981. Seismic study of fractured sedimentary rocks. In: Trans. 26th Geophys. Symp., Leipzig, pp. 116–123.

ISRM, 1978. Suggested methods for the quantitative description of discontinuities in rock masses. International Society for Rock Mechanics, Commission on Standardization of Laboratory and Field Tests. Int. J. Rock Mech. Min. Sci. Geomech. Abstr. 15, 319–368.

Ivanova, V., 1998. Geological and Stochastic Modeling of Fracture Systems in Rocks. PhD thesis, Massachusetts Institute of Technology, Cambridge, MA.

Jamscikov, W. S., Schkuratnik, W. L., Bobrov, A. B., 1985. O kolicestvennoi ozenke mikrotrecinobatocti gornich porod ultrasvukovim velocimmetriceskim metodom. Akad. Nauk SSSR, Sibirskoe otdelenie, 4, 110–114.

Kachanov, M., 1980. Continuum model of medium with cracks. J. Eng. Mech. Div. ASCE 106, 1039–1051.

Kawamoto, T., Ichikawa, Y., Kyoya, T., 1988. Deformation and fracturing behavior of discontinuous rock mass and damage mechanics theory. Int. J. Num. Analy. Meth. Geomech. 12, 1–30.

Kim, B.H., Cai, M., Kaiser, P.K., Yang, H.S., 2007. Estimation of block sizes for rock masses with non-persistent joints. Rock Mech. Rock. Eng. 40, 169–192.

Kulatilake, P.H.S.W., 1985a. Fitting Fisher distributions to discontinuity orientation data. J. Geol. Educ. 33, 266–269.

Kulatilake, P.H.S.W., 1986. Bivariate normal distribution fitting on discontinuity orientation clusters. Math. Geol. 18, 181–195.

Kulatilake, P.H.S.W., 1993. Application of probability and statistics in joint network modeling in three dimensions. In: Proc. Conf. on Probabilistic Methods in Geotech. Eng., Lanberra, Australia, pp. 63–87.

Kulatilake, P.H.S.W., Wu, T.H., 1984c. Estimation of mean trace length of discontinuities. Rock Mech. Rock. Eng. 17, 215–232.

Kulatilake, P.H.S.W., Wu, T.H., Wathugala, D.N., 1990. Probabilistic modeling of joint orientation. Int. J. Num. Analy. Meth. Geomech. 14, 325–350.

Laslett, G.M., 1982. Censoring and edge effects in areal and line transect sampling of rock joint traces. Math. Geol. 14, 125–140.

Mahtab, M.A., Yegulalp, T.M., 1984. A similarity test for grouping orientation data in rock mechanics. In: Proc. 25th U.S. Symp. on Rock Mech, pp. 495–502.

Mahtab, M.A., Bolstad, D.D., Alldredge, J.R., Shanley, R.J., 1972. Analysis of fracture orientations for input to structural models of discontinuous rock. US Bur. Mines Rep. Invest. 7669, 76.

Mauldon, M., 1994. Intersection probabilities of impersistent joints. Int. J. Rock Mech. Min. Sci. Geomech. Abstr. 31, 107–115.

Mauldon, M., 1998. Estimating mean fracture trace length and density from observations in convex windows. Rock Mech. Rock. Eng. 31, 201–216.

Mauldon, M., Rohrbaugh Jr., M.B., Dunne, W.M., Lawdermilk, W., 1999. Fracture intensity estimates using circular scanlines. In: Proc. 37th US Rock Mech. Symp.—Rock Mech. for Industry, Balkema, Rotterdam, pp. 777–784.

Meyer, T., 1999. Geologic Stochastic Modeling of Rock Fracture Systems Related to Crustal Faults. MS thesis, Massachusetts Institute of Technology, Cambridge, MA.

Miller, S.M., 1983. A statistical method to evaluate homogeneity of structural populations. Math. Geol. 15, 317–328.

Mostyn, G.R., Li, K.S., 1993. Probabilistic slope analysis—state-of-play. In: Proc. Conf. on Probabilistic Methods in Geotech. Eng., Canberra, Australia, pp. 89–109.

Oda, M., 1982. Fabric tensor for discontinuous geological materials. Soils Fdns. 22, 96–108.

Pahl, P.J., 1981. Estimating the mean length of discontinuity traces. Int. J. Rock Mech. Min. Sci. Geomech. Abstr. 18, 221–228.

Palmström, A., 1974. Characterization of Jointing Density and the Quality of Rock Masses (in Norwegian). Internal report, A.B. Berdal, Norway, 26 p.

Palmström, A., 1982. The volumetric joint count—a useful and simple measure of the degree of rock jointing. In: Proc. 41st Int. Congress Int. Ass. Eng. Geol., Delphi, vol. 5. pp. 221–228.

Palmström, A., 1995. RMi—A Rock Mass Classification System for Rock Engineering Purposes. PhD Thesis, University of Oslo.

Palmström, A., 2000. Recent developments in rock support estimates by the RMi. J. Rock Mech. Tunneling Technol. 6 (1), 1–19.

Palmström, A., 2002. Measurement and characterization of rock mass jointing. In: Sharma, V.M., Saxena, K.R. (Eds.), In-Situ Characterization of Rocks. Balkema, Lisse, pp. 49–98.

Palmström, A., 2005. Measurements of and correlations between block size and rock quality designation (RQD). Tunnels Underground Space Technol. 20, 362–377.

Parker, P., Cowan, R., 1976. Some properties of line segment processes. J. Appl. Prob. 13, 96–107.

Petit, J.-P., Massonnat, G., Pueo, F., Rawnsley, K., 1994. Rapport de forme des fractures de mode 1 dans les roches stratifiees: Une etude de cas dans le Bassin Permian de Lodeve (France). Bulletin du Centre de Recherches Elf Exploration Production 18, 211–229.

Price, N.J., 1966. Fault and Joint Development in Brittle and Semi-Brittle Rock. Pergamon, Oxford, UK.

Priest, S.D., 1993. Discontinuity Analysis for Rock Engineering. Chapman & Hall, London.

Priest, S.D., Hudson, J., 1976. Discontinuity spacing in rock. Int. J. Rock Mech. Min. Sci. Geomech. Abstr. 13, 135–148.

Priest, S.D., Hudson, J., 1981. Estimation of discontinuity spacing and trace length using scanline surveys. Int. J. Rock Mech. Min. Sci. Geomech. Abstr. 18, 183–197.

Rives, T., Razack, M., Petit, J.-P., Rawnsley, K.D., 1992. Joint spacing: analogue and numerical simulations. J. Struct. Geol. 14, 925–937.

Robertson, A., 1970. The interpretation of geologic factors for use in slope theory. In: Proc. Symp. on the Theoretical Background to the Planning of Open Pit Mines, Johannesburg, South Africa, pp. 55–71.

Savic, A.J., Koptev, W.J., Nikitin, W.N., Jascenko, Z.D., 1969. Seismoakusticeskie metody izucenia massivov skalnych porod, Izdat. Nedra, Moskva.

Schön, J.H., 1996. Physical Properties of Rocks—Fundamentals and Principles of Petrophysics. Pergamon, Oxford, UK.

Sen, Z., 1993. RQD-fracture frequency chart based on a Weibull distribution. Int. J. Rock Mech. Min. Sci. Geomech. Abstr. 30, 555–557.

Sen, Z., Kazi, A., 1984. Discontinuity spacing and RQD estimates from finite length scanlines. Int. J. Rock Mech. Min. Sci. Geomech. Abstr. 21, 203–212.

Shanley, R.J., Mahtab, M.A., 1976. Delineation and analysis of centers in orientation data. Math. Geol. 8, 9–23.

Sjögren, B., Øvsthus, A., Sandberg, J., 1979. Seismic classification of rock mass qualities. Geophys. Prospect. 27, 409–442.

Song, J.-J., 2006. Estimation of areal frequency and mean trace length of discontinuities observed in non-planar surfaces. Rock Mech. Rock. Eng. 39 (2), 131–146.

Song, J.-J., Lee, C.-I., 2001. Estimation of joint length distribution using window sampling. Int. J. Rock Mech. Min. Sci. 38 (4), 519–528.

Spencer, E.W., 1969. Introduction to the Structure of the Earth. McGraw-Hill, New York.

Terzaghi, R., 1965. Sources of error in joint surveys. Geotechnique 5, 287–304.

Titley, S.R., Thompson, R.C., Haynes, F.M., Manske, S.L., Robison, L.C., White, J.L., 1986. Evaluation of fractures and alteration in the Sierrita-Esperanza Hydrothermal System, Pima County, Arizona. Econ. Geol. 81, 343–370.

Villaescusa, E., Brown, E.T., 1992. Maximum likelihood estimation of joint size from trace length measurements. Rock Mech. Rock. Eng. 25, 67–87.

Wallis, P.F., King, M.S., 1980. Discontinuity spacing in a crystalline rock. Int. J. Rock Mech. Min. Sci. Geomech. Abstr. 17, 63–66.

Warburton, P.M., 1980a. A stereological interpretation of joint trace data. Int. J. Rock Mech. Min. Sci. Geomech. Abstr. 17, 181–190.

Warburton, P.M., 1980b. Stereological interpretation of joint trace data: influence of joint shape and implications for geological surveys. Int. J. Rock Mech. Min. Sci. Geomech. Abstr. 17, 305–316.

Wathugala, D.N., 1991. Stochastic Three Dimensional Joint Geometry Modeling and Verification. Ph.D. Dissertation, University of Arizona, Tucson.

Watkins, H., Bond, C.E., Healy, D., Butler, R.W.H., 2015. Appraisal of fracture sampling methods and a new workflow to characterise heterogeneous fracture networks at outcrop. J. Struct. Geol. 72, 67–82.

Wyllie, D.C., Mah, C.W., 2004. Rock Slope Engineering, fourth ed. SPON Press, Taylor & Francis Group, London and New York.

Zanbak, C., 1977. Statistical interpretation of discontinuity contour diagrams. Int. J. Rock Mech. Min. Sci. Geomech. Abstr. 14, 111–120.

Zhang, L., 1999. Analysis and Design of Drilled Shafts in Rock. PhD thesis, Massachusetts Institute of Technology, Cambridge, MA.

Zhang, L., Ding, X., 2010. Variance of non-parametric rock fracture mean trace length estimator. Int. J. Rock Mech. Min. Sci. 47, 1222–1228.

Zhang, L., Einstein, H.H., 1998. Estimating the mean trace length of rock discontinuities. Rock Mech. Rock. Eng. 31, 217–235.

Zhang, L., Einstein, H.H., 2000. Estimating the intensity of rock discontinuities. Int. J. Rock Mech. Min. Sci. 37, 819–837.

Zhang, L., Einstein, H.H., 2010. The planar shape of rock joints. Rock Mech. Rock. Eng. 43, 55–68.

Zhang, L., Einstein, H.H., Dershowitz, W.S., 2002. Stereological relationship between trace length distribution and size distribution of elliptical discontinuities. Geotechnique 52, 419–433.

Chapter 5

Rock Masses

5.1 INTRODUCTION

The classification and index properties of intact rock and the characterization of rock discontinuities have been described in Chapters 3 and 4, respectively. Field rock masses usually contain both intact rock and discontinuities. It is the properties of the rock mass (the combination of intact rock and discontinuities) that should be used in the design of a rock structure. This chapter describes different rock mass classification systems that are useful in the estimation of rock mass properties. The correlations between different classification indices are also presented. Finally, the classification of weathering of rocks is discussed.

5.2 CLASSIFICATION OF ROCK MASSES

Numerous rock mass classification systems have been developed, including Terzaghi's Rock Load Height Classification (Terzaghi, 1946), Lauffer's Classification (Lauffer, 1958), Deere's Rock Quality Designation (RQD) (Deere, 1964), the Rock Structure Rating (RSR) concept (Wickham et al., 1972, 1974), the Rock Mass Rating (RMR) system (Bieniawski, 1973, 1976, 1989), the Modified Rock Mass Rating (MRMR) system for mining (Laubscher, 1977, 1984, 1990), the Q-System (Barton et al., 1974), the Rock Mass index (RMi) (Palmström, 1995, 1996a,b), and the Geological Strength Index (GSI) system (Hoek and Brown, 1997). Most of these classification systems were primarily developed for the design of underground excavations. However, five of the above classification systems have been used extensively in the estimation of rock mass properties. These five classification systems are the RQD, the RMR, the Q-System, the RMi, and the GSI.

5.2.1 Rock Quality Designation (RQD)

The RQD was introduced by Deere (1964) as an index assessing rock quality quantitatively. Table 5.1 shows the relationship between the RQD index and the rock mass quality. RQD can be determined directly by logging boring cores or indirectly by using different correlations such as the correlation between RQD and discontinuity frequency λ, and the correlation between RQD and

Engineering Properties of Rocks. http://dx.doi.org/10.1016/B978-0-12-802833-9.00005-5

TABLE 5.1 Correlation Between RQD and Rock Mass Quality

RQD (%)	Rock Mass Quality
<25	Very poor
25–50	Poor
50–75	Fair
75–90	Good
90–100	Excellent

seismic velocities. The different procedures for determining RQD have been discussed in detail in Chapter 4.

Although the RQD is a simple and inexpensive index, when considered alone it is not sufficient to provide an adequate description of a rock mass because it disregards discontinuity orientation, discontinuity condition, type of discontinuity filling and other features. As will be seen later in this chapter, RQD is only one of the important parameters for determining RMR, Q and GSI.

5.2.2 Rock Mass Rating (RMR)

The RMR or the Geomechanics Classification System, proposed by Bieniawski (1973), was initially developed for tunnels. In recent years, it has been applied to the preliminary design of rock slopes and foundations as well as to the estimation of the in-situ deformation modulus and strength of rock masses. The RMR uses six parameters that can be determined in the field (see Table 5.2):

- unconfined compressive strength of the intact rock
- RQD
- spacing of discontinuities
- condition of discontinuities
- ground water conditions
- orientation of discontinuities

All but the intact rock strength are normally determined in the standard geological investigations and are entered on an input data sheet. Table 5.3 shows the guidelines for assessing the discontinuity condition. The unconfined compressive strength of intact rock is determined in accordance with standard laboratory procedures but can be estimated in situ from the point-load strength index.

Rating adjustments for discontinuity orientation are summarized for underground excavations, foundations and slopes in Part B of Table 5.2. A more detailed explanation of these rating adjustments for dam foundations is given in Table 5.4, after ASCE (1996).

TABLE 5.2 Geomechanics Classification of Jointed Rock Masses

A. Classification Parameters and Their Rating

	Parameter		Range of Values						
1	Strength of intact rock	Point-load strength index (MPa)	>10	4–10	2–4	1–2	For this low range, unconfined compressive test is preferred		
		Unconfined compressive strength (MPa)	>250	100–250	50–100	25–50	5–25	1–5	<1
		Rating	15	12	7	4	2	1	0
2	Drill core quality RQD (%)		90–100	75–90	50–75	25–50	<25		
	Rating		20	17	13	8	3		
3	Spacing of discontinuities (m)		>2	0.6–2	0.2–0.6	0.06–0.2	<0.06		
	Rating		20	15	10	8	5		
4	Conditions of discontinuities		Very rough surfaces, Not continuous, No separation, Unweathered wall rock	Slightly rough surfaces, separation <1 mm, Slightly weathered walls	Slightly rough surfaces, separation <1 mm, Highly weathered walls	Slickensided surfaces or Gouge <5 mm thick or Separation 1–5 mm continuous	Soft gouge >5 mm thick or Separation >5 mm Continuous		
	Rating		30	25	20	10	0		

Continued

TABLE 5.2 Geomechanics Classification of Jointed Rock Masses—cont'd

A. Classification Parameters and Their Rating

	Parameter		Range of Values				
5	Ground water	Inflow per 10 m tunnel length (l/min)	None or	<10 or	10–25 or	25–125 or	>125 or
		Ratio of joint water pressure to major principal stress	0 or	<0.1 or	0.1–0.2 or	0.2–0.5 or	>0.5 or
		General conditions	Completely dry	Damp	Wet	Dripping	Flowing
	Rating		15	10	7	4	0

B. Rating adjustment for joint orientations

Strike and dip orientations of discontinuities		Very favorable	Favorable	Fair	Unfavorable	Very Unfavorable
Ratings	Tunnels and mines	0	−2	−5	−10	−12
	Foundations	0	−2	−7	−15	−25
	Slopes	0	−5	−25	−50	−60

Continued

TABLE 5.2 Geomechanics Classification of Jointed Rock Masses—cont'd

C. Rock mass classes and corresponding design parameters and engineering properties

Class No.	I	II	III	IV	V
RMR	100–81	80–61	60–41	40–21	<20
Description	Very Good	Good	Fair	Poor	Very poor
Average stand-up time	20 years for 15 m span	1 year for 10 m span	1 week for 5 m span	10 h for 2.5 m span	30 min for 1 m span
Cohesion of rock mass (MPa)	>0.4	0.3–0.4	0.2–0.3	0.1–0.2	<0.1
Internal friction angle of rock mass (°)	>45	35–45	25–35	15–25	<15
Deformation modulus (GPa)[a]	>56	56–18	18–5.6	5.6–1.8	<1.8

[a]*Deformation modulus values are from Serafim and Pereira (1983).*
Based on Bieniawski, Z.T., 1989. Engineering Rock Mass Classifications. John Wiley, Rotterdam.

TABLE 5.3 Guidelines for Classifying Discontinuity Condition

Parameter		Range of Values				
Discontinuity length (persistence/continuity)	Rating	6	4	2	1	0
	Measurement (m)	<1	1–3	3–10	10–20	>20
Separation (aperture)	Rating	6	5	4	1	0
	Measurement (mm)	None	<0.1	0.1–1	1–5	>5
Roughness	Rating	6	5	3	1	0
	Description	Very rough	Rough	Slight	Smooth	Slickensided
Infilling (gouge)	Rating	6	4	2	2	0
	Description and Measurement (mm)	None	Hard filling <5	Hard filling >5	Soft filling <5	Soft filling >5
Degree of weathering	Rating	6	5	3	1	0
	Description	None	Slight	Moderate	High	Decomposed

Note: Some conditions are mutually exclusive. For example, if infilling is present, it is irrelevant what the roughness may be, since its effect will be overshadowed by the influence of the gouge. In such cases, use Table 5.2 directly.
Based on Bieniawski, Z.T., 1989. Engineering Rock Mass Classifications. John Wiley, Rotterdam.

TABLE 5.4 Ratings for Discontinuity Orientations for Dam Foundations and Tunneling

A. Dam Foundations

Dip 0°–10°	Dip 10°–30°		Dip 30°–60°	Dip 60°–90°
	Dip direction			
	Upstream	Downstream		
Very favorable	Unfavorable	Fair	Favorable	Very unfavorable

B. Tunneling

Strike perpendicular to tunnel axis

Drive with dip		Drive against dip		Strike parallel to tunnel axis		Irrespective of strike
Dip 45°–90°	Dip 20°–45°	Dip 45°–90°	Dip 20°–45°	Dip 45°–90°	Dip 20°–45°	Dip 0°–20°
Very favorable	Favorable	Fair	Unfavorable	Very unfavorable	Fair	Fair

Based on ASCE, 1996. Rock Foundations: Technical Engineering and Design Guides as Adapted from the US Army Corps of Engineers. No. 16, ASCE Press, New York, NY.

The six separate ratings are summed to give an overall RMR, with a higher RMR indicating a better quality rock. Based on the observed RMR value, the rock mass is classified into five classes named as very good, good, fair, poor and very poor, as shown in Part C of Table 5.2. Also shown in Part C of Table 5.2 is an interpretation of these five classes in terms of roof stand-up time, cohesion, internal friction angle and deformation modulus for the rock mass.

It is noted that Table 5.2 shows the 1989 version of the RMR system. In many cases, the RMR data may be based on the 1976 version of the RMR system. RMR_{76} can be converted to RMR_{89} by adding a value of 5.

Seismic velocity measurements can also be used to estimate RMR values. Based on the data of limestones, mudstones, marls and shales at a dam site in Wadi Mujib, Jordan, El-Naqa (1996) obtained the following empirical correlation between RMR and P-wave velocity:

$$RMR = 59.8 \left(\frac{v_{pF}}{v_{p0}}\right)^{0.26} \quad (r = 0.84) \tag{5.1}$$

where v_{pF} is the P-wave velocity of the in situ rock mass; v_{p0} is the P-wave velocity of the corresponding intact rock; and r is the correlation coefficient.

Cha et al. (2006) obtained the following simple linear relation between RMR and shear-wave velocity measured using a refraction microtremor technique, in order to evaluate the rock condition for the design of a proposed railway tunnel at a site consisting of different types of granite (granite and felsite) and volcanic rocks (dacitic tuff, andesitic tuff, and andesite):

$$RMR = 36.2v_s - 10 \tag{5.2}$$

where v_s is the shear-wave velocity of the rock mass in km/s.

Banks (2005) derived an empirical relation between the basic RMR and the slope angles in mature, natural rock outcrops:

$$basic\,RMR = 0.4S + 52 \tag{5.3}$$

where the basic RMR is the RMR without the adjustment to account for the influence that discontinuity orientations may have on the particular application; and S is the slope angle in mature, natural outcrops in degrees.

5.2.3   Q-System

The Q-system, proposed by Barton et al. (1974), was developed specifically for the design of tunnel support systems. As the RMR system, the Q-system has been expanded to provide preliminary estimates of rock mass properties. The Q-system incorporates the following six parameters and the equation for obtaining rock mass quality Q:

- RQD
- number of discontinuity sets

- roughness of the most unfavorable discontinuity
- degree of alteration or filling along the weakest discontinuity
- water inflow
- stress condition.

$$Q = \frac{\text{RQD}}{J_n} \times \frac{J_r}{J_a} \times \frac{J_w}{\text{SRF}} \qquad (5.4)$$

where RQD is the Rock Quality Designation; J_n is the joint set number; J_r is the joint roughness number; J_a is the joint alteration number; J_w is the joint water reduction number; and SRF is the stress reduction number.

The meaning of the parameters used to determine the value of Q in Eq. (5.4) can be seen from the following comments by Barton et al. (1974):

The first quotient (RQD/J_n), representing the structure of the rock mass, is a crude measure of the block or particle size, with the two extreme values (100/0.5 and 10/20) differing by a factor of 400. If the quotient is interpreted in units of centimeters, the extreme "particle sizes" of 200 to 0.5 cm are seen to be crude but fairly realistic approximations. Probably the largest blocks should be several times this size and the smallest fragments less than half the size. (Clay particles are of course excluded).

The second quotient (J_r/J_a) represents the roughness and frictional characteristics of the joint walls or filling materials. This quotient is weighted in favor of rough, unaltered joints in direct contact. It is to be expected that such surfaces will be close to peak strength, that they will dilate strongly when sheared, and they will therefore be especially favorable to tunnel stability.

When rock joints have thin clay mineral coatings and fillings, the strength is reduced significantly. Nevertheless, rock wall contact after small shear displacements have occurred may be a very important factor for preserving the excavation from ultimate failure.

Where no rock wall contact exists, the conditions are extremely unfavorable to tunnel stability. The "friction angles" (given in Table 5.5) are a little below the residual strength values for most clays, and are possibly downgraded by the fact that these clay bands or fillings may tend to consolidate during shear, at least if normal consolidation or if softening and swelling has occurred. The swelling pressure of montmorillonite may also be a factor here.

The third quotient (J_w/SRF) consists of two stress parameters. SRF is a measure of: (1) loosening load in the case of an excavation through shear zones and clay bearing rock, (2) rock stress in competent rock, and (3) squeezing loads in plastic incompetent rocks. It can be regarded as a total stress parameter. The parameter J_w is a measure of water pressure, which has an adverse effect on the shear strength of joints due to a reduction in effective normal

TABLE 5.5 The Q-System and Associated Parameters RQD, J_n, J_r, J_a, SRF and J_w

Rock Quality Designation (RQD) (%)		
Very poor	0–25	Note:
Poor	25–50	(i) Where RQD is reported or measured to be <10 a nominal value of 10 is used to evaluate Q in Eq. (5.4)
Fair	50–75	
Good	75–90	(ii) Take RQD to be nearest 5%
Excellent	90–100	

Joint Set Number J_n		
Massive, none or few joints	0.5–1.0	Note:
One joint set	2	(i) For intersections use $(3.0 \times J_n)$
One joint set plus random	3	(ii) For portals use $(2.0 \times J_n)$
Two joint sets	4	
Two joint sets plus random	6	
Three joint sets	9	
Three joint sets plus random	12	
Four or more joint sets, random, heavily jointed, "sugar cube", etc.	15	
Crushed rock, earthlike	20	

Joint Roughness Number J_r		
(a) *Rock wall contact and* (b) *Rock wall contact before 10 cm shear*		
Discontinuous joint	4	Note:
Rough or irregular, undulating	3	(i) Add 1.0 if the mean spacing of the relevant joint set is greater than 3 m
Smooth, undulating	2	(ii) $J_r = 0.5$ can be used for planar slickensided joints having lineations, provided the lineations are favorably orientated
Slickensided, undulating		
Rough and irregular, planar	1.5	
Smooth or irregular	1.5	
Slickensided, planar	1	
	0.5	
(c) *No rock wall contact when sheared*		
Zone containing clay minerals thick enough to prevent rock wall contact	1	(Nominal)
Sandy, gravelly or crushed zone thick enough to prevent rock wall contact	1	(Nominal)

Continued

TABLE 5.5 The Q-System and Associated Parameters RQD, J_n, J_v J_a, SRF and J_w—cont'd

	Joint Alternation Number J_a	
		Approximate residual angle of friction (deg)
(a) *Rock wall contact*		
A. Tightly healed, hard, non-softening, impermeable filling, i.e. quartz or epidote	0.75	–
B. Unaltered joint walls, surface staining only	1	25–35
C. Unaltered joint walls. Non-softening mineral coatings, sandy particles, clay-free disintegrated rock, etc.	2	25–30
D. Silty or sandy clay coatings, small clay fraction (non-softening)	3	20–25
E. Softening or low friction clay mineral coatings, ie, kaolinite, mica. Also chlorite, talc, gypsum and graphite, etc, and small quantities of swelling clays (discontinuous coatings, 1–2 mm or less in thickness)	4	8–16
(b) *Rock wall contact before 10 cm shear*		
F. Sandy particles, clay free disintegrated rock, etc	4	25–30
G. Strongly over-consolidated, non-softening clay mineral fillings (continuous, <5 mm in thickness)	6	16–24
H. Medium or low over-consolidation, softening, clay mineral fillings (continuous, <5 mm in thickness)	8	12–16

Continued

TABLE 5.5 The Q-System and Associated Parameters RQD, J_n, J_r, J_a, SRF and J_w —cont'd

J. Swelling clay fillings, ie, montmorillonite (continuous, <5 mm in thickness). Value of Ja depends on percentage of swelling clay-sized particles and access to water, etc.	8–12	6–12
(c) *No rock wall contact when sheared*		
K. Zones or bands of disintegrated or crushed rock and clay (see G, H, J for description of clay condition)	6, 8 or 8–12	6–24
L. Zones or bands of silty or sandy clay, small clay fraction (non-softening)	5	–
M. Thick, continuous zones or bands of clay (see G, H, J for description of clay condition)	10,13 or 13–20	6–24

Joint Water Reduction Factor J_w

		Approximate water pressure (kPa)
A. Dry excavations or minor inflow, ie, <5 L/min locally	1	<100
B. Medium inflow or pressure occasional outwash of joint fillings	0.66	100–250
C. Large inflow or high pressure in competent rock with unfilled joints	0.5	250–1000
D. Large inflow or high pressure, considerable occasional outwash of joint fillings	0.33	250–1000
E. Exceptionally high inflow or water pressure at blasting, decaying with time	0.1	>1000
F. Exceptionally high inflow or water pressure continuing without decay	0.1–0.05	>1000

Continued

TABLE 5.5 The Q-System and Associated Parameters RQD, J_n, J_r, J_a, SRF and J_w—cont'd

Note:
(i) Factors C–F are crude estimates. Increase J_w if drainage measures are installed
(ii) Special problems caused by ice formation are not considered

Stress Reduction Factor, SRF		
(a) *Weakness zones intersecting excavation, which may cause loosening of rock mass when tunnel is excavated*		
A. Multiple occurrences of weakness zones containing clay or chemically disintegrated rock, very loose surrounding rock (any depth)	10	Note: (i) Reduce these values by 25–50% if the relevant shear zones only influence but do not intersect the excavation
B. Single weakness zones containing clay or chemically disintegrated rock (depth of excavation <50 m)	5	
C. Single weakness zones containing clay or chemically disintegrated rock (depth of excavation >50 m)	2.5	
D. Multiple shear zones in competent rock (clay free), loose surrounding rock (any depth)	7.5	
E. Single shear zones in competent rock (clay free, depth of excavation <50 m)	5	
F. Single shear zones in competent rock (clay free, depth of excavation >50 m)	2.5	
G. Loose open joints, heavily jointed, or "sugar cube" etc (any depth)	5	

Continued

TABLE 5.5 The Q-System and Associated Parameters RQD, J_n, J_r, J_a, SRF and J_w—cont'd

(b) *Competent rock, rock stress problems*	Strength/stress ratios			
	σ_c/σ_1	σ_1/σ_1		
H. Low stress, near surface	>200	>13	2.5	(ii) If stress field is strongly anisotropic: when $5<\sigma_1/\sigma_3<10$, reduce σ_c and σ_t to $0.8\sigma_c$ and $0.8\sigma_t$; when $\sigma_1/\sigma_3>10$, reduce σ_c and σ_t to $0.6\sigma_c$ and $0.6\sigma_t$. Where $\sigma_c=$unconfined compressive strength, $\sigma_t=$tensile strength, σ_1 and $\sigma_3=$major and minor principal stresses
J. Medium stress	200–10	13–0.66	1	
K. High stress, very tight structure (usually favorable to stability, maybe unfavorable to wall stability)	10–5	0.66–0.33	0.5–2.0	
L. Mild rock burst (massive rock)	5–2.5	0.33–0.16	5–10	
M. Heavy rock burst (massive rock)	<2.5	<0.16	10–20	
(c) *Swelling rock; chemical swelling activity depending on presence of water*				(iii) Few case records available where depth of crown below surface is less than span width. Suggest SRF increase from 2.5 to 5 for such cases (see H)
P. Mild swelling rock pressure		5–10		
R. Heavy swelling rock pressure		10–15		

Based on Barton, N., Lien, R., Lunde, J., 1974. Engineering classification of rock masses for the design of tunnel support. Rock Mech. 6, 189–236.

stress. Water may, in addition, cause softening and possible out-wash in the case of clay-filled joints. It has proved impossible to combine these two parameters in terms of inter-block effective stress, because paradoxically a high value of effective normal stress may sometimes signify less stable conditions than a low value, despite the higher shear strength. The quotient (J_w/SRF) is a complicated empirical factor describing the "active stress".

So the Q may be considered a function of three parameters which are approximate measures of:

(i) Block size (RQD/J_n): It represents the overall structure of rock masses.
(ii) Inter block shear strength (J_r/J_a): It represents the roughness and frictional characteristics of the joint walls or filling materials.
(iii) Active stress (J_w/SRF): It is an empirical factor describing the active stress.

Table 5.5 provides the necessary guidance for assigning values to the six parameters. Depending on the six assigned parameter values reflecting the rock mass quality, Q can vary between 0.001 and 1000. Rock quality is divided into nine classes ranging from exceptionally poor (for Q from 0.001 to 0.01) to exceptionally good (for Q from 400 to 1000) as shown in Table 5.6.

Based on data from hard rock tunneling projects in several countries, Barton (1991) proposed a correlation between Q and P-wave velocity:

$$Q = 10^{v_p - 3.5} \qquad (5.5)$$

where v_p is the P-wave velocity of the in situ rock mass in km/s.

TABLE 5.6 Classification of Rock Mass Based on Q-Values

Group	Q	Classification
1	1000–400	Exceptionally good
	400–100	Extremely good
	100–40	Very good
	40–10	Good
2	10–4	Fair
	4–1	Poor
	1–0.1	Very poor
3	0.1–0.01	Extremely poor
	0.01–0.001	Exceptionally poor

Based on Barton, N., Lien, R., Lunde, J., 1974. Engineering classification of rock masses for the design of tunnel support. Rock Mech. 6, 189–236.

Barton (2002) extended the above relation between Q and v_p to rocks that could be weaker or stronger than the assumed "hard" rock by introducing the normalized Q:

$$Q_c = Q \times \frac{\sigma_c}{100} \tag{5.6}$$

where Q_c is the normalized Q; σ_c is the unconfined compressive strength of the intact rock; and 100 MPa is the σ_c assumed for the hard rock norm. The generalized Q_c–v_p correlation is

$$Q_c = 10^{v_p-3.5} \tag{5.7}$$

Based on the data of limestones, mudstones, marls and shales at a dam site in Wadi Mujib, Jordan, El-Naqa (1996) obtained the following empirical correlation between Q and P-wave velocity:

$$\ln Q = 2.61 \left(\frac{v_{pF}}{v_{p0}}\right)^{0.97} \quad (r = 0.78) \tag{5.8}$$

where v_{pF} is the P-wave velocity of the in situ rock mass; v_{p0} is the P-wave velocity of the corresponding intact rock; and r is the correlation coefficient.

5.2.4   Rock Mass Index (RMi)

Palmström (1995, 1996a, b) proposed the RMi to characterize the strength of rock masses as a construction material. RMi considers the reduction in strength of the rock mass caused by discontinuities and is expressed as:

$$RMi = \sigma_c \times JP \tag{5.9}$$

where σ_c is the unconfined compressive strength of the intact rock measured on 50 mm diameter specimens; and JP is a jointing parameter which is a reduction factor representing the combined effect of block volume (V_b) and joint (discontinuity) condition (jC). The joint condition factor jC is related to the joint roughness (jR), the joint alteration (jA), and the joint size and termination (jL) (Fig. 5.1).

JP can be determined using the following expression which was derived based on calibrations of test results (Palmström, 1995, 1996a,b):

$$JP = 0.2\sqrt{jC} \times V_b^D \tag{5.10}$$

where V_b is given in m³; and D is related to jC by:

$$D = 0.37 jC^{-0.2} \tag{5.11}$$

The jC in Eqs. (5.10) and (5.11) can be determined by:

$$jC = (jR/jA)jL \tag{5.12}$$

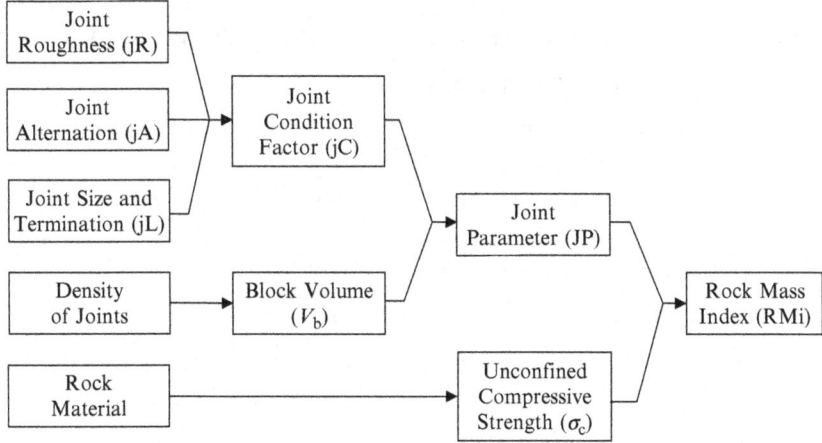

FIG. 5.1   Parameters used in RMi. *(Based on Palmström, A., 1995. RMi—A Rock Mass Classification System for Rock Engineering Purposes. PhD Thesis, University of Oslo.)*

The ratings of iR, jA and jL can be determined using Tables 5.7–5.9, respectively. The factors jA and jR are similar to the joint roughness number (J_r) and the joint alternation number (J_a) in the Q-system, respectively. The factor iL was introduced in the RMi system to represent the scale effect of the joints.

TABLE 5.7 Ratings of Joint Roughness Factor (jR) Found From Smoothness and Waviness

Small-Scale Smoothness of Joint Surface	Large-Scale Waviness of Joint Plane				
	Planar	Slightly Undulating	Undulating	Strongly Undulating	Stepped or Interlocking
Very rough	2	3	4	6	6
Rough	1.5[a]	2	4	4.5	6
Smooth	1	1.5	2	3	4
Polished or slickensided**	0.5	1	1.5	2	3

For filled joints, jR = 1; for irregular joints, jR = 5 is suggested.
**For slickensided surfaces, the ratings apply to possible movement along the lineations.*
[a]*The ratings in **bold** are similar to J_r in the Q-system.*
Based on Palmström, A., 1995. RMi—A Rock Mass Classification System for Rock Engineering Purposes. PhD Thesis, University of Oslo; Palmström, A., 2000. Recent developments in rock support estimates by the RMi. J. Rock Mech. Tunn. Technol. 6(1), 1–19.

TABLE 5.8 Characterization and Rating of Joint Alternation Factor (jA)

A. Contact Between Rock Wall Surfaces

Joint Wall Character		Condition	jA
Clean joints	Healed or welded joints	Softening, impermeable filling (quartz, epidote, etc.)	0.75
	Fresh rock walls	No coating or filling on joint surface, except for staining	1
	Altered joint walls	One grade higher alteration than the rock	2
		Two grades higher alteration than the rock	4
Coating or thin filling	Sand, silt, calcite etc.	Coating of friction materials without clay	3
	Clay, chlorite, talc etc.	Coating of softening and cohesive minerals	4

B. Filled joints with partial or no contact between rock wall surfaces

Type of filling material	Description	jA	
		Partial wall contact (thin filling <5 mm)	No wall contact (thick filling or gouge)
Sand, silt calcite, etc. (non-softening)	Filling of friction materials without clay	4	8
Compacted filling of clay, chlorite, talc, etc.	"Hard" filling of softening and cohesive materials	6	10
Medium to low overconsolidated clay, chlorite, talc, etc.	Medium to low over-consolidation of filling	8	12
Filling material exhibits swelling properties	Filling material exhibits clear swelling properties	8–12	12–20

Based on Palmström, A., 1995. RMi—A Rock Mass Classification System for Rock Engineering Purposes. PhD Thesis, University of Oslo; Palmström, A., 2000. Recent developments in rock support estimates by the RMi. J. Rock Mech. Tunn. Technol. 6(1), 1–19.

TABLE 5.9 Ratings of Joint Size and Termination Factor (jL)

			jL	
Type	Length (m)	Description	Continuous Joints	Discontinuous Joints[a]
Bedding/ foliation partings	<0.5	Very short	3	6
Joint	0.1–1.0	Short/small	2	4
	1–10	Medium	1	2
	10–30	Long/large	0.75	1.5
Filled joint, seam or shear[b]	>30	Very long/ large	0.5	1

[a]*Discontinuous joints end in massive rock.*
[b]*Often occurs as a single discontinuity, and should in these cases be treated separately.*
Based on Palmström, A., 1995. RMi—A Rock Mass Classification System for Rock Engineering Purposes. PhD Thesis, University of Oslo; Palmström, A., 2000. Recent developments in rock support estimates by the RMi. J. Rock Mech. Tunn. Technol. 6(1), 1–19.

The block volume (V_b) can be determined using the methods presented in Chapter 4.

It is noted that significant scale effects are generally involved when the tested rock volume is enlarged from laboratory size to field size. For the method described above for determining RMi, the RMi is related to large samples where the scale effect has been included in JP. The joint size factor (jL) is also a scale variable. However, for massive rock masses where the jointing parameter JP ≈ 1, the scale effect for the unconfined compressive strength (σ_c) has not been accounted for, as σ_c is related to the 50 mm sample size. For a large field sample with a diameter D_b, the unconfined compressive strength (σ_{cf}) may be determined from:

$$\sigma_{cf} = \sigma_{c50}(0.05/D_b)^{0.2} \tag{5.13}$$

where σ_{c50} is the unconfined compressive strength for a 50 mm diameter sample; and D_b is the field sample diameter measured in meter.

Eq. (5.13) is valid for sample diameters up to several meters, and may, therefore, be applied for massive rock masses. The equivalent block diameter (D_b) in Eq. (5.13) may be found from $D_b = (V_b)^{1/3}$ or, in cases where a pronounced joint set occurs, from $D_b = s$, where s is the spacing of this joint set as described in

Chapter 4. If the block shape is considered, the equivalent block diameter can be determined by:

$$D_b = \frac{27}{\beta} \sqrt[3]{V_b} \qquad (5.14)$$

where β is the block shape factor which has been discussed in detail in Chapter 4.

Based on the RMi values, rock masses can be divided into seven categories as shown in Table 5.10.

5.2.5 Geological Strength Index (GSI)

Hoek and Brown (1997) introduced the GSI, both for hard and weak rock masses. Experienced field engineers and geologists generally show a liking for a simple, fast, yet reliable classification which is based on visual inspection of geological conditions. Hoek and Brown (1997) proposed such a practical classification for estimating GSI based on visual inspection alone (see Fig. 5.2). In this classification, there are five main qualitative classifications of rock mass structures:

- **(i)** Intact/Massive
- **(ii)** Blocky
- **(iii)** Very blocky
- **(iv)** Blocky/Disturbed
- **(v)** Disintegrated

TABLE 5.10 Classification of Rock Masses According to RMi

RMi Value	Description of RMi	Description of Rock Mass Strength
<0.001	Extremely low	Extremely weak
0.001–0.01	Very low	Very weak
0.01–0.1	Low	Weak
0.1–1	Moderate	Medium
1–10	High	Strong
10–100	Very high	Very strong
>100	Extremely high	Extremely strong

Based on Palmström, A., 1995. RMi—A Rock Mass Classification System for Rock Engineering Purposes. PhD Thesis, University of Oslo; Palmström, A., 2000. Recent developments in rock support estimates by the RMi. J. Rock Mech. Tunn. Technol. 6(1), 1–19.

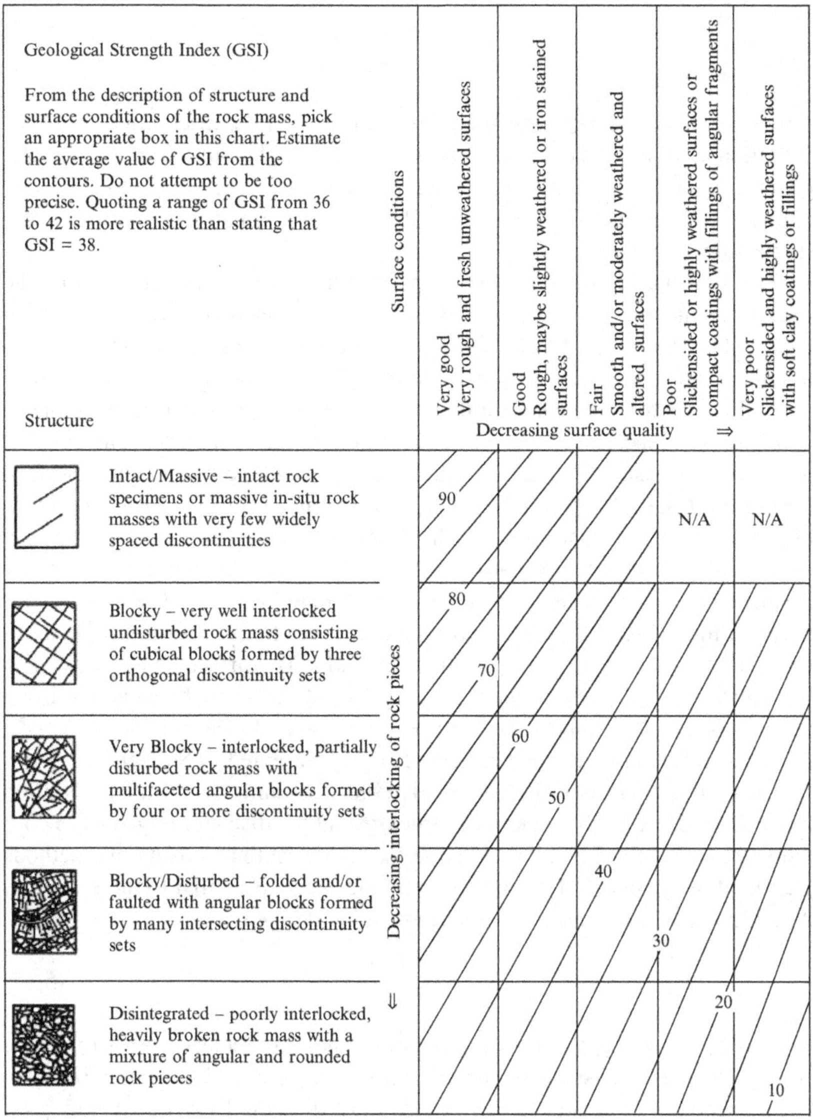

Geological Strength Index (GSI) — From the description of structure and surface conditions of the rock mass, pick an appropriate box in this chart. Estimate the average value of GSI from the contours. Do not attempt to be too precise. Quoting a range of GSI from 36 to 42 is more realistic than stating that GSI = 38.	Surface conditions → Decreasing surface quality ⇒				
Structure	Very good — Very rough and fresh unweathered surfaces	Good — Rough, maybe slightly weathered or iron stained surfaces	Fair — Smooth and/or moderately weathered and altered surfaces	Poor — Slickensided or highly weathered surfaces or compact coatings with fillings of angular fragments	Very poor — Slickensided and highly weathered surfaces with soft clay coatings or fillings
Intact/Massive – intact rock specimens or massive in-situ rock masses with very few widely spaced discontinuities	90			N/A	N/A
Blocky – very well interlocked undisturbed rock mass consisting of cubical blocks formed by three orthogonal discontinuity sets	80	70			
Very Blocky – interlocked, partially disturbed rock mass with multifaceted angular blocks formed by four or more discontinuity sets		60	50		
Blocky/Disturbed – folded and/or faulted with angular blocks formed by many intersecting discontinuity sets			40	30	
Disintegrated – poorly interlocked, heavily broken rock mass with a mixture of angular and rounded rock pieces				20	10

(Left axis: Decreasing interlocking of rock pieces ⇓)

FIG. 5.2 Characterization of rock masses on the basis of interlocking and joint alteration. *(Based on Hoek, E., Brown, E.T., 1997. Practical estimates of rock mass strength. Int. J. Rock Mech. Min. Sci. 34, 1165–1186.)*

Further, discontinuities are classified into five surface conditions which are similar to discontinuity conditions in RMR as described earlier:

(i) Very good
(ii) Good

(iii) Fair
(iv) Poor
 (v) Very poor

Based on the actual rock structure classification and the discontinuity surface condition, a block in the 5×5 matrix of Fig. 5.2 can be picked up and the corresponding GSI value can then be read from the figure. According to Hoek and Brown (1997), a range of values of GSI should be estimated in preference to a single value.

The GSI chart based on visual inspection has been commonly used by the rock mechanics community since it was developed. However, due to the lack of measurable parameters for describing the rock mass structures and the discontinuity surface conditions, it is possible for different persons to estimate different GSI values from the chart for the same rock mass, particularly for engineers with limited experience. Therefore, researchers have attempted to develop quantitative measures of the rock mass structures and the discontinuity surface conditions (Cai et al., 2004; Hoek et al., 2013; Sonmez and Ulusay, 1999, 2002). Fig. 5.3 shows the quantitative GSI chart proposed by Sonmez and Ulusay (1999, 2002). The structure rating, SR, based on the volumetric discontinuity frequency λ_v, is introduced to describe the rock mass structure. The surface condition rating, SCR, estimated from roughness, weathering and infilling conditions, is introduced to describe the discontinuity surface conditions. The volumetric discontinuity frequency λ_v and discontinuity roughness and infilling conditions can be evaluated as described in Chapter 4. The weathering condition can be evaluated using the methods to be presented in Section 5.4.

Cai et al. (2004) used the quantitative block volume V_b and joint condition factor J_C to describe the rock mass structure and the discontinuity surface condition, respectively. The block volume V_b can be estimated using the methods presented in Chapter 4. The joint condition factor J_C is similar to the joint condition factor jC used in the RMi system and is defined as:

$$J_C = \frac{J_W J_S}{J_A} \tag{5.15}$$

where J_W, J_S, and J_A are the large-scale waviness (in meters from 1 to 10 m), small-scale smoothness (in centimeter from 1 to 20 cm) and joint alteration factor, respectively. J_W and J_S can be determined using Tables 5.11 and 5.12, respectively, while J_A can be determined by treating it as jA and using Table 5.8. With the obtained V_b and J_C, GSI can be determined using the quantified GSI system chart in Fig. 5.4 or the following equation (Cai and Kaiser, 2006):

$$GSI = \frac{26.5 + 8.79 \ln J_C + 0.9 \ln V_b}{1 + 0.015 \ln J_C - 0.0253 \ln V_b} \tag{5.16}$$

where J_C is dimensionless and V_b is in cm^3.

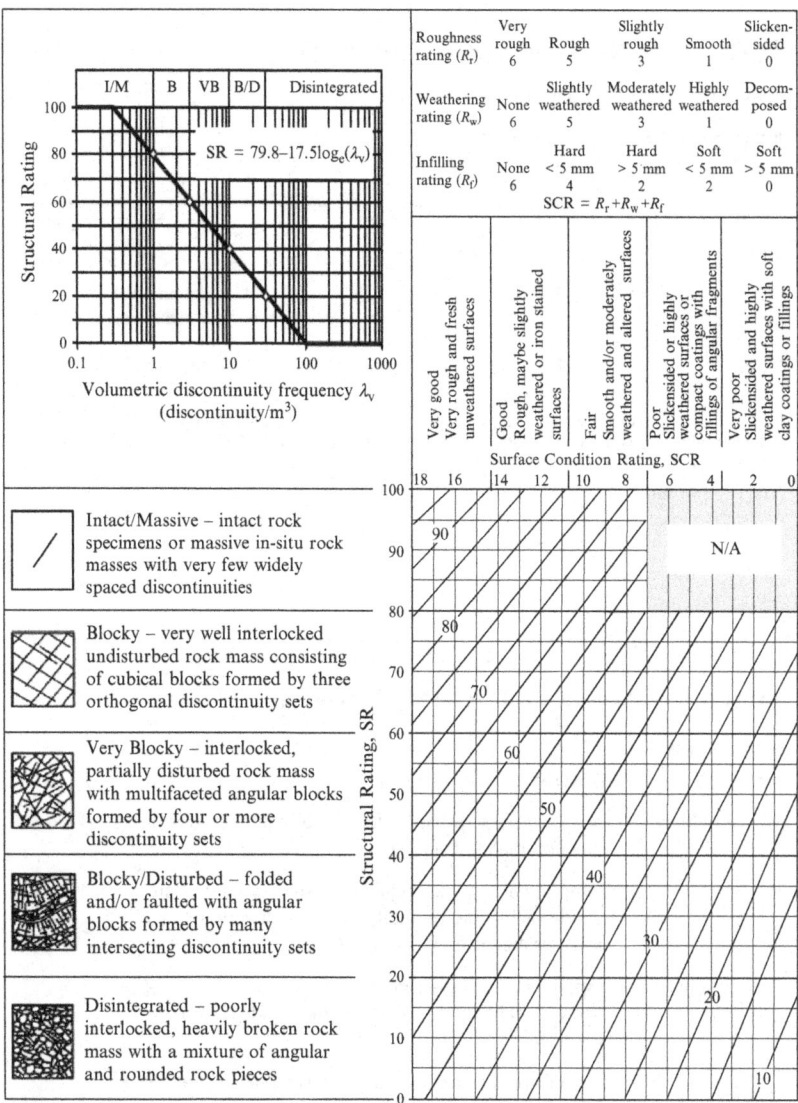

FIG. 5.3 Quantification of GSI chart by Sonmez and Ulusay (1999, 2002).

Hoek et al. (2013) quantified the GSI chart on the basis of joint condition and RQD, and proposed the following simple expression for estimating GSI:

$$GSI = 1.5JCond_{89} + RQD/2 \qquad (5.17)$$

where $JCond_{89}$ is the joint condition rating defined by Bieniawski (1989) and can be determined using Table 5.2 or 5.3; and RQD can be determined using the different methods presented in Chapter 4.

TABLE 5.11 Ratings of Large-Scale Waviness J_W

Waviness Terms	Undulation	J_W
Interlocking (large-scale)	Extremely low	3
Stepped	Very low	2.5
Large undulation	Low	2
Small to moderate undulation	Moderate	1.5
Planar	High	1

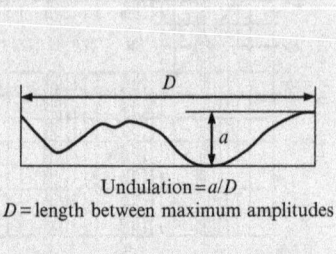

Undulation $= a/D$
$D =$ length between maximum amplitudes

Based on Cai, M., Kaiser, P. K., Uno, H., Tasaka, Y., Minami, M., 2004. Estimation of rock mass deformation modulus and strength of jointed hard rock masses using the GSI System. Int. J. Rock Mech. Min. Sci. 41, 3–19.

TABLE 5.12 Ratings of Small-Scale Smoothness J_S

Smoothness terms	Description	J_S
Very rough	Near vertical steps and ridges occur with interlocking effect on the joint surface	3
Rough	Some ridge and side-angle are evident; asperities are clearly visible; discontinuity surface feels very abrasive (rougher than sandpaper grade 30)	2
Slightly rough	Asperities on the discontinuity surfaces are distinguishable and can be felt (like sandpaper grade 30–300)	1.5
Smooth	Surface appear smooth and feels so to touch (smoother than sandpaper grade 300)	1
Polished	Visual evidence of polishing exists. This is often seen in coating of chlorite and specially talc	0.75
Slickensided	Polished and striated surface that results from sliding along a fault surface or other movement surface	0.6–1.5

Based on Cai, M., Kaiser, P. K., Uno, H., Tasaka, Y., Minami, M., 2004. Estimation of rock mass deformation modulus and strength of jointed hard rock masses using the GSI System. Int. J. Rock Mech. Min. Sci. 41, 3–19.

Discontinuity or Block Wall Condition

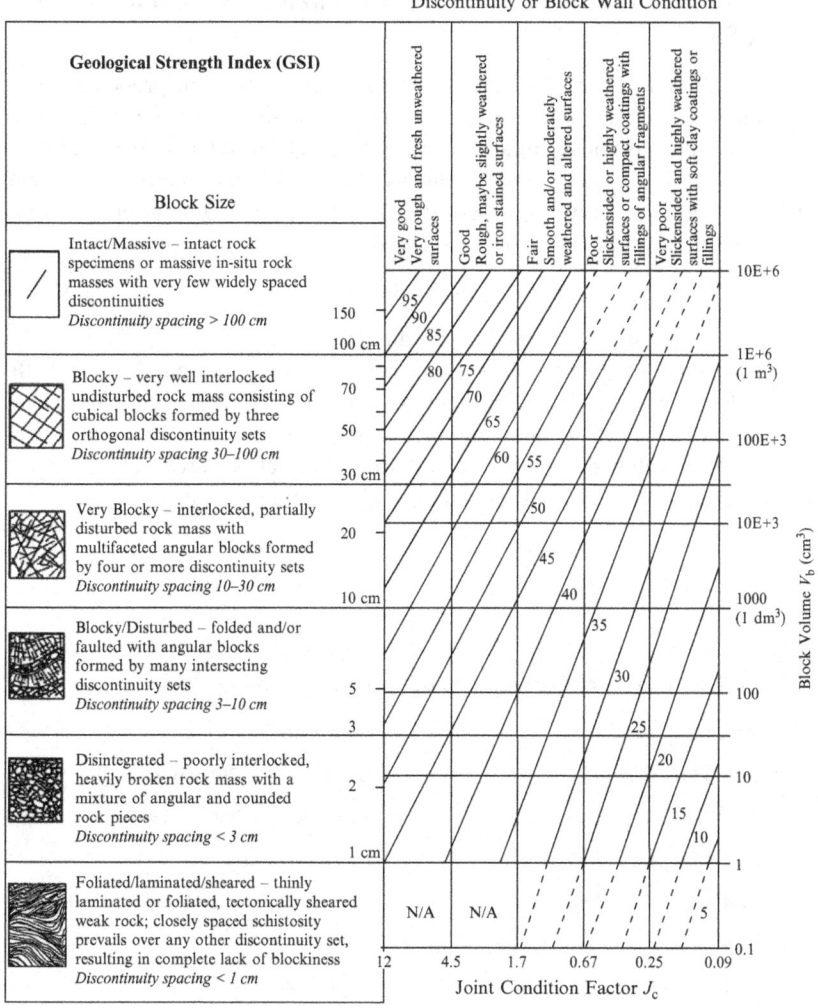

FIG. 5.4   Quantification of GSI chart by Cai et al. (2004).

When the joint roughness number J_r and the joint alteration number J_a in the Q-system (see Table 5.5) are used to describe the joint condition, the following simple expression can be used to estimate GSI (Hoek et al., 2013):

$$\text{GSI} = \frac{52(J_r/J_a)}{1+J_r/J_a} + \text{RQD}/2 \tag{5.18}$$

5.3 CORRELATIONS BETWEEN DIFFERENT CLASSIFICATION INDICES

Since the Q and RMR systems are based on essentially the same properties, they are highly correlated and can be predicted one from the other. Various researchers have proposed relationships between Q and RMR in the following general form (Abad et al., 1984; Bieniawski, 1976, 1989; Cameron-Clarke and Budavari, 1981; Goel et al., 1996; Hashemi et al., 2010; Kumar et al., 2004; Rutledge and Preston, 1978; Tuğrul, 1998):

$$RMR = A \ln Q + B \qquad (5.19)$$

where A is typically in the range 5–15; and B in the range 35–60 (Table 5.13).

Noting that the Q and RMR systems are not truly equivalent (eg, the RMR system does not consider the stress condition of the rock mass, while the Q system does not consider discontinuity orientation and intact rock strength), Goel et al. (1996) developed a new type of empirical correlation between RMR and Q by introducing two new rock mass indices RCR and N:

$$RCR = 8 \ln N + 30 \quad (r = 0.92) \qquad (5.20)$$

TABLE 5.13 Empirical Correlations Between RMR and Q

Correlation	Reference
$RMR = 9.0 \ln Q + 44$	Bieniawski (1976), Jethwa et al. (1982)
$RMR = 5.9 \ln Q + 43$	Rutledge and Preston (1978)
$RMR = 5.4 \ln Q + 55$	Moreno (1980)
$RMR = 4.6 \ln Q + 56$ (Drill core)	Cameron-Clarke and Budavari (1981)
$RMR = 5.0 \ln Q + 61$ (In situ results)	
$RMR = 10.5 \ln Q + 42$	Abad et al. (1984)
$RMR = 8.7 \ln Q + 38$	Kaiser et al. (1986)
$RMR = 9.0 \ln Q + 49$	Al-Harthi (1993)
$RMR = 7.0 \ln Q + 41$ (Bore cores)	El-Naqa (1994)
$RMR = 7.0 \ln Q + 44$ (Scanlines)	
$RMR = 15 \ln Q + 50$	Barton (1995)
$RMR = 7.0 \ln Q + 36$	Tuğrul (1998)
$RMR = 6.4 \ln Q + 50$	Kumar et al. (2004)
$RMR = 5.4 \ln Q + 40$	Hashemi et al. (2010)

where RCR is the rock condition rating defined as RMR without rating for discontinuity orientation and intact rock strength; N is the rock mass number defined as Q with SRF as 1; and r is the correlation coefficient.

Kumar et al. (2004) and Hashemi et al. (2010) also derived correlations between RCR and N similar to Eq. (5.20):

$$\text{RCR} = 8 \ln N + 42.7 \quad \text{(Kumar et al., 2004)} \tag{5.21}$$

$$\text{RCR} = 6 \ln N + 33.8 \quad \text{(Hashemi et al., 2010)} \tag{5.22}$$

Researchers also derived the correlations between RMi and RMR or Q, such as:

$$\text{RMR} = 5.4 \ln \text{RMi} + 54.4 \quad \text{(Kumar et al., 2004)} \tag{5.23}$$

$$\text{RMR} = 7.5 \ln \text{RMi} + 36.8 \quad \text{(Hashemi et al., 2010)} \tag{5.24}$$

$$\text{RMi} = 0.5 \ln Q^{0.93} \, \text{(Kumar et al., 2004)} \tag{5.25}$$

$$\text{RMi} = 1.5 \ln Q^{0.72} \quad \text{(Kumar et al., 2004)} \tag{5.26}$$

$$\text{RMi} = 1.08 \ln Q^{0.49} \quad \text{(Hashemi et al., 2010)} \tag{5.27}$$

GSI can also be estimated from RMR and Q (Hoek and Brown, 1997). When using Bieniawski's 1989 RMR (see Part A of Table 5.2) to estimate the value of GSI, the rock mass should be assumed to be completely dry and a rating of 15 assigned to the groundwater value. Very favorable discontinuity orientations should be assumed and the Adjustment for Discontinuity Orientation value set to zero. The minimum value which can be obtained for the 1989 classifications is 23. The estimated RMR can be used to estimate the value of GSI as follows:

$$\text{GSI} = \text{RMR} - 5 \tag{5.28}$$

Hashemi et al. (2010) derived the following relation between GSI and Bieniawski's, 1989 RMR:

$$\text{GSI} = 0.7 \text{RMR} + 22.3 \tag{5.29}$$

The Q value of Barton et al. (1974) can be used to estimate the value of GSI as follows:

$$\text{GSI} = 9 \ln Q + 44 \tag{5.30}$$

where Q is calculated from Eq. (5.4) by setting a value of 1 for both J_w (discontinuity water reduction factor) and SRF (stress reduction factor).

5.4 CLASSIFICATION OF WEATHERING OF ROCK

Weathering is the process of alteration of rock brought about by physical disintegration, chemical decomposition and biological activity. Weathering leads to change of the engineering properties of a rock at varying degrees depending on the stages of weathering. The early stages of weathering usually are represented by discoloration of the rock material, which increases from slightly to highly discolored as the degree of weathering increases. As weathering proceeds, the rock material becomes increasingly decomposed and/or disintegrated until a soil is formed. Various classification schemes have been proposed for classifying the weathering grades of rock masses, based on the presence or absence of discoloration in rock material, the rock to soil ratio, and the presence or absence of relict rock fabric in the groups which are predominantly soil (Bell, 1987). Classification of weathered rocks helps in better understanding their engineering behavior, allows samples to be grouped for description and for geotechnical models to be developed, and ensures the best use of the geotechnical data determined in that index properties can be related to engineering properties (Anon, 1995).

There exist different classification schemes for classifying the weathering grades of rock masses. Table 5.14 shows the general weathering categories and grades suggested by ISRM (1978), which may be modified to suit

TABLE 5.14 Weathering Grade of Rock Mass

Term	Description	Grade
Fresh rock	No visible sign of rock material weathering; perhaps slight discoloration on major discontinuity surfaces	I
Slightly weathered	Discoloration indicates weathering of rock material and discontinuity surfaces. All the rock material may be discolored by weathering and the external surface may be somewhat weaker than in its fresh condition	II
Moderately weathered	Less than half of the rock material is decomposed and/or disintegrated to soil. Fresh or discolored rock is present either as continuous framework or as corestones	III
Highly weathered	More than half of the rock material is decomposed and/or disintegrated to soil. Fresh or discolored rock is present either as discontinuous framework or as corestones	IV
Completely weathered	All rock material is decomposed and/or disintegrated to soil. The original mass structure is still largely intact	V
Residual soil	All rock material is converted to soil. The mass structure and material fabric are destroyed. There is a large change in volume, but the soil has not been significantly transported	VI

Based on ISRM, 1978. Suggested methods for the quantitative description of discontinuities in rock masses. International Society for Rock Mechanics, Commission on Standardization of Laboratory and Field Tests. Int. J. Rock Mech. Min. Sci. Geomech. Abstr. 15, 319–368.

particular situations. Fig. 5.5 shows different classifications of weathering grades of rock masses. Some grades of weathering may not be seen in a given rock mass, and, in some cases, a particular grade may be present to a very small extent.

The classifications of weathering grades presented in Table 5.14 and Fig. 5.5 are qualitative and based on subjective criteria. Quantitative classifications of weathering grades using index properties are also developed by different researchers. Irfan and Dearman (1978) suggested that quick absorption and point load strength tests could be used to determine a weathering grade for granite as illustrated in Table 5.15.

Researchers advocated that Schmidt hammer could be used as an index tool over the full range of weathering. Hencher and Martin (1982) proposed non-overlapping in situ N hammer value ranges (>45, 25–45, 0–25, no rebound) for differentiating weathered states (from Grades II to V) of igneous rocks of Hong Kong; and Karpuz and Pasamehmetoglu (1997) proposed non-overlapping L hammer value ranges (54–61, 39–54, 28–39, 18–28 and <18) for classifying weathered states (from Grades I to V) of Ankara andesite. However, researchers reported overlapping ranges of Schmidt hammer rebound values in adjacent and even gap weathering grades (Basu et al., 2009; Ebuk, 1991; Irfan, 1996; Irfan and Powell, 1985). The degree of such overlapping is increases at higher weathering grades because increasing degree of weathering intensifies heterogeneity of rocks resulting in larger scatter in impact rebound values (Aydin and Basu, 2005).

Hachinohe et al. (1999), Seiki and Aydan (2003), Aydan et al. (2014) and other researchers used needle penetration index (NPI) to determine the degree of weathering. Hachinohe et al. (1999) used the residual strength ratio R_s (%) as defined below to describe the degree of weathering:

$$R_s = \left(\text{NPI}/\text{NPI}_{fp}\right) \times 100 \qquad (5.31)$$

where NPI is the measured value for the part to be evaluated; and NPI_{fp} is the average value for the fresh (unweathered) part of each drill core. They determined the R_s values for Tertiary sandstone and mudstone from the bedrock of marine terraces in Boso Peninsula, Japan. The R_s decreases with increasing weathering degree and longer weathering time. Aydan et al. (2014) also evaluated the weathering degree of soft rocks by using the NPI/NPI_{fp} ratio.

Table 5.16 lists the measured total porosity and dry density of granitic rocks in the Northwest of Turkey at different weathering stages (Arel and Onalp, 2004).

Table 5.17 shows the relationship between weathering and RQD for rocks at the Gilgel Gibe hydropower project site located in the western part of Ethiopia (Ayalew et al., 2002).

Fig. 5.6 shows the variation of unconfined compressive strength (UCS) with weathering grade of different rocks.

Schematic profile	Love (1951), Little (1961)	Vargas (1951)	Sowers (1954, 1963)	Chandler (1969)	Geological Soci. Eng. Group (1970)	Deere and Patton (1971)
	Igneous rocks	Ignics, basaltics and sandstones	Igneous and metamorphic rocks	Marl and limolites	Igneous rocks	Igneous and metamorphic rocks
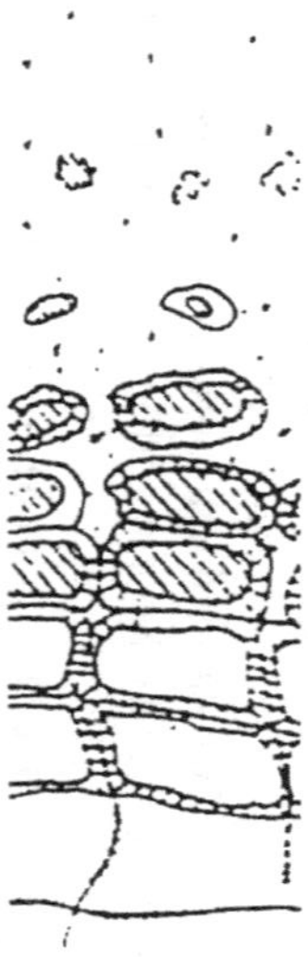	VI Soil	Residual soil	Upper zone	V Completely weathered	VI Residual soil	IA Horizon / IB Horizon (Residual soil)
	V Completely weathered	Young residual soil	Intermediate zone	IV (Partially weathered)	V Completely weathered	IC Horizon (saprolite) (Residual soil)
	IV Highly weathered	Disintegrated soil layers	Partially weathered zone	III (Partially weathered)	IV Highly weathered	IA Saprolite-weathered rock transition (Transition zone)
	III Moderately weathered			II (Partially weathered)	III Moderately weathered	IB Partially weathered (Transition zone)
	II Slightly weathered				II Weakly weathered	
					IB Softly weathered	
	I Fresh rock	Fresh rock	Unweathered rock	I Unweathered rock	IA Fresh rock	Fresh rock

FIG. 5.5 Different classifications of weathering grades. (Based on Oteo, C.S., 2002. In situ characterization of rocks. In V.M. Sharma and K.R. Saxena (Eds.). In-Situ Characterization of Rocks. Balkema, Lisse, pp. 1–48.)

TABLE 5.15 Weathering Grades for Granite

Term	Quick Absorption (%)	Bulk Density (Mg/m³)	Point Load Strength (MPa)	Unconfined Compressive Strength (MPa)
Fresh	<0.2	2.61	>10	>250
Partially stained[a]	0.2–1.0	2.56–2.61	6–10	150–250
Completely stained[a]	1.0–2.0	2.51–2.56	4–6	100–150
Moderately weathered	2.0–10.0	2.05–2.51	0.1–4	2.5–100
Highly/ completely weathered	>10.0	<2.05	<0.1	<2.5

[a]Slightly weathered.
Based on Irfan, T.Y., Dearman, W.R., 1978. Engineering classification and index properties of weathered granite. Bull. Int. Assoc. Eng. Geol. 17, 79–90.

TABLE 5.16 Total Porosity and Dry Density of Granitic Rocks at Different Weathering Grades

Grade	Term	Total Porosity n (%)	Dry Density ρ_d (Mg/m³)
I	Fresh rock	3.48	2.63
II	Slightly weathered	3.57	2.59
III	Moderately weathered	4.65	2.46
IV	Highly weathered	5.42	2.38
V	Completely weathered	9.08	2.30
VI	Residual soil	15.5	2.00

Based on Arel, E., Onalp, A., 2004. Diagnosis of the transition from rock to soil in a granodiorite. J. Geotech. Geoenvir. Eng., ASCE, 130, 968–974.

TABLE 5.17 The Relationship Between Weathering and RQD

Grade	Term	RQD (%)
I	Discolored (Fresh rock)	66–100
II	Slightly weathered	41–65
III	Moderately weathered	16–40
IV	Highly weathered	9–15
V	Decomposed (Completely weathered)	0–8

Based on Ayalew, L., Reik, G., Busch, W., 2002. Characterizing weathered rock masses—a geostatistical approach. Int. J. Rock Mech. Min. Sci. 39, 105–114.

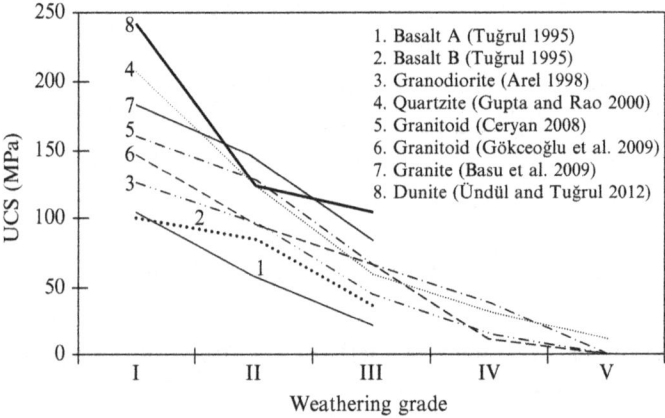

FIG. 5.6 Change of unconfined compressive strength (UCS) with weathering grade.

REFERENCES

Abad, J., Caleda, B., Chacon, E., Gutierez, V., Hidlgo, E., 1984. Application of geomechanical classification to predict the convergence of coal mine galleries and to design their supports. In: 5th Int. Cong. on Rock Mech., Epp. 15–19.

Al-Harthi, A.A., 1993. Application of CSIR and NGI classification systems along tunnel no. 3 at Al-Dela Descant, Asir Province, Saudi Arabia. In: Cripps, J.C., Coulthard, J.M., Culshaw, M.G., Forster, A., Hencher, S.R., Moon, C.F. (Eds.), The Engineering Geology of Weak Rock. Balkema, Rotterdam, pp. 323–328.

Anon, 1995. The description and classification of weathered rocks for engineering purposes. Q. J. Eng. Geol. 28, 207–242. Working Party Report.

Arel, E., Onalp, A., 2004. Diagnosis of the transition from rock to soil in a granodiorite. J. Geotech. Geoenviron. Eng. ASCE 130, 968–974.

ASCE, 1996. Rock Foundations: Technical Engineering and Design Guides as Adapted from the US Army Corps of Engineers, No. 16. ASCE Press, New York, NY.

Ayalew, L., Reik, G., Busch, W., 2002. Characterizing weathered rock masses—a geostatistical approach. Int. J. Rock Mech. Min. Sci. 39, 105–114.

Aydan, Ö., Sato, A., Yagi, M., 2014. The inference of geo-mechanical properties of soft rocks and their degradation from needle penetration tests. Rock Mech. Rock. Eng. 47, 1867–1890.

Aydin, A., Basu, A., 2005. The Schmidt hammer in rock material characterization. Eng. Geol. 81, 1–14.

Banks, D., 2005. Rock mass ratings (RMRs) predicted from slope angles of natural rock outcrops. Int. J. Rock Mech. Min. Sci. 42, 440–449.

Barton, N., 1991. Geotechnical design. World Tunn., 410–416.

Barton, N., 1995. Permanent support for tunnels using NMT—Special Lecture. In: Proc. Symp. of KRMS (Korea Rock Mechanics Society) and KSEG (Korea Society of Engineering Geology) pp. 1–26.

Barton, N., 2002. Some new Q value correlations to assist in site characterization and tunnel design. Int. J. Rock Mech. Min. Sci. Geomech. Abstr. 39, 185–216.

Barton, N., Lien, R., Lunde, J., 1974. Engineering classification of rock masses for the design of tunnel support. Rock Mech. 6, 189–236.

Basu, A., Celestino, T.B., Bortolucci, A.A., 2009. Evaluation of rock mechanical behaviors under uniaxial compression with reference to assessed weathering grades. Rock Mech. Rock. Eng. 42, 73–93.

Bell, F.G., 1987. Properties and behavior of the ground. In: Bell, F.G. (Ed.), Ground Engineer's Reference Book. Butterworths, London, UK.

Bieniawski, Z.T., 1973. Engineering classification of jointed rock masses. Trans South African Inst. Civil Eng. 15, 335–344.

Bieniawski, Z.T., 1976. Rock mass classification in rock engineering. In: Bieniawski, Z.T. (Ed.), Proc. Symp. on Exploration for Rock Eng. vol. 1. Balkema, Cape Town, pp. 97–106.

Bieniawski, Z.T., 1989. Engineering Rock Mass Classifications. John Wiley, Rotterdam.

Cai, M., Kaiser, P.K., 2006. Visualization of rock mass classification systems. Geotech. Geol. Eng. 24, 1089–1102.

Cai, M., Kaiser, P.K., Uno, H., Tasaka, Y., Minami, M., 2004. Estimation of rock mass deformation modulus and strength of jointed hard rock masses using the GSI System. Int. J. Rock Mech. Min. Sci. 41, 3–19.

Cameron-Clarke, I.S., Budavari, S., 1981. Correlation of rock mass classification parameters obtained from bore core and in-situ observations. Eng. Geol. 17, 19–53.

Ceryan, S., 2008. New chemical weathering indices for estimating the mechanical properties of rocks: a case study from the Kürtün granodiorite, NE Turkey. Turk. J. Earth Sci. 17, 187–207.

Cha, Y.H., Kang, J.S., Jo, C.-H., 2006. Application of linear-array microtremor surveys for rock mass classification in urban tunnel design. Explor. Geophys. 37, 108–113.

Deere, D.U., 1964. Technical description of rock cores for engineering purposes. Rock Mech. Rock. Eng. 1, 107–116.

Ebuk, E.J., 1991. The Influence of Fabric on the Shear Strength Characteristics of Weathered Granites. PhD Thesis, The University of Leeds, UK.

El-Naqa, A., 1994. Rock mass characterization of Wadi Mujib Damsite, Central Jordan. Eng. Geol. 38, 81–93.

El-Naqa, A., 1996. Assessment of geotechnical characterization of a rock mass using a seismic geophysical technique. Geotech. Geol. Eng. 14, 291–305.

Goel, R.K., Jethwa, J.L., Paithankar, A.G., 1996. Correlation between Barton's Q and Bieniaswki's RMR—a new approach. Int. J. Rock Mech. Min. Sci. Geomech. Abstr. 33, 179–181.

Gupta, A.S., Rao, S.K., 2000. Weathering effects on the strength and deformational behaviour of crystalline rocks under uniaxial compression state. Eng. Geol. 56, 257–274.

Hachinohe, S., Hiraki, N., Suzuki, T., 1999. Rates of weathering and temporal changes in strength of bedrock of marine terraces in Boso Peninsula, Japan. Eng. Geol. 55, 29–43.

Hashemi, M., Moghaddas, S., Ajalloeian, R., 2010. Application of rock mass characterization for determining the mechanical properties of rock mass: a comparative study. Rock Mech. Rock. Eng. 43, 305–320.

Hencher, S.R., Martin, R.P., 1982. The description and classification of weathered rocks in Hong Kong for engineering purposes. In: Proc. 7th SE Asian Geotech. Conf., Hong Kong, pp. 125–142.

Hoek, E., Brown, E.T., 1997. Practical estimates of rock mass strength. Int. J. Rock Mech. Min. Sci. 34, 1165–1186.

Hoek, E., Carter, T.G., Diederichs, M.S., 2013. Quantification of the geological strength index chart. In: 47th US Rock Mech./Geomech. Symp., San Francisco, CA, USA, June 23–26, 2013, Paper ARMA 13-672.

Irfan, T.Y., 1996. Mineralogy, fabric properties and classification of weathered granites in Hong Kong. Q. J. Eng. Geol. 29, 5–35.

Irfan, T.Y., Dearman, W.R., 1978. Engineering classification and index properties of weathered granite. Bull. Int. Assoc. Eng. Geol. 17, 79–90.

Irfan, T.Y., Powell, G.E., 1985. Engineering geological investigations for pile foundation on a deeply weathered granitic rock in Hong Kong. Bull. Assoc. Eng. Geol. 32, 67–80.

ISRM, 1978. Suggested methods for the quantitative description of discontinuities in rock masses. International Society for Rock Mechanics, Commission on Standardization of Laboratory and Field Tests. Int. J. Rock Mech. Min. Sci. Geomech. Abstr. 15, 319–368.

Jethwa, J.L., Dube, A.K., Singh, B., Mithal, R.S., 1982. Evaluation of methods for tunnel support design in squeezing rock conditions. In: Proc. 4th Int. Cong. Int. Assoc. Eng. Geol. vol. 5. Balkema, Rotterdam, pp. 125–134.

Kaiser, P.K., Mackay, C., Gale, A.D., 1986. Evaluation of rock classifications at B.C. Rail Tumbler Ridge Tunnels. Rock Mech. Rock. Eng. 19, 205–234.

Karpuz, C., Pasamehmetoglu, A.G., 1997. Field characterization of weathered Ankara andesites. Eng. Geol. 46, 1–17.

Kumar, N., Samadhiya, N.K., Anbalagan, R., 2004. Application of rock mass classification system for tunneling in Himalaya, India. Int. J. Rock Mech. Min. Sci. 41 (3), 531. SINOROCK2004 Symposium, Paper 3B 14.

Laubscher, D.H., 1977. Geomechanics classification of jointed rock masses—mining applications. Trans. Inst. Min. Metall. 86, 1–8.

Laubscher, D.H., 1984. Design aspects and effectiveness of support system in different mining conditions. Trans. Inst. Min. Metall. 93, A70–A81.

Laubscher, D.H., 1990. A geomechanics classification system for rating of rock mass in mine design. J. South Africa Inst. Min. Metall. 90 (10), 257–273.

Lauffer, H., 1958. Gebirgsklassifizierung für den Stollenbau. Geol. Bauwesen 24, 46–51.

Moreno, T.E., 1980. Application de las classificaciones geomechnicas a los tuneles de parjares. II Cursode Sostenimientos Activosen Galeriasy Tunnels. Foundation Gomez-Parto, Madrid.

Palmström, A., 1995. RMi—A Rock Mass Classification System for Rock Engineering Purposes. PhD Thesis, University of Oslo.

Palmström, A., 1996a. Characterizing rock masses by the RMi for use in practical rock engineering, Part 1: the development of the Rock Mass Index (RMi). Tunn. Undergr. Space Tech. 11 (2), 175–188.

Palmström, A., 1996b. Characterizing rock masses by the RMi for use in practical rock engineering, Part 2: some practical applications of the Rock Mass Index (RMi). Tunn. Undergr. Space Tech. 11 (3), 287–303.

Rutledge, J.C., Preston, R.L., 1978. Experience with engineering classifications of rock. In: Proc. Int. Tunneling Symp., Tokyo, pp. A3.1–A3.7.

Seiki, T., Aydan, Ö., 2003. Deterioration of Oya tuff and its mechanical property change as building stone. In: Proc. Int. Symp. on Industrial Minerals and Building Stones, Istanbul, Turkey, September 2003pp. 329–336.

Serafim, J.L., Pereira, J.P., 1983. Considerations of the geomechanics classification of Bieniawski. In: Proceedings International Symposium Engineering Geology and Underground Construction, vol. 1, Lisbon, pp, II33–II42.

Sonmez, H., Ulusay, R., 1999. Modifications to the geological strength index (GSI) and their applicability to stability of slopes. Int. J. Rock Mech. Min. Sci. 36, 743–760.

Sonmez, H., Ulusay, R., 2002. A discussion on the Hoek-Brown failure criterion and suggested modification to the criterion verified by slope stability case studies. Yerbilimleri (Earthsciences) 26, 77–99.

Terzaghi, K., 1946. Rock defects and loads on tunnel supports. In: Proctor, R.V., White, T.L. (Eds.), Rock Tunneling With Steel Supports, vol. 1. Commercial Shearing and Stamping Company, Youngstown, OH, pp. 17–99.

Tuğrul, A., 1995. The Effects of Weathering on the Engineering Properties of Basalts in the Niksar Region. PhD Thesis, Istanbul University.

Tuğrul, A., 1998. The application of rock mass classification systems to underground excavation in weak limestone, Ataturk dam, Turkey. Eng. Geol. 50, 337–345.

Ündül, Ö., Tuğrul, A., 2012. The influence of weathering on the engineering properties of dunites. Rock Mech. Rock. Eng. 45, 225–239.

Wickham, G.E., Tiedemann, H.R., Skinner, E.H., 1972. Support determination based on geologic predictions. In: Lane, K.S., Garfield, L.A. (Eds.), Proc. North American Rapid Excav. Tunneling Conf., Chicago, pp. 43–46.

Wickham, G.E., Tiedemann, H.R., Skinner, E.H., 1974. Ground support prediction model—RSR concept. In: Proc. 2nd North American Rapid Excav. Tunneling Conf., San Francisco. vol. 1. American Institute of Mining, Metallurgical and Petroleum Engineers (AIME), New York, pp. 691–707.

Chapter 6

Deformability

6.1 INTRODUCTION

Evaluation of the deformability of rock masses is an important task in rock mechanics and rock engineering because it is used in the design of different structures in or on rock, from underground openings to foundations (Deere et al., 1967; Dershowitz et al., 1979; Bieniawski, 1978; Wyllie, 1999; Zhang and Einstein, 2004). This chapter first discusses the deformability of intact rock and rock discontinuities and then presents the different methods for determining the deformation modulus and Poisson's ratio of rock masses. The effect of different factors such as scale, pressure, and temperature on rock deformability and the anisotropy of rock deformability are also discussed.

The presence of discontinuities has long been recognized as an important factor that influences the deformability of rock masses. Compared to intact rock, jointed rock masses show increased deformability. The discontinuities also induce some degree of anisotropy to the deformability of rock masses. Therefore, the determination of rock mass deformability should consider not only the deformability of the intact rock but also that of the discontinuities.

Since a rock mass seldom behaves as an ideal elastic material, its modulus is dependent upon the proportion of the stress-strain response considered. Fig. 6.1 shows a stress-strain curve typical of an in situ rock mass containing discontinuities with the various types of modulus that can be obtained. Although the curve, as shown, is representative of a jointed rock mass, it is also typical of intact rock except that the upper part of the curve tends to be concaved downward at stress levels approaching failure. As can be seen in Fig. 6.1, there are at least four portions of the stress-strain curve that can be used for determining in situ rock mass modulus: the initial tangent modulus, the elastic modulus, the recovery modulus, and the deformation modulus (ASCE, 1996; ASTM, 2004):

(a) *Initial tangent modulus.* The initial tangent modulus is determined from the slope of a line constructed tangent to the initial concave upward section of the stress-strain curve (ie, line 1 in Fig. 6.1). The initial curved section reflects the effects of discontinuity closure in in situ tests and micro-crack closure in tests on small laboratory specimens.

(b) *Elastic modulus.* Upon closure of discontinuities/micro-cracks, the stress-strain curve becomes essentially linear. The elastic modulus, frequently referred to as the modulus of elasticity, is derived from the slope of this

Engineering Properties of Rocks. http://dx.doi.org/10.1016/B978-0-12-802833-9.00006-7

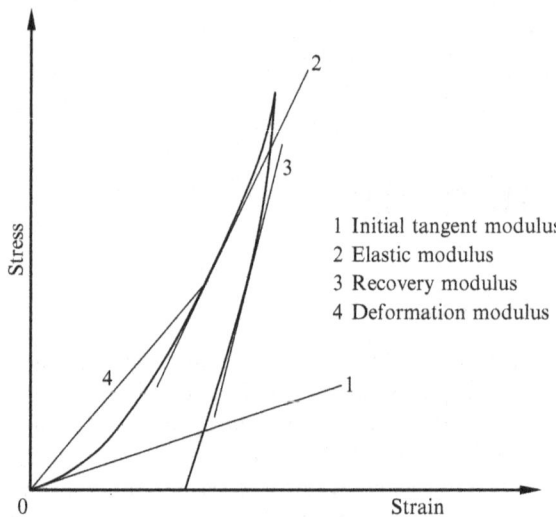

FIG. 6.1   Stress-strain curve typical of in situ rock mass with various moduli that can be obtained. *(Based on ASCE, 1996. Rock Foundations: Technical Engineering and Design Guides as Adapted from the US Army Corps of Engineers. No. 16, ASCE Press, New York, NY; ASTM, 2004. Annual Book of ASTM Standards, Vol. 4.08, Soil and Rock. American Society for Testing and Materials, West Conshohocken, Philadelphia, PA.)*

linear (or near linear) portion of the curve (ie, line 2 in Fig. 6.1). In some cases, the elastic modulus is derived from the slope of a line constructed tangent to the stress-strain curve at some specified stress level. The stress level is usually specified as 50% of the maximum or peak stress.

(c) *Recovery modulus.* The recovery modulus is obtained from the slope of a line constructed tangent to the segment of the unloading stress-strain curve (ie, line 3 in Fig. 6.1). As such, the recovery modulus is primarily derived from in situ tests where test specimens are seldom stressed to failure.

(d) *Deformation modulus or Secant modulus.* The deformation modulus is determined from the slope of the secant line established between zero and some specified stress level (ie, line 4 in Fig. 6.1). The stress level is usually specified as 50% of the maximum or peak stress.

Since the actual jointed rock masses do not behave elastically, deformation modulus is usually used in practice.

6.2   DEFORMABILITY OF INTACT ROCK

This section first presents the range and typical values of the elastic modulus of intact rock. Then the different empirical correlations for determining the elastic modulus of intact rock are discussed. Finally, the range and typical values of the Poisson's ratio of intact rock are presented.

6.2.1   Elastic Modulus of Intact Rock

Fig. 3.1 in Chapter 3 shows the range of the elastic modulus of different rocks. The typical values of the elastic modulus of different rocks can also be found from Table 6.1.

The common method for determining the elastic modulus of intact rock is to perform unconfined compression tests on rock core samples obtained from drilling using a diamond core barrel (ISRM, 1979; ASTM, 2004). Since standard sample preparation and testing is time consuming and expensive, different empirical correlations are also often used to estimate the elastic modulus.

TABLE 6.1 Typical Values of Elastic Modulus of Intact Rocks

Rock Type	No. of Values	No. of Rock Types	Elastic Modulus (GPa)			Standard Deviation
			Maximum	*Minimum*	*Mean*	
Granite	26	26	100	6.41	52.7	24.5
Diorite	3	3	112	17.1	51.4	42.7
Gabbro	3	3	84.1	67.6	75.8	6.69
Diabase	7	7	104	69.0	88.3	12.3
Basalt	12	12	84.1	29.0	56.1	17.9
Quartzite	7	7	88.3	36.5	66.1	16.0
Marble	14	13	73.8	4.00	42.6	17.2
Gneiss	13	13	82.1	28.5	61.1	15.9
Slate	11	2	26.1	2.41	9.58	6.62
Schist	13	12	69.0	5.93	34.3	21.9
Phyllite	3	3	17.3	8.62	11.8	3.93
Sandstone	27	19	39.2	0.62	14.7	8.21
Siltstone	5	5	32.8	2.62	16.5	11.4
Shale	30	14	38.6	0.007	9.79	10.0
Limestone	30	30	89.6	4.48	39.3	25.7
Dolostone	17	16	78.6	5.72	29.1	23.7

Based on AASHTO, 1989. Standard Specifications for Highway Bridges. 14th edition, American Association of State Highway and Transportation Officials, Washington DC.

6.2.1.1   Elastic Modulus Versus Unconfined Compressive Strength

With the modulus ratio (MR) shown in Fig. 3.1 of Chapter 3, the elastic modulus (E) of an intact rock can be determined when its unconfined compressive strength (σ_c) is known:

$$E = \text{MR} \times \sigma_c \tag{6.1}$$

where E and σ_c are both in MPa.

The MR values can also be obtained from Table 6.2, which was proposed by Hoek and Diederichs (2006) based in part on the correlations in Fig. 3.1 and on additional correlations from Palmström and Singh (2001).

Palchik (2011) examined the MR values for 11 heterogeneous carbonate (dolomites, limestones, and chalks) rock formations in different regions of Israel and found that MR is closely related to the maximum axial strain $\varepsilon_{a,max}$ at σ_c and their relation can be expressed by (Fig. 6.2):

$$\text{MR} = \frac{2k}{\varepsilon_{a,max}\left(1 + e^{-\varepsilon_{a,max}}\right)} \tag{6.2}$$

where k is a conversion coefficient equal to 100; and $\varepsilon_{a,max}$ is in %.

There are also other types of relations between the elastic modulus E and the unconfined compressive strength σ_c for intact rock, such as those listed in Table 6.3.

Palchik (2011) performed linear multiple regression analysis on experimental data and found that the elastic modulus E for some rock formations can be reasonably presented as a composite function of density ρ and unconfined compressive strength σ_c:

$$E = a + b\rho + c\sigma_c \tag{6.3}$$

where E is in GPa; σ_c is in MPa; and a, b, and c are regression coefficients which are $a = -88.4$, -147.9, and -40.6, $b = 52.1$, 75.2, and 24.3, and $c = 0.042$, 0.046, and 0.162 for Nekorot limestone, Aminadav dolomite, and Bina limestone, respectively.

6.2.1.2   Elastic Modulus Versus Porosity

As expected, the elastic modulus decreases as the porosity increases. Leite and Ferland (2001) derived a linear empirical correlation between elastic modulus E and porosity n based on test results of artificial porous rocks:

$$E = 10.10 - 0.109n \quad \left(r^2 = 0.74\right) \tag{6.4}$$

where E is in GPa; n is in %; and r^2 is the determination coefficient.

Lashkaripour (2002) derived a negative exponential relationship between elastic modulus E and porosity n based on the test results of claystone, clay shale, mudstone, mud shale, siltstone, and silt shale:

$$E = 37.9e^{-0.863n} \quad \left(r^2 = 0.68\right) \tag{6.5}$$

where E, n, and r^2 are as defined earlier.

TABLE 6.2 Values of Modulus Ratio (MR) for Different Rocks

Rock Type	Class	Group	Texture			
			Coarse	Medium	Fine	Very Fine
Sedimentary	Clastic		Conglomerates 300–400	Sandstones 230–350	Siltstones 350–400	Claystones 200–300
			Breccias 230–350		Greywackes 350	Shales 150–250[a]
						Marls 150–200
	Nonclastic	Carbonates	Crystalline limestones 400–600	Sparitic limestones 600–800	Micritic limestones 800–1000	Dolomites 350–500
		Evaporites		Gypsum (350)[b]	Anhydrite (350)[b]	
		Organic				Chalk 1000 +
Metamorphic	Nonfoliated		Marble 700–1000	Hornfels 400–700	Quartzites 300–450	
				Metasandstone 200–300		
	Slightly foliated		Migmatite 350–400	Amphibolites 400–500	Gneiss 300–750[a]	
	Foliated[a]			Schists 250–1100[a]	Phyllites/Mica Schist 300–800[a]	Slates 400–600[a]

Continued

TABLE 6.2 Values of Modulus Ratio (MR) for Different Rocks—cont'd

Rock Type	Class	Group	Texture			
			Coarse	Medium	Fine	Very Fine
Igneous	Plutonic	Light	Granite[c] 300–550 Granodiorote[c] 400–450	Diorite[c] 300–350		
		Dark	Gabbro 400–500 Norite 350–400	Dolerite 300–400		
	Hypabyssal		Porphyries (400)[b]		Diabase 300–350	Peridotite 250–300
	Volcanic	Lava		Rhyolite 300–500	Dacite 350–450	
				Andesite 300–500	Basalt 250–450	
		Pyroclastic	Agglomerate 400–600	Volcanic breccias (500)[b]	Tuff 200–400	

[a]Highly anisotropic rocks: the values of MR will be significantly different if normal strain and/or loading occurs parallel (high MR) or perpendicular (low MR) to a weakness plane. Uniaxial test loading direction should be equivalent to field application.
[b]No data available, estimated on the basis of geologic logic.
[c]Felsic Granitoids: coarse grained or altered (high MR), fine grained (low MR).
Based on Hoek, E., Diederichs, M.S., 2006. Empirical estimation of rock mass modulus. Int. J. Rock Mech. Min. Sci., 36, 203–215.

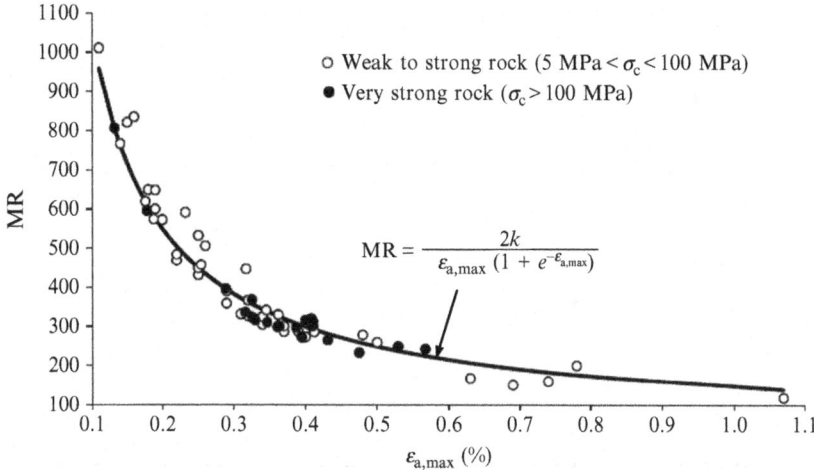

FIG. 6.2 Relation between modulus ratio (MR) and maximum axial strain $\varepsilon_{a,max}$. *(From Palchik, V., 2011. On the ratios between elastic modulus and uniaxial compressive strength of heterogeneous carbonate rocks. Rock Mech. Rock Eng., 44, 121–128.)*

TABLE 6.3 Other Types of Relations Between Elastic Modulus E and Unconfined Compressive Strength σ_c

Relation	r^2	Rock Type	Reference
$E = 0.103\sigma_c^{1.086}$	0.81	Mudrock	Lashkaripour (2002)
$E = 0.531\sigma_c + 9.567$	0.71	Shale and dolomite	Shalabi et al. (2007)
$E = 0.243\sigma_c - 0.555$		Commonly used in oil industry	Rabbani et al. (2012)
$E = 0.199\sigma_c - 3.970$	0.81	Dolomite at Taormina	Pappalardo (2015)
$E = 0.166\sigma_c - 1.301$	0.92	Dolomite at Castelmola	

Notes: E is in GPa; σ_c is in MPa; and r^2 is the determination coefficient.

Using the experimental data of gypsum, Yilmaz and Yuksek (2009) derived the following logarithmic relation between elastic modulus E and porosity n:

$$E = -39.1 \ln(n) + 100.3 \quad (r^2 = 0.83) \tag{6.6}$$

where E is in GPa; n is in %; and r^2 is the determination coefficient.

Fig. 6.3 shows the variation of elastic modulus E with porosity n for dolomite and limestone (Palchik and Hatzor, 2002). The relationship between E and

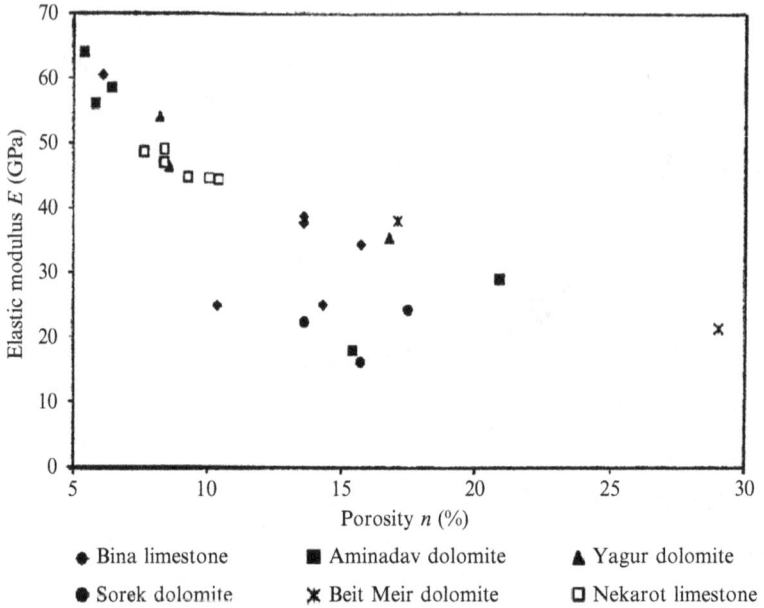

FIG. 6.3 Variation of elastic modulus E with porosity n for dolomites and limestones. *(From Palchik, V., Hatzor, Y.H., 2002. Crack damage stress as a composite function of porosity and elastic matrix stiffness in dolomites and limestones. Eng. Geol., 63, 233–245.)*

n can be described approximately by a negative exponential function or a log-arithmic function.

6.2.1.3   Elastic Modulus Versus Density

Based on an extensive study of different types of rocks (basalt, diabase, dolo-mite, gneiss, granite, limestone, marble, quartzite, rock salt, sandstone, schist, siltstone, and tuff), Deere and Miller (1966) derived the following simple linear relation between elastic modulus E and dry density ρ_d:

$$E = 64.64\rho_d - 115.4 \quad (r^2 = 0.61) \tag{6.7}$$

where E is in GPa; ρ_d is in g/cm^3; and r^2 is the determination coefficient.

6.2.1.4   Elastic Modulus Versus Dynamic Elastic Modulus

The propagation velocity of elastic waves measured on intact rock is often used to determine the *dynamic* elastic properties:

$$\nu_{dyn} = \frac{(v_p/v_s)^2 - 2}{2\left[(v_p/v_s)^2 - 1\right]} \tag{6.8}$$

$$E_{dyn} = \frac{\rho v_p^2 \left(1 - 2\nu_{dyn}\right)\left(1 + \nu_{dyn}\right)}{1 - \nu_{dyn}} \qquad (6.9)$$

$$G_{dyn} = \rho v_s^2 \qquad (6.10)$$

$$E_{dyn} = 2G_{dyn}\left(1 + \nu_{dyn}\right) \qquad (6.11)$$

where ν_{dyn} is the dynamic Poisson's ratio; v_p is the velocity of the P-wave; v_s is the velocity of the S-wave; G_{dyn} is the dynamic shear modulus; ρ is the density; and E_{dyn} is the dynamic elastic modulus.

The dynamic elastic modulus calculated from Eqs. (6.8)–(6.11) is usually larger than the (static) elastic modulus mainly because of the lower strain magnitude in the dynamic testing than in the static testing (Zimmer, 2003). (To be simple, "elastic modulus" will mean "static elastic modulus" in later discussion). Fig. 6.4 shows the ratio of dynamic elastic modulus to elastic modulus compiled by Stacey et al. (1987). The ratio varies between about 1 and 3, and can be used for a quick estimation of the elastic modulus when the dynamic elastic modulus is known.

The elastic modulus can also be estimated from the dynamic elastic modulus using the closed-form empirical correlations in Table 6.4. It is noted that different correlations may give very different elastic modulus values. To obtain reliable results for a specific site, a series of tests should be carried out to calibrate the correlations to be used for the site.

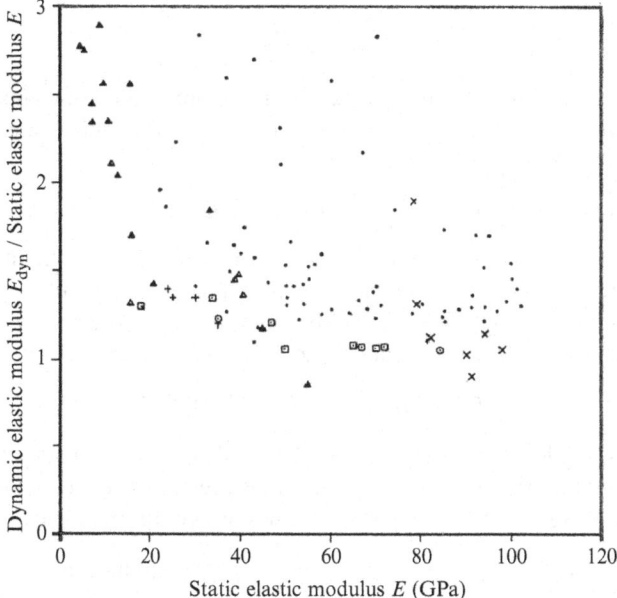

FIG. 6.4 Comparison of static and dynamic elastic modulus. *(From Stacey, T.R., van Veerden, W.L., Vogler, U.W., 1987. Properties of intact rock. In: Bell, F.G., Ground Engineer's Reference Book. Butterworths, London, UK.)*

TABLE 6.4 Relations Between Static Elastic Modulus E and Dynamic
Elastic Modulus E_{dyn}

Relation	Rock Type	Reference
$E = 1.137E_{dyn} - 9.685$	Granite	Belikov et al. (1970)
$E = 1.263E_{dyn} - 29.5$	Igneous and metamorphic rocks	King (1983)
$E = 0.64E_{dyn} - 0.32$	Different rocks	Eissa and Kazi (1988)
$E = 0.69E_{dyn} + 6.40$	Granite	McCann and Entwisle (1992)
$E = 0.48E_{dyn} - 3.26 \ (r^2 = 0.82)$	Crystalline rocks	
$E = 0.0158E_{dyn}^{2.74}$	Shale	Ohen (2003)
$E = 0.4145E_{dyn} - 1.059$		Rabbani et al. (2012)

Notes: Both E and E_{dyn} are in GPa; and r^2 is the determination coefficient.

The research by Asef and Najibi (2013) indicates that the dynamic to static elastic modulus ratio (E_{dyn}/E) decreases when the confining pressure is higher, and the trend can be expressed by:

$$\frac{E_{dyn}}{E} = aP^{-b} \tag{6.12}$$

where P is the confining pressure; and a and b are two coefficients. For the Sarvak limestone tested by Asef and Najibi (2013), a and b are equal to 4.295 and 0.337, respectively.

6.2.1.5 Elastic Modulus Versus Wave Velocity

The wave velocity can also be used directly to estimate the elastic modulus of intact rock. Based on best fitting analysis of test data for dolomite, marble and limestone, Yasar and Erdoğan (2004b) derived the following simple linear correlation between elastic modulus E and P-wave velocity v_p:

$$E = 10.67v_p - 18.71 \ (r^2 = 0.86) \tag{6.13}$$

where E is in GPa; v_p is in km/s; and r^2 is the determination coefficient.
Pappalardo (2015) also derived simple linear relations between elastic modulus E and P-wave velocity v_p for dolostones at two different sites:

$$E = 6.623v_p - 22.64 \ (r^2 = 0.78) \ \text{(Taormina site)} \tag{6.14}$$

$$E = 5.076v_p - 15.72 \ (r^2 = 0.80) \ \text{(Castelmola site)} \tag{6.15}$$

where σ_c, n, and r^2 are as defined earlier.

Yilmaz and Yuksek (2009) derived the following exponential relation between elastic modulus and P-wave velocity based on the experimental data of gypsum:

$$E = 6.8545e^{0.5561v_p} \quad (r^2 = 0.83) \tag{6.16}$$

where E, v_p, and r^2 are as defined earlier.

6.2.1.6 Elastic Modulus Versus Point Load Index

Using the experimental data of gypsum, Yilmaz and Yuksek (2009) derived the following linear relation between elastic modulus and point load index:

$$E = 14.12I_{s(50)} - 2.745 \quad (r^2 = 0.56) \tag{6.17}$$

where E is the elastic modulus in GPa; $I_{s(50)}$ is the point load index in MPa; and r^2 is the determination coefficient.

6.2.1.7 Elastic Modulus Versus Schmidt Hammer Rebound Number

Table 6.5 lists a number of empirical correlations between the elastic modulus and the Schmidt hammer rebound number. It is important to note whether the rebound number is obtained from L- or N-type hammer so that the corresponding correlation(s) are used.

6.2.1.8 Elastic Modulus Versus Needle Penetration Index

The test results of different types of rocks show that the relation between the elastic modulus and the needle penetration index (NPI) can be simply described by a linear function (Fig. 6.5):

$$E = A \times \text{NPI} \tag{6.18}$$

where A is a coefficient; and E and NPI are in GPa and N/mm, respectively. The value of A ranges from 0.015 to 0.12, with an average of 0.05. At $A = 0.05$, the determination coefficient r^2 is 0.62 (Aydan et al., 2014).

6.2.1.9 Elastic Modulus Versus Shore Sclerscope Hardness

Deere and Miller (1966) performed an extensive study on different types of rocks (basalt, diabase, dolomite, gneiss, granite, limestone, marble, quartzite, rock salt, sandstone, schist, siltstone, and tuff) and derived the following empirical relations between elastic modulus and Shore Sclerscope hardness:

$$E = 0.739H + 11.51 \quad (r^2 = 0.56) \tag{6.19a}$$

$$E = 0.268\rho_d H + 12.62 \quad (r^2 = 0.64) \tag{6.19b}$$

where E is the elastic modulus in GPa; ρ_d is the dry density in g/cm^3; H is the Shore Sclerscope hardness; and r^2 is the determination coefficient.

TABLE 6.5 Correlations Between Elastic Modulus E and Schmidt Hammer Rebound Number R_n

Correlation	r^2	Rock Type	Reference
$E = 1.786\rho_d R_{n(L)} - 29.58$	0.53	28 Lithological units, 3 base rock types	Deere and Miller (1966)
$E = 0.601\rho_d R_{n(L)} - 20.27$	0.72		
$E = 0.0069 \times 10^{[1.061\log(\rho R_{n(L)}) + 1.861]}$		25 Lithological units	Aufmuth (1973)
$E = 0.192\rho^2 R_{n(L)} - 12.71$		20 Lithological units	Beverly et al. (1979)
$E = 1.940 R_{n(L)} - 33.92$	0.78	Marble, limestone, dolomite	Sachpazis (1990)
$E = e^{cR_{n(L)} + d}$, c and d are coefficients depending on rock type	0.77–0.92	Mica-sachist, prasinite, serpentinite, gabbro, mudstone	Xu et al. (1990)
$E = 0.00013 R_{n(N)}^{3.09074}$	0.99	Chalk, limestone, sandstone, marble, syenite, granite	Katz et al. (2000)
$E = e^{0.054 R_{n(L)} + 1.146}$	0.90	Gypsum	Yilmaz and Sendir (2002)
$E = 0.47 R_{n(L)} - 6.25$	0.85	Andesita, tuff, Basalt	Dincer et al. (2004)
$E = 6.999 e^{0.0345 R_{n(L)}}$	0.79	Gypsum	Yilmaz and Yuksek (2009)

Notes: E is in GPa; ρ is the rock density in g/cm^3; $R_{n(L)}$ and $R_{n(N)}$ are, respectively, the L- and N-type Schmidt hammer rebound numbers (see Chapter 3 for detailed description of Schmidt hammer rebound tests); and r^2 is the determination coefficient.

Based on the experimental data of shale, Shalabi et al. (2007) also derived a similar linear relation between elastic modulus and Shore Sclerscope hardness for shale rock:

$$E = 0.971H - 26.91 \quad (r^2 = 0.85) \tag{6.20}$$

where E, H, and r^2 are as defined earlier.

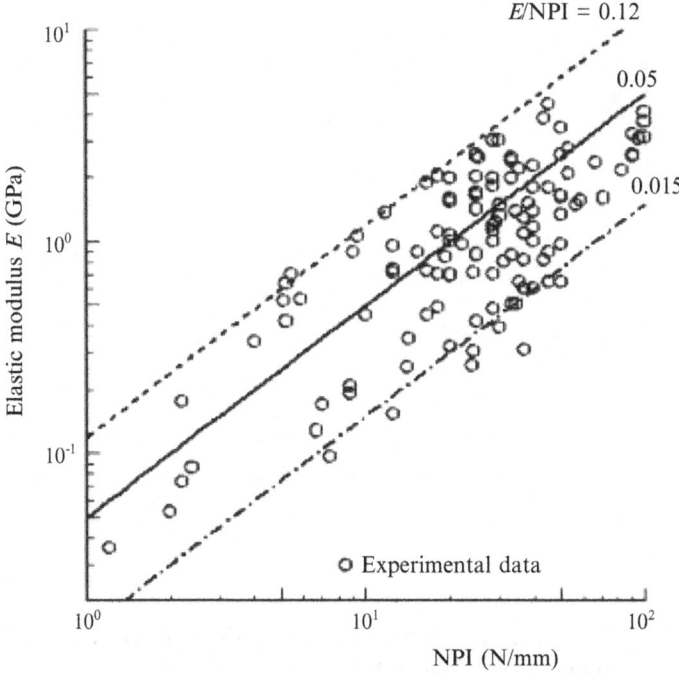

FIG. 6.5 Variation of elastic modulus E with needle penetration index NPI for various rock types. *(Based on Aydan, Ö., Sato, A., Yagi, M., 2014. The inference of geo-mechanical properties of soft rocks and their degradation from needle penetration tests. Rock Mech. Rock Eng., 47, 1867–1890.)*

6.2.1.10 Effect of Water Content on Elastic Modulus

Water content has a great effect on the deformability of intact rock. The elastic modulus of intact rock decreases as the water content increases. For example, the experimental data of the massive gypsum of the Hafik formation in the Sivas basin show that the elastic modulus decreases with the water content approximately following the relation below (Yilmaz and Yuksek, 2009):

$$E = -13.94 \ln(w) + 43.71 \quad (r^2 = 0.84) \tag{6.21}$$

where E is the elastic modulus in GPa; w is the water content in %; and r^2 is the determination coefficient.

Using about the same data (Fig. 6.6), Yilmaz (2010) derived the following relation between elastic modulus and water content:

$$E = 13.23 e^{-0.4701w} + 9.3 \quad (r^2 = 0.92) \tag{6.22}$$

where E, w, and r^2 are as defined earlier.

The effect of water on the elastic modulus can also be clearly seen from the ratio of the elastic modulus at saturated condition, $E_{\text{saturated}}$, to that at dry condition, E_{dry}, for different rocks (Table 6.6). In general, the elastic modulus at saturated condition is 20–80% of that at dry condition.

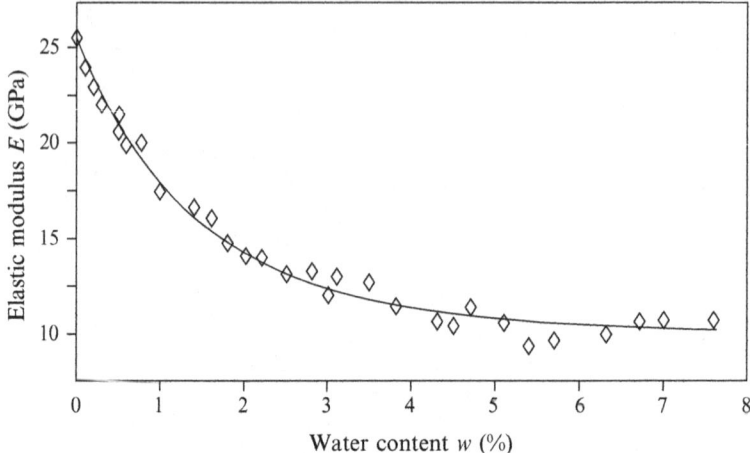

FIG. 6.6   Influence of water content *w* on elastic modulus *E* for gypsum. *(From Yilmaz, I., 2010. Influence of water content on the strength and deformability of gypsum. Int. J. Rock Mech. Min. Sci., 47, 342–347.)*

TABLE 6.6   Ratio of Elastic Modulus at Saturated Condition $E_{saturated}$ to that at Dry Condition E_{dry} for Different Rocks

$E_{saturated}/E_{dry}$	Rock	Reference
0.76	British sandstone	Vasarhelyi (2003)
0.66	Miocene limestone	Vasarhelyi (2005)
0.68	Jastrzębie sandstone	Kwasniewski and Oitaben (2009)
0.34	Anna mudstone	
0.54	Gypsum	Yilmaz (2010)
0.79	Andesite	Karakul and Ulusay (2013)
0.19	Ignimbrite	
0.32	Marl	

6.2.1.11   Effect of Temperature on Elastic Modulus

Temperature also affects the deformability of intact rock. Fig. 6.7 shows the normalized elastic modulus (E/E_0) values at various temperatures for several granites and one mudstone, where E_0 is the elastic modulus at the lowest test temperature which is 30°C for the British granite, Salisbury granite, Remiremont granite, Senones granite, and Indian granite, 20°C for the Ningbo granite, and 25°C for the mudstone. For the granites, when the temperature is below

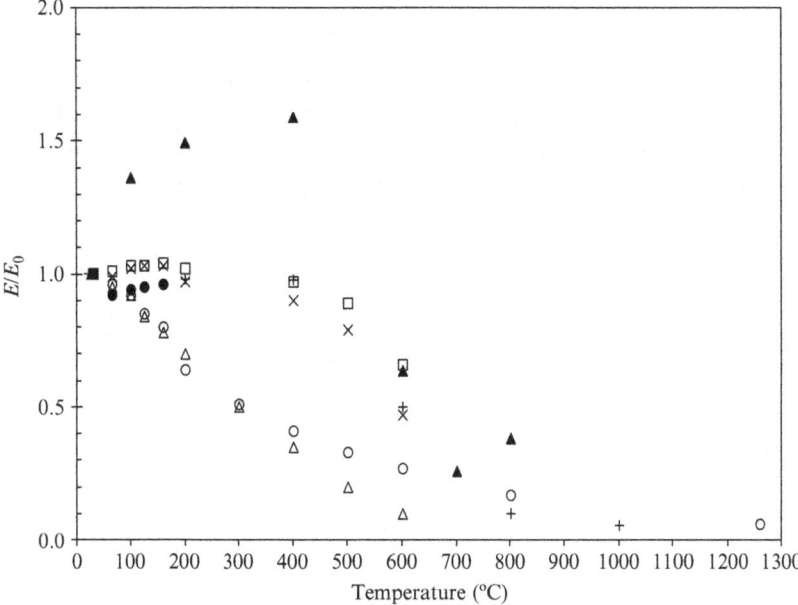

FIG. 6.7   Variation of normalized elastic modulus (E/E_0) with temperature (E_0 is the elastic modulus of rock at the lowest test temperature for each data set). Δ, British granite (McLaren and Titchel, 1981); ○, Salisbury granite (Heuze, 1983); □, Remiremont granite (Homand-Etienne and Houpert, 1989); ×, Senones granite (Homand-Etienne and Houpert, 1989); ●, Indian granite (Dwivedi et al., 2008); +, Ningbo granite (Chen et al., 2012); ▲, Mudstone (Zhang et al., 2014).

200°C, the elastic modulus may slightly increase, stay about the same or decrease with higher temperature. After the temperature goes above 200°C, the elastic modulus decreases with higher temperature for all of the granites. For the mudstone, however, the elastic modulus increases substantially (more than 50%) when the temperature is raised up to 400°C. After the temperature is above 400°C, the elastic modulus decreases when the temperature is higher.

6.2.2   Poisson's Ratio of Intact Rock

Table 6.7 lists the typical values of (static) Poisson's ratio of different intact rocks from AASHTO (1989). Fig. 6.8, compiled by Gercek (2007), shows the typical ranges of Poisson's ratio for more types of intact rocks. It is noted that the ranges shown in Table 6.7 and Fig. 6.8 for the same rock type may be slightly different simply because of the different data sources used.

There are also different empirical relations which can be used to estimate the Poisson's ratio. For example, after the dynamic Poisson's ratio ν_{dyn} is determined using Eq. (6.8) based on wave velocity measurements, the (static) Poisson's ratio ν can be estimated by (Rabbani et al., 2012):

$$\nu = 0.7\nu_{dyn} \tag{6.23}$$

TABLE 6.7 Typical Values of Poisson's Ratio of Intact Rocks

Rock Type	No. of Values	No. of Rock Types	Poisson's Ratio			Standard Deviation
			Maximum	*Minimum*	*Mean*	
Basalt	11	11	0.32	0.16	0.23	0.05
Diabase	6	6	0.38	0.20	0.29	0.06
Dolostone	5	5	0.35	0.14	0.29	0.08
Gabbro	3	3	0.20	0.16	0.18	0.02
Gneiss	11	11	0.40	0.09	0.22	0.09
Granite	22	22	0.39	0.09	0.20	0.08
Limestone	19	19	0.33	0.12	0.23	0.06
Marble	5	5	0.40	0.17	0.28	0.08
Quartzite	6	6	0.22	0.08	0.14	0.05
Sandstone	12	9	0.46	0.08	0.20	0.11
Schist	12	11	0.31	0.02	0.12	0.08
Shale	3	3	0.18	0.03	0.09	0.06
Siltstone	3	3	0.23	0.09	0.18	0.06

Based on AASHTO, 1989. Standard Specifications for Highway Bridges. 14th edition, American Association of State Highway and Transportation Officials, Washington DC.

Based on the experimental data, Shalabi et al. (2007) derived the following simple expressions relating the Poisson's ratio to Shore Sclerscope hardness and unconfined compressive strength, respectively:

$$\nu = -0.00365H + 0.383 \quad (r^2 = 0.66) \tag{6.24}$$

$$\nu = -0.00324\sigma_c + 0.293 \quad (r^2 = 0.76) \tag{6.25}$$

where ν is the Poisson's ratio; H is the Shore Sclerscope hardness; σ_c is the unconfined compressive strength in MPa; and r^2 is the determination coefficient.

The pressure and temperature may affect the Poisson's ratio of intact rock. As Table 6.8 shows, the Poisson's ratio of all three granites at a confining pressure of 500 MPa decreases when the temperature increases from 25 to 200°C. The two Llano granites show increase in Poisson's ratio at higher confining pressure, but the Woodbury granite shows the reverse effect of confining pressure.

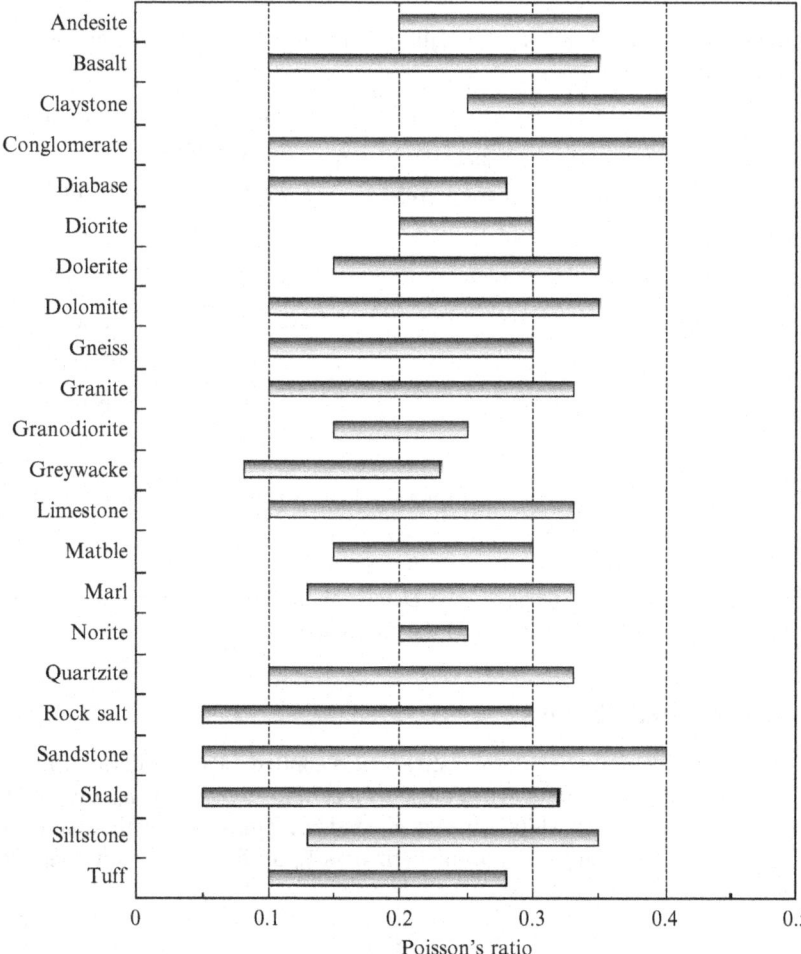

FIG. 6.8 Typical ranges of Poisson's ratio values of different intact rocks. *(Based on Gercek, H., 2007. Poisson's ratio values for rocks. Int. J. Rock Mech. Min. Sci., 44, 1–13.)*

6.3   DEFORMABILITY OF ROCK DISCONTINUITIES

The behavior of jointed rock masses is dominated by the behavior of discontinuities in the rock mass. To consider the effect of discontinuities on the deformability of rock masses, the deformability of rock discontinuities should be known first.

The deformation properties of individual rock discontinuities can be described by normal stiffness k_n and shear stiffness k_s. These refer to the rate of change of normal stress and shear stress with respect to normal displacement

TABLE 6.8 Variation of Poisson's Ratio of Granites With Confining Pressure and Temperature

Rock	Confining Pressure (MPa)	Poisson's Ratio at Different Temperatures	
		25°C	200°C
Woodbury grabite	50	0.257	0.260
	500	0.242	0.237
Llano granite (pink)	50	0.306	0.302
	500	0.317	0.312
Llano granite (gray)	50	0.225	0.236
	500	0.253	0240

Based on Clark, S.P., 1966. Handbook of Physical Constants. Geological Society of America, New York.

and shear displacement, respectively. Details about the definition and determination of k_n and k_s are presented in the following.

6.3.1   Normal Stiffness

If an effective compressive normal stress σ'_n is applied on a rock discontinuity, it would cause the discontinuity to close by a certain amount, say u_n. Fig. 6.9A shows a typical relationship between σ'_n and u_n. The slope of the curve in Fig. 6.9A gives the tangential normal stiffness k_n of the discontinuity and, at any stress level, is defined as:

$$k_n = \frac{\Delta \sigma'_n}{\Delta u_n} \qquad (6.26)$$

where Δ denotes an increment.

It is noted that k_n is small when σ'_n is small but rapidly increases as the discontinuity closes. There is actually a limit of discontinuity closure and $\sigma'_n \to \infty$ as this limit (u_{nc}) is reached. The relation between σ'_n and u_n can be expressed by the following hyperbolic function (Goodman et al., 1968; Bandis et al., 1983):

$$\sigma'_n = \frac{\alpha u_n}{u_{nc} - u_n} \qquad (6.27)$$

where α is an empirical constant; and u_{nc} is the limit of discontinuity closure.

Differentiating Eq. (6.27), the expression for k_n can be obtained as:

$$k_n = \frac{d\sigma'_n}{du_n} = \frac{\alpha u_{nc}}{(u_{nc} - u_n)^2} \qquad (6.28)$$

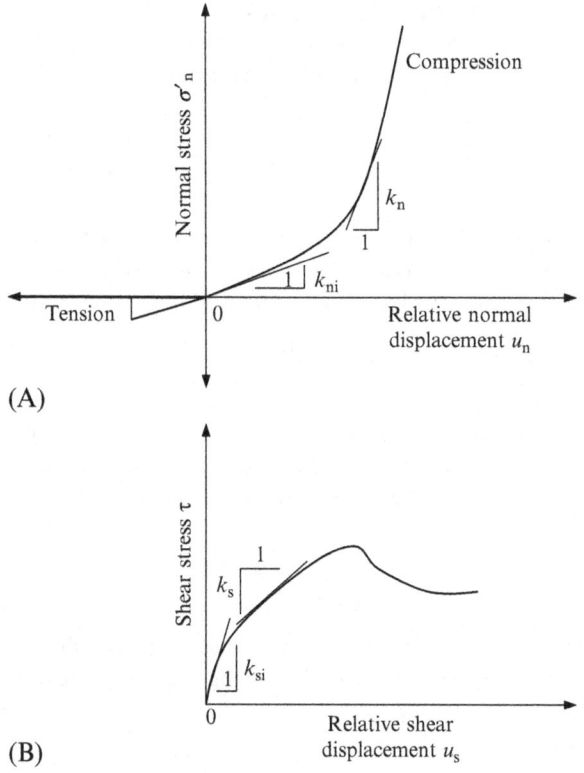

FIG. 6.9 Typical stress-relative displacement relationship: (A) σ'_n versus u_n; and (B) τ versus u_s.

When $u_n = 0$, the initial tangential normal stiffness k_{ni} can be obtained as:

$$k_{ni} = \frac{\alpha}{u_{nc}} \tag{6.29}$$

Combining Eqs. (6.28) and (6.29) gives:

$$k_n = k_{ni} \frac{u_{nc}^2}{(u_{nc} - u_n)^2} \tag{6.30}$$

Solving Eq. (6.27) for u_n and inserting it in equation (6.30) gives:

$$k_n = k_{ni} \left(1 + \frac{\sigma'_n}{k_{ni} u_{nc}}\right)^2 \tag{6.31}$$

It is noted that Eq. (6.31) is valid for compressive normal stress only. It is usual to assume that discontinuities do not offer any resistance to tensile normal stresses implying $k_n = 0$ if σ'_n is tensile.

To determine the normal stiffness k_n at a normal stress σ'_n, one has to know the initial normal stiffness k_{ni} and the limit of discontinuity closure u_{nc}. According to Bandis et al. (1983), the initial normal stiffness k_{ni} in MPa/mm can be estimated from:

$$k_{ni} \approx -7.15 + 1.75 JRC + 0.02 \left(\frac{JCS}{e}\right) \qquad (6.32)$$

where JRC is the joint (discontinuity) roughness coefficient; JCS is the joint (discontinuity) wall compressive strength in MPa; and e is the discontinuity aperture in mm at the beginning of loading which can be estimated from (Bandis et al., 1983):

$$e \approx JRC \left(\frac{0.04\sigma_c}{JCS} - 0.02\right) \qquad (6.33)$$

where σ_c is the unconfined compressive strength of the rock material.

The discontinuity aperture at the beginning of loading, e, can be estimated from the corresponding hydraulic aperture e_h of the discontinuity:

$$e \approx \sqrt{e_h \cdot JRC^{2.5}} \qquad (6.34)$$

The hydraulic aperture e_h can be determined as described in Chapter 8.

By analyzing experimental data for discontinuities with different values of JRC, Bandis et al. (1983) obtained the following expression for determining the limit of discontinuity closure u_{nc}:

$$u_{nc} \approx A + B(JRC) + C \left(\frac{JCS}{e}\right)^D \qquad (6.35)$$

where JRC, JCS, and e are the same as defined earlier; and A, B, C, and D are empirical parameters which can be estimated from Table 6.9.

Eq. (6.35) is only applicable to unfilled, interlocked discontinuities for which JRC is 5–15, JCS 22–182 MPa, and e 0.1–0.06 mm (Bandis et al.,

TABLE 6.9 Empirical Parameters A, B, C, and D in Eq. (6.35)

Parameter	Load Cycle 1	Load Cycle 2	Load Cycle 3
A	−0.2960	−0.1005	−0.1032
B	−0.0056	−0.0073	−0.0074
C	2.2410	1.0082	1.1350
D	−0.2450	−0.2301	−0.2510

Based on Bandis, S.C., Lumsden, A.C., Barton, N.R., 1983. Fundamentals of rock joint deformation. Int. J. Rock Mech. Min. Sci. Geomech. Abstr., 20, 249–268; Priest, S.D., 1993. Discontinuity Analysis for Rock Engineering. Chapman & Hall, London.

1983; Priest, 1993). Bandis et al. (1983) also obtained the following expression for determining u_{nc} without the data of JRC:

$$u_{nc} \approx R \left(\frac{JCS}{e} \right)^S \tag{6.36}$$

where JCS and e are the same as defined earlier; and R and S are empirical parameters which can be estimated from Table 6.10.

The discontinuity roughness coefficient JRC provides an angular measure of the geometrical roughness of the discontinuity surface in the approximate range of 0 (smooth) to 20 (very rough). The JRC can be estimated in a number of ways. Barton and Choubey (1977) presented a selection of scaled typical roughness profiles (Fig. 6.10), which facilitate the estimation of JRC for real discontinuities by visual matching. Barton (1987) published a table relating J_r (discontinuity roughness number in the Q classification system) to JRC (Fig. 6.11). Barton and Bandis (1990) suggested that JRC can also be estimated from a simple tilt shear test in which a pair of matching discontinuity surfaces are tilted until one slides over the other. The JRC can be back-figured from the tilt angle α (Fig. 6.12) using the following equation:

$$JRC = \frac{\alpha - \phi_r}{\log \left(\frac{JCS}{\sigma'_n} \right)} \tag{6.37}$$

where σ'_n is the normal stress on the discontinuity plane; and ϕ_r is the residual friction angle of the discontinuity which can be estimated from:

$$\phi_r = (\phi_b - 20) + 20 \frac{R_{n(L), disc}}{R_{n(L), rock}} \tag{6.38}$$

where ϕ_b is the basic friction angle of the rock material; and $R_{n(L),disc}$ and $R_{n(L),rock}$ are the rebound numbers from the L-type Schmidt hammer tests, respectively, on the discontinuity surface and the fresh rock surface. If the discontinuity surfaces are unweathered, ϕ_r can be simply taken equal to ϕ_b. The basic friction angle ϕ_b can be determined from direct shear tests or tilt tests

TABLE 6.10 Empirical Parameters R and S in Eq. (6.36)

Parameter	Load Cycle 1	Load Cycle 2	Load Cycle 3
R	8.57	4.46	6.41
S	−0.68	−0.65	−0.72

Based on Bandis, S.C., Lumsden, A.C., Barton, N.R., 1983. Fundamentals of rock joint deformation. Int. J. Rock Mech. Min. Sci. Geomech. Abstr., 20, 249–268; Priest, S.D., 1993. Discontinuity Analysis for Rock Engineering. Chapman & Hall, London.

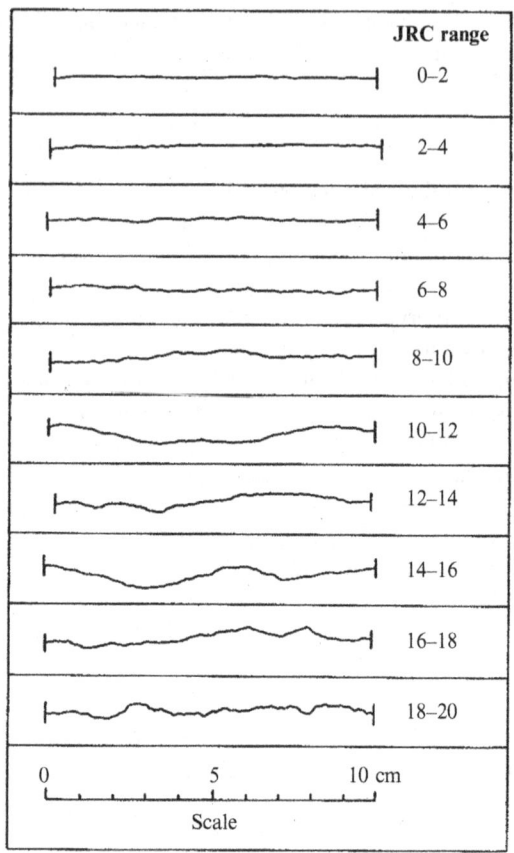

FIG. 6.10   Typical discontinuity roughness profiles and associated JRC values. *(Based on Barton, N., Choubey, V., 1977. The shear strength of rock joints in theory and practice. Rock Mech., 10, 1–54.)*

on saw-cut rock surfaces. The values of ϕ_b depend on the rock type and water content. Table 6.11 lists the basic friction angle values for different types of rocks.

The basic friction angle can also be estimated using tilt testing of diamond core samples (Stimpson, 1981). The tilt test involves attaching two pieces of core to a horizontal base, ensuring that the core samples are in contact with one another and are not free to slide. A third piece of core is then placed on top of the first two pieces and the base is rotated about a horizontal axis until sliding of the upper piece of core along the two line contacts with the lower pieces of core begins. The following equation can then be used to estimate the basic friction angle:

$$\phi_b = \arctan\left(1.155\tan\alpha\right) \tag{6.39}$$

Profile	J_r	JRC (200 mm)	JRC (1 m)
Rough	4	20	11
Smooth	3	14	9
Slickensided	2	11	8
Stepped			
Rough	3	14	9
Smooth	2	11	8
Slickensided	1.5	7	6
Undulating			
Rough	1.5	2.5	2.3
Smooth	1.0	1.5	0.9
Slickensided	0.5	0.5	0.4
Planar			

FIG. 6.11   Relationship between J_r in Q-system and JRC for 200 mm and 1 m samples. *(Based on Barton, N., 1987. Predicting the Behavior of Underground Openings in Rock. Manuel Rocha Memorial Lecture, Lisbon, Oslo, Norwegian Geotech. Inst.)*

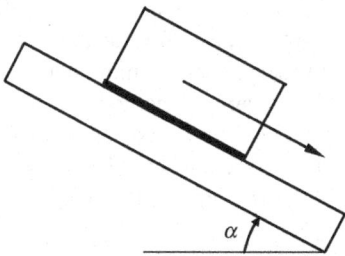

FIG. 6.12   Tilt test to measure the tilt angle α. *(Based on Barton, N., Bandis, S.C., 1990. Review of predictive capabilities of JRC-JCS model in engineering practice. In: Barton, N., Stephansson, O. (Eds.), Proc. Int. Symp. on Rock Joints. Loen, Norway, Balkema, Rotterdam, pp. 603–610.)*

TABLE 6.11 Basic Friction Angles ϕ_b for Different Rocks

Rock Family	Rock Type	ϕ_b Dry (Degrees)	ϕ_b Wet (Degrees)
Sedimentary	Conglomerate	35	
	Chalk	32	
	Limestone	31–37	27–35
	Mudstone	31–33	27–31
	Sandstone	26–35	25–34
	Shale		27
	Siltstone	31–33	27–31
Igneous	Basalt	35–38	31–36
	Dolerite	36	32
	Coarse-grained granite	31–35	31–33
	Fine-grained granite	31–35	29–31
	Porphyry	31	31
Metamorphic	Amphibolite	32	
	Gneiss	26–29	23–26
	Schist	25–30	21
	Slate	25–30	21

Based on Barton, N., Choubey, V., 1977. The shear strength of rock joints in theory and practice. Rock Mech., 10, 1–54; Alejano, L.R., González, J., Muralha, J., 2012. Comparison of different techniques of tilt testing and basic friction angle variability assessment. Rock Mech. Rock Eng., 45, 1023–1035.

where ϕ_b is the basic friction angle for the upper piece of core; and α is the tilt angle at which sliding commences.

Alejano et al. (2012) did a good comparison of the different types of tilt tests by applying them to measure the basic friction angle of various types of rocks and the reader can refer to their paper for more details.

The nail brush is one of the simple methods for recording surface profiles. Tse and Cruden (1979) presented a method for estimating JRC based on digitization of the discontinuity surface into a total of M data points spaced at a constant small distance Δx along the profile. If y_i is the amplitude of the ith data point measured above (y_i^+) and below (y_i^-) the center line, the root mean square Z_2 of the first derivative of the roughness profile is given by:

$$Z_2 = \sqrt{\frac{\sum_{i=1}^{M}(y_{i+1}-y_i)^2}{M(\Delta x)^2}} \qquad (6.40)$$

By digitizing the ten typical roughness profiles presented in Fig. 6.10 and then conducting a series of regression analyses, Tse and Cruden (1979) found that there is a strong correlation between JRC and Z_2. On this basis, they proposed the following expression for estimating JRC:

$$JRC \approx 32.2 + 32.47\log Z_2 \qquad (6.41)$$

The increasing availability of image analysis hardware and low-cost digitizing pads makes the method of Tse and Cruden (1979) a valuable objective alternative for the assessment of JRC. This approach should be used with caution, however, since Bandis et al. (1981) have shown that both JRC and JCS decrease with increasing scale. The idea of applying statistical and probabilistic analysis of surface profiles to the calculation of JRC has recently been examined and extended by several authors, notably McWilliams et al. (1990), Roberds et al. (1990), and Yu and Vayssade (1990). These last authors, noting that the value of JRC is dependent upon the sampling interval along the profile, proposed the following extension to Eq. (6.41):

$$JRC \approx AZ_2 - B \qquad (6.42)$$

where the constants A and B depend on the sampling interval Δx, taking values of 60.32 and 7.51, respectively, for an interval of 0.25 mm, 61.79 and 3.47 for an interval of 0.5 mm, and 64.22 and 2.31 for an interval of 1.0 mm. Lee et al. (1990), applying the concept of fractals to discontinuity surface profiles, obtained an empirical relation linking the fractal dimension D to the JRC value, as follows:

$$JRC = -0.87804 + 37.7844\left(\frac{D-1}{0.015}\right) - 16.9304\left(\frac{D-1}{0.015}\right)^2 \qquad (6.43)$$

Unfortunately, Lee et al. (1990) did not explain adequately how the fractal dimension D should be determined in practice. Odling (1994) proposed a method for determining the fractal dimension D, in which the roughness of a discontinuity surface is represented by the structure function S. For a discontinuity surface profile, S is defined as:

$$S(\Delta x) = \frac{\sum_{i=1}^{M}(y_{i+1}-y_i)^2}{M} \qquad (6.44)$$

where M is the number of data points at a sampling interval Δx, and y_i is the amplitude of the ith data point measured above (y_i^+) and below (y_i^-) the center line. The structure function is thus simply the mean square height difference of points on the profile at horizontal separations of Δx. The structure function is related to the Hurst exponent H (Voss, 1988; Poon et al., 1992):

$$S(\Delta x) = A(\Delta x)^{2H} \qquad (6.45)$$

Thus, if a log-log plot of $S(\Delta x)$ versus Δx gives an acceptably straight line, the slope of this line gives $2H$. A is an amplitude parameter and is equivalent to the mean square height difference at a sampling interval of 1 unit, and is therefore dependent on the units of measurement. From H, the fractal dimension D can be determined from the following equation (Voss, 1988):

$$D = E - H \qquad (6.46)$$

where E is the Euclidean dimension of embedding medium and $E = 2$ for surface profiles.

If the discontinuity is unweathered, JCS is equal to the unconfined compressive strength of the rock material, σ_c, which can be determined using the typical values and correlations presented in Section 7.2. If there has been softening or other forms of weathering along the discontinuity, JCS will be smaller than σ_c and must be estimated in some way. Suggested methods for estimating JCS are published by ISRM (1978). Barton and Choubey (1977) explained how the Schmidt hammer rebound test can be used to estimate JCS with the following empirical expression:

$$\log \text{JCS} \approx 0.88 \gamma R_{n(L)} + 1.01 \qquad (6.47)$$

where γ is the unit weight of the rock material in MN/m^3; $R_{n(L)}$ is the rebound number from the L-type Schmidt hammer test on the discontinuity surface; and JCS is in the unit of MPa from 20 to 300 MPa. Although the Schmidt hammer is notoriously unreliable, particularly for heterogeneous materials, it is one of the few methods available for estimating the surface strength of a material (see Chapter 3 for more detailed description of Schmidt hammer tests).

6.3.2   Shear Stiffness

If a shear stress τ is applied on the discontinuity, there will be a relative shear displacement u_s on the discontinuity. Fig. 6.9B shows a typical relationship between τ and u_s. It is now possible to define a tangential shear stiffness k_s exactly in the same way as for the normal stiffness k_n, ie,

$$k_s = \frac{\Delta \tau}{\Delta u_s} \qquad (6.48)$$

The relation between shear stress τ and relative shear displacement u_s can be expressed by the following hyperbolic function (Duncan and Chang, 1970; Bandis et al., 1983; Priest, 1993):

$$\tau = \left(\frac{1}{k_{si} u_s} + \frac{R_f}{\tau_f} \right)^{-1} \tag{6.49}$$

where k_{si} and τ_f are, respectively, the initial tangent shear stiffness and the shear strength of the discontinuity; and R_f is the failure ratio given by τ_f/τ_{ult} in which τ_{ult} is the ultimate shear stress at large shear displacement.

Differentiating Eq. (6.49), the expression for k_s can be obtained as:

$$k_s = \frac{d\tau}{du_s} = k_{si} \left(1 + \frac{R_f k_{si} u_s}{\tau_f} \right)^{-2} \tag{6.50}$$

Solving Eq. (6.49) for u_s and inserting it in Eq. (6.50) gives:

$$k_s = k_{si} \left(1 - \frac{R_f \tau}{\tau_f} \right)^2 \tag{6.51}$$

To determine the shear stiffness k_s at a shear stress τ, one has to know the initial shear stiffness k_{si}, the shear strength τ_f, and the failure ratio R_f. Bandis et al. (1983) found that the initial shear stiffness k_{si} increases with normal stress σ'_n and can be estimated from:

$$k_{si} \approx k_j \left(\sigma'_n \right)^{n_j} \tag{6.52}$$

where k_j and n_j are empirical constants termed the stiffness number and the stiffness exponent, respectively. Based on test results of dolerite, limestone, sandstone, and slate at normal stresses from 0.23 to 2.36 MPa, Bandis et al. (1983) found that n_j is in the range of 0.615–1.118 MPa^2/mm with an average of about 0.761 and R_f in the range of 0.652–0.887 with an average of about 0.783. The stiffness number k_j in MPa/mm was found to vary with JRC and could be estimated from:

$$k_j \approx -17.19 + 3.86 JRC \quad (\text{for } JRC > 4.5) \tag{6.53}$$

The shear strength τ_f of a discontinuity at a normal stress σ'_n can be determined by:

$$\tau_f = \sigma'_n \tan \left[JRC \log \left(\frac{JCS}{\sigma'_n} \right) + \phi_r \right] \tag{6.54}$$

where JRC, JCS, and ϕ_r are the same as defined earlier. For more detailed discussion of shear strength τ_f, the reader can refer to Chapter 7.

6.3.3 Dilation of Discontinuities

It is noted that for discontinuities (especially rough discontinuities), an increment of shear stress can produce an increment of relative displacement in the normal direction and vice versa an increment of normal stress can produce

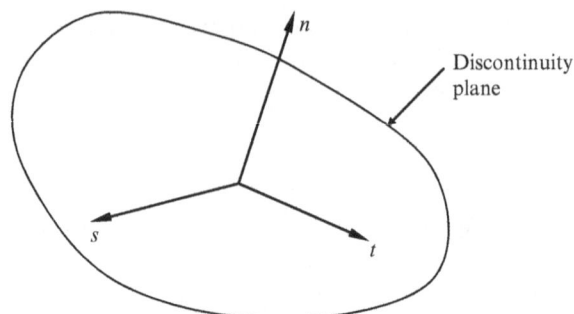

FIG. 6.13   A local coordinate system s, t, n.

an increment of relative displacement in the shear direction. This behavior is called dilation of discontinuities. If the relative shear displacement is broken into two components (along two perpendicular coordinate axes s and t on the discontinuity plane; see Fig. 6.13), the general constitutive relation for a discontinuity including the dilation behavior can be expressed as:

$$\left\{ \begin{array}{c} u_s \\ u_t \\ u_n \end{array} \right\} = \left[\begin{array}{ccc} C_{ss} & C_{st} & C_{sn} \\ C_{ts} & C_{tt} & C_{tn} \\ C_{ns} & C_{nt} & C_{nn} \end{array} \right] \left\{ \begin{array}{c} \tau_s \\ \tau_t \\ \sigma'_n \end{array} \right\} \qquad (6.55)$$

where the subscripts s and t represent two orthogonal directions in the discontinuity plane; the subscript n represents the direction normal to the discontinuity plane; u_s and u_t are the shear displacements in directions s and t, respectively; u_n is the closure displacement; τ_s and τ_t are the shear stresses in directions s and t, respectively; σ'_n is the effective normal stress; and $[C_{ij}]$ $(i,j = s, t, n)$ is the compliance matrix of the discontinuity. Elements of the compliance matrix can be found experimentally by holding two of the stresses constant (for example at zero) and then monitoring the three relative displacement components associated with changes in the third stress component (Priest, 1993).

For simplicity, the following assumptions are often made for the behavior of a single discontinuity:

(1) Deformation behavior is the same in all directions in the discontinuity plane. Thus $C_{ss} = C_{tt}$, $C_{st} = C_{ts}$, $C_{sn} = C_{tn}$, and $C_{ns} = C_{nt}$.
(2) The dilation (coupling) effect is neglected, ie, C_{ij} $(i \neq j)$ in Eq. (6.37) are zero.

With the above two assumptions, Eq. (6.55) can be simplified to:

$$\left\{ \begin{array}{c} u_s \\ u_t \\ u_n \end{array} \right\} = \left[\begin{array}{ccc} C_{ss} & 0 & 0 \\ 0 & C_{ss} & 0 \\ 0 & 0 & C_{nn} \end{array} \right] \left\{ \begin{array}{c} \tau_s \\ \tau_t \\ \sigma'_n \end{array} \right\} = \left[\begin{array}{ccc} \dfrac{1}{k_s} & 0 & 0 \\ 0 & \dfrac{1}{k_s} & 0 \\ 0 & 0 & \dfrac{1}{k_n} \end{array} \right] \left\{ \begin{array}{c} \tau_s \\ \tau_t \\ \sigma'_n \end{array} \right\} \qquad (6.56)$$

where k_s and k_n are the discontinuity shear and normal stiffness, respectively, as described in previous subsections.

6.4 DEFORMABILITY OF ROCK MASS

6.4.1 Empirical Methods for Estimating Rock Mass Deformation Modulus

A number of empirical methods have been developed for estimating the deformation modulus of rock masses. The commonly used include the correlations between the deformation modulus and various rock quality indices such as RQD, RMR, GSI, and Q. The definition of RQD, RMR, GSI, and Q and the methods for determining them have been discussed in Chapter 5.

6.4.1.1 Methods Relating Deformation Modulus With RQD

Based on field studies at Dworshak Dam, Deere et al. (1967) suggested that RQD be used for determining the rock mass deformation modulus. By adding further data from other sites, Coon and Merritt (1970) developed a relation between RQD and E_m/E_r, where E_m and E_r are the deformation modulus of the rock mass and the intact rock, respectively (Fig. 6.14).

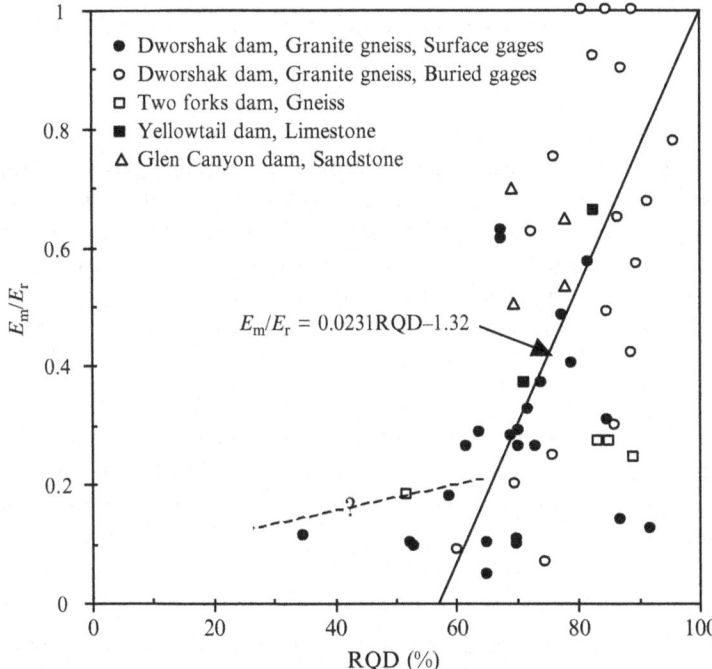

FIG. 6.14 Variation of E_m/E_r with RQD. *(Based on Coon, R.F., Merritt, A.H., 1970. Predicting in situ modulus of deformation using rock quality indices. Determination of the in Situ Modulus of Deformation of Rock, ASTM STP 477, pp. 154–173.)*

Gardner (1987) proposed the following relation for estimating the rock mass deformation modulus E_m from the intact rock modulus E_r by using a reduction factor α_E which is a function of RQD:

$$E_m = \alpha_E E_r \tag{6.57a}$$

$$\alpha_E = 0.0231(\text{RQD}) - 1.32 \geq 0.15 \tag{6.57b}$$

This method is adopted by the American Association of State Highway and Transportation Officials in the *Standard Specification for Highway Bridges* (AASHTO, 1989). For RQD > 57%, Eq. (6.57) is the same as the relation of Coon and Merritt (1970). For RQD < 57%, Eq. (6.57) gives $E_m/E_r = 0.15$.

It is noted that the RQD—E_m/E_r relations of Coon and Merritt (1970) and Gardner (1987) have the following limitations (Zhang and Einstein, 2004):

(1) The range of RQD < 60% is not covered and only an arbitrary value of E_m/E_r can be selected in this range.
(2) For RQD = 100%, E_m is assumed to be equal to E_r. This is obviously unsafe in design practice because RQD = 100% does not mean that the rock is intact. There may be discontinuities in rock masses with RQD = 100% and thus E_m may be smaller than E_r even when RQD = 100%.

Zhang and Einstein (2004) added further data collected from the published literature to cover the entire range $0 \leq \text{RQD} \leq 100\%$ (Fig. 6.15). It can be seen that the data in Fig. 6.15 shows a large scatter, which may be caused by many different factors as discussed in the following.

Testing Methods

The data in Fig. 6.15 were obtained with different testing methods. For example, Deere et al. (1967) used plate load tests while Ebisu et al. (1992) used borehole jacking tests. Different testing methods may give different values of deformation modulus even for the same rock mass. According to Bieniawski (1978), even a single testing method, such as the flat jack test, can lead to a wide scatter in the results even where the rock mass is very uniform.

Directional Effect

Most rock masses are anisotropic and do not have a single deformation modulus. RQD also varies with direction through a fractured rock mass. The dependence of both RQD and deformation modulus on direction adds to the scatter of the data.

Discontinuity Conditions

RQD does not consider the discontinuity conditions, such as the aperture and fillers. However, the discontinuity conditions have a great effect on the rock mass deformation modulus. Fig. 6.16 shows the variation of E_m/E_r with the

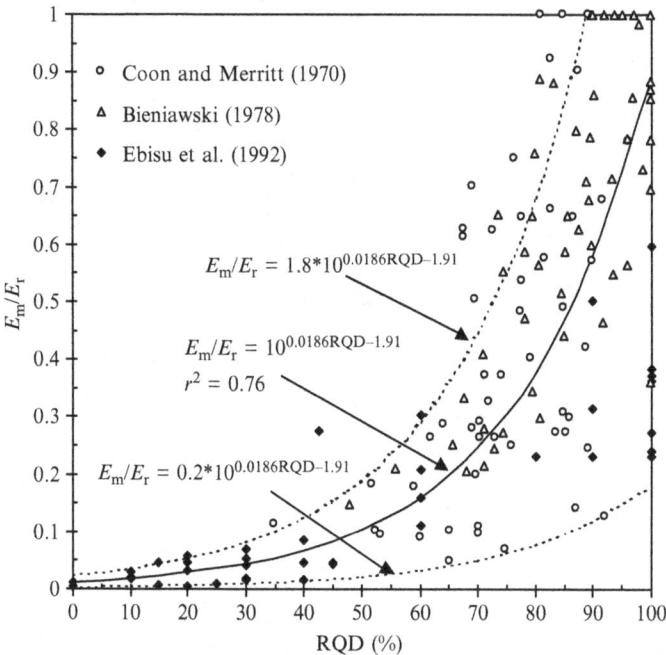

FIG. 6.15 E_m/E_r—RQD data and proposed E_m/E_r—RQD relations. *(From Zhang, L., Einstein, H.H., 2004. Estimating the deformation modulus of rock masses. Int. J. Rock Mech. Min. Sci., 41, 337–341.)*

average discontinuity spacing s for different values of k_n/E_r using the Kulhawy (1978) model (see Section 6.4.2). It can be seen that k_n/E_r which represents the discontinuity conditions has a great effect on the rock mass deformation modulus.

Kayabasi et al. (2003) derived the following relation from a database of 57 test values showing the influence of weathering of discontinuities on the rock mass deformation modulus:

$$E_m = 0.1423 \left[\frac{E_r(1+0.01\text{RQD})}{\text{WD}} \right]^{1.1747} \tag{6.58}$$

where WD is the weathering degree of discontinuities. By adding 58 new test values to the database of Kayabasi et al. (2003), Gokceoglu et al. (2003) derived the following relation based on regresion analysis:

$$E_m = 0.001 \left[\frac{(E_r/\sigma_c)(1+0.01\text{RQD})}{\text{WD}} \right]^{1.5528} \tag{6.59}$$

The new relation considers the effect of the unconfined compressive strength of intact rock on the rock mass deformation modulus.

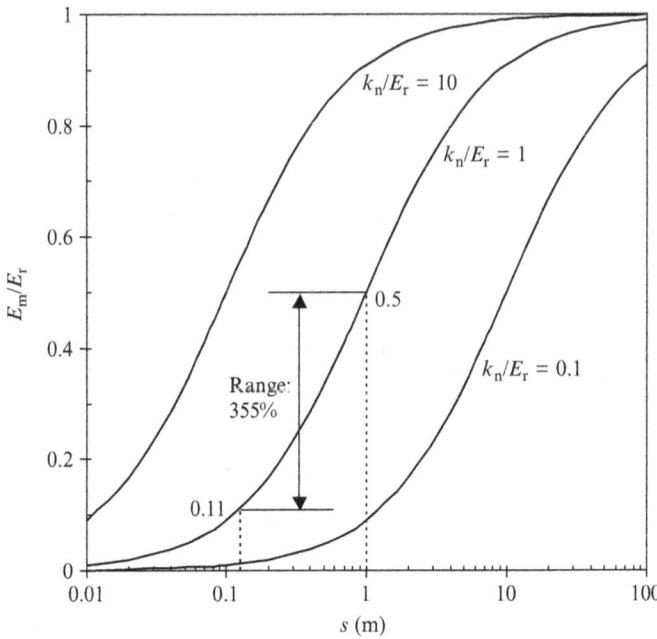

FIG. 6.16   Variation of E_m/E_r with average discontinuity spacing s for different values of k_n/E_r. *(From Zhang, L., Einstein, H.H., 2004. Estimating the deformation modulus of rock masses. Int. J. Rock Mech. Min. Sci., 41, 337–341.)*

Insensitivity of RQD to Discontinuity Frequency

RQD used in Fig. 6.15 is defined in terms of the percentage of intact pieces of rock (or discontinuity spacings) greater than a threshold value t of 0.1 m. According to Harrison (1999), the adoption of a threshold value t of 0.1 m leads to the insensitivity of RQD to the change of discontinuity frequency λ or mean discontinuity spacing s. As discussed in Chapter 4, for a negative exponential distribution of discontinuity spacings, the theoretical RQD can be related to the discontinuity frequency λ by Eq. (4.21). Fig. 6.17 shows the variation of RQD with λ. It can be seen that, for a threshold value t of 0.1 m, when discontinuity frequency λ increases from 1 m^{-1} to 8 m^{-1} (ie, the mean discontinuity spacing s decreases from 1 m to 0.125 m), RQD only decreases from 99.5% to 80.9%, which is a range of only 23%. However, when the mean discontinuity spacing s decreases from 1 m to 0.125 m, the rock mass deformation modulus will vary over a large range. As shown in Fig. 6.16, with $k_n/E_r = 1$, E_m/E_r changes from 0.5 to 0.11 when s decreases from 1 m to 0.125 m. Harrison (1999) showed that the sensitivity of RQD to the mean discontinuity spacing s is closely related to the adopted threshold value t. For example, if a threshold value t of 0.5 m is used, the corresponding RQD will change from 91.0% to 9.2% when λ increases from 1 m^{-1} to 8 m^{-1}, which is a range of 889% (Fig. 6.17).

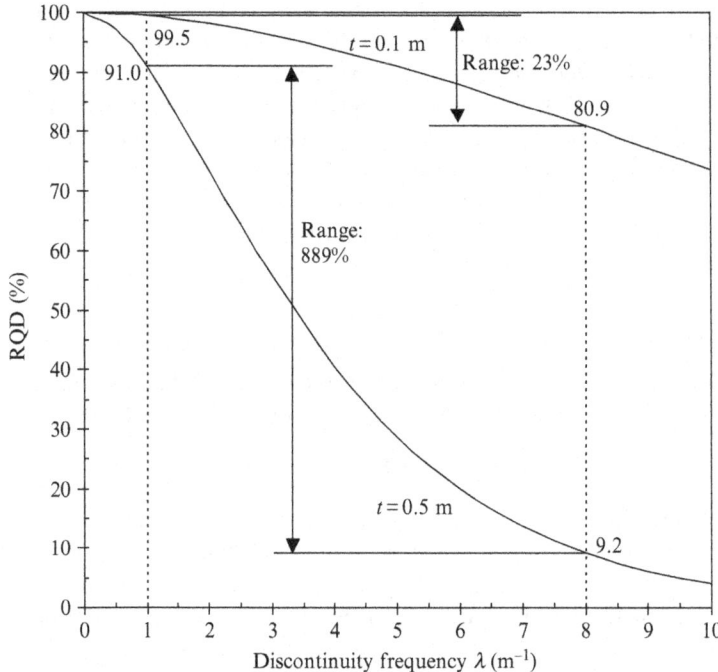

FIG. 6.17 RQD—discontinuity frequency relations for threshold values of 0.1 and 0.5 m. *(From Zhang, L., Einstein, H.H., 2004. Estimating the deformation modulus of rock masses. Int. J. Rock Mech. Min. Sci., 41, 337–341.)*

Considering the data shown in Fig. 6.15, Zhang and Einstein (2004) proposed the following relations between the rock mass deformation modulus and RQD:

Lower bound:

$$E_m/E_r = 0.2 \times 10^{0.0186RQD-1.91} \tag{6.60a}$$

Upper bound:

$$E_m/E_r = 1.8 \times 10^{0.0186RQD-1.91} \tag{6.60b}$$

Mean:

$$E_m/E_r = 10^{0.0186RQD-1.91} \tag{6.60c}$$

The mean relation between E_m/E_r and RQD was obtained by regression analysis of the data in Fig. 6.15. The coefficient of regression, r^2, is 0.76. The upper bound could be put somewhat higher but it was selected to be conservative.

RQD is a directionally dependent parameter and its value may change significantly at different orientations. Therefore, it is important to specify the orientation of the RQD when estimating the rock mass deformation modulus E_m using the E_m/E_r—RQD relationship. To reduce the directional dependence of RQD, Eq. (4.44) or (4.45) in Chapter 4 can be used to estimate RQD from the volumetric discontinuity frequency λ_v. The other option is to do core boring, scanline sampling and/or wave velocity measurements at different directions and then evaluate the overall RQD of the rock mass.

6.4.1.2   Methods Relating Deformation Modulus With RMR or GSI

Bieniawski (1978) studied seven projects and suggested the following correlation for estimating rock mass deformation modulus E_m from RMR:

$$E_m = 2RMR - 100 \quad \text{(GPa)} \tag{6.61}$$

The obvious deficiency of this equation is that it gives negative modulus values when RMR is smaller than 50. Additional studies carried out on rock masses with qualities ranging from poor to very good indicated that the rock mass deformation modulus E_m could be related to RMR by (Serafim and Pereira, 1983):

$$E_m = 10^{(RMR-10)/40} \quad \text{(GPa)} \tag{6.62}$$

It is noted that Eqs. (6.61) and (6.62) were developed before 1989 and the RMR in them is RMR_{76}, which is equal to RMR_{89} - 5. For simplicity, if not specifically stated, the RMR will simply mean RMR_{89} in later discussion.

Eq. (6.62) has been found to work well for good-quality rocks. However, for poor-quality rocks, it appears to predict deformation modulus values which are too high (Hoek and Brown, 1997). Based on practical observations and back analysis of excavation behavior in poor-quality rock masses, Hoek and Brown (1997) modified equation (6.62) for unconfined compressive strength of intact rock $\sigma_c < 100$ MPa as follows:

$$E_m = \sqrt{\frac{\sigma_c}{100}} 10^{(GSI-10)/40} \quad \text{(GPa)} \tag{6.63}$$

Note that GSI (Geological Strength Index) has been substituted for RMR in Eq. (6.63).

Johnston et al. (1980) also found that Eq. (6.62) overestimates the rock mass deformation modulus for poor-quality rocks. They reported that the results of various in situ load tests in moderately weathered Melbourne mudstone of σ_c in the range 2–3 MPa yielded a rock mass deformation modulus of about 0.5 GPa for estimated RMR of about 70 (note that the RMR here is RMR_{76}). If Eq. (6.63) with $\sigma_c = 2.5$ MPa and $GSI = RMR_{76} = 70$ is used, E_m of

5.0 GPa is obtained which is much closer to the measured value of about 0.5 GPa than the value of 31.6 GPa calculated using Eq. (6.62).

Read et al. (1999) proposed the following simple relationship for estimating the rock mass deformation modulus E_m from RMR:

$$E_m = 0.1(\text{RMR}/10)^3 \quad (\text{GPa}) \tag{6.64}$$

Using a database including 115 data values obtained from in situ plate loading and dilatometer tests, Gokceoglu et al. (2003) obtained the following correlations based on regression analyses:

$$E_m = 0.0736e^{0.0755\text{RMR}} \quad (\text{GPa}) \tag{6.65}$$

$$E_m = 0.1451e^{0.0654\text{GSI}} \quad (\text{GPa}) \tag{6.66}$$

Hoek (2004) presented the following simplified correlation for estimating the rock mass deformation modulus E_m from GSI:

$$E_m = 0.33e^{0.064\text{GSI}} \quad (\text{GPa}) \tag{6.67}$$

Based on data from a large number of in situ measurements in China and Taiwan, Hoek and Diederichs (2006) derived the following relationship between rock mass deformation modulus E_m and GSI:

$$E_m = 100\left(\frac{1 - D/2}{1 + e^{(75 + 25D - \text{GSI})/11}}\right) \quad (\text{GPa}) \tag{6.68}$$

where D is the disturbance factor indicating the degree of disturbance due to blast damage and stress relaxation, which ranges from 0 for undisturbed in situ rock masses to 1 for very disturbed rock masses.

Based on the analysis of a database of 150 data sets using the genetic programming approach, Beiki et al. (2010) derived the following two relationships for estimating the rock mass deformation modulus E_m:

$$E_m = \tan\left(\ln(\text{GSI})\right)\log(\sigma_c)(\text{RQD})^{1/3} \quad (\text{GPa}) \tag{6.69}$$

$$E_m = \tan\left(\sqrt{1.56 + (\ln(\text{GSI}))^2}\right)(\sigma_c)^{1/3} \quad (\text{GPa}) \tag{6.70}$$

Fig. 6.18 shows a comparison of some of the above correlations with the test data from different researchers. The wide range of the estimated values from the empirical correlations can be clearly seen.

There are also empirical correlations between the ratio of the rock mass deformation modulus E_m to the intact rock deformation modulus E_r and RMR or GSI. The following are some of them.

Nicholson and Bieniawski (1990):

$$\frac{E_m}{E_r} = 0.0028\text{RMR}^2 + 0.9e^{\text{RMR}/22.82} \tag{6.71}$$

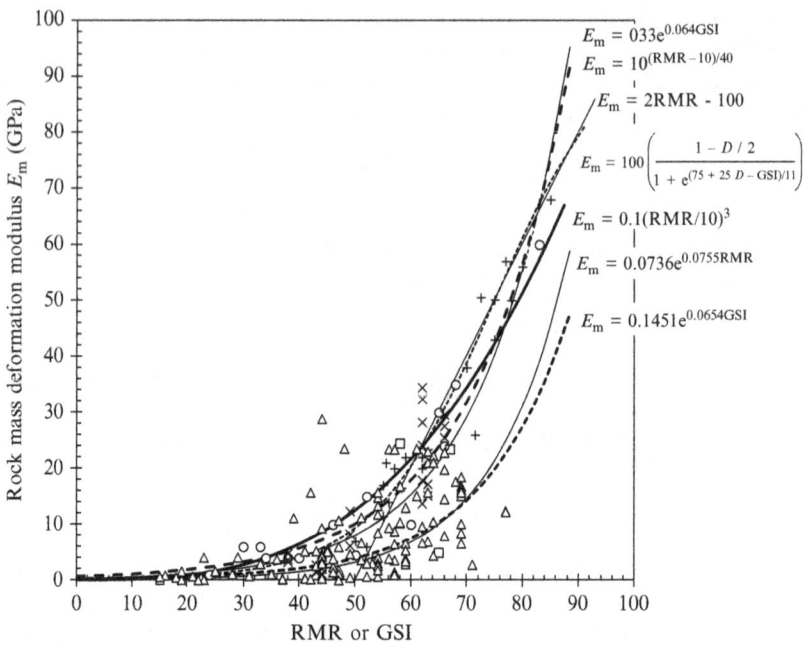

FIG. 6.18 Correlation between deformation modulus E_m and RMR or GSI ($D=0$ is used). +, Bieniawski (1978); ○, Serafim and Pereira (1983); ×, Stephans and Banks (1989); □, Schultz (1996); Δ, Gokceoglu et al. (2003).

Mitri et al. (1994):

$$\frac{E_m}{E_r} = \frac{1 - \cos\left(\pi \times RMR/100\right)}{2} \tag{6.72}$$

Sonmez et al. (2004):

$$\frac{E_m}{E_r} = (s^a)^{0.4}; s = e^{\frac{GSI-100}{9}}; a = \frac{1}{2} + \frac{e^{-GSI/15} - e^{-20/3}}{6} \tag{6.73}$$

Ramamurthy (2004):

$$\frac{E_m}{E_r} = e^{(RMR-100)/17.4} \tag{6.74}$$

Hoek and Diederichs (2006):

$$\frac{E_m}{E_r} = 0.02 + \frac{1 - D/2}{1 + e^{(60 + 15D - GSI)/11}} \tag{6.75}$$

Sonmez et al. (2006):

$$\frac{E_m}{E_r} = 10^{[((RMR-100)((100-RMR)/4000\exp(-RMR/100))]} \tag{6.76}$$

6.4.1.3   Methods Relating Deformation Modulus With Q

Barton et al. (1980) suggested the following relationships between rock mass deformation modulus E_m and Q:
Lower bound:

$$E_m = 10 \log Q \quad \text{(GPa)} \tag{6.77a}$$

Upper bound:

$$E_m = 40 \log Q \quad \text{(GPa)} \tag{6.77b}$$

Mean:

$$E_m = 25 \log Q \quad \text{(GPa)} \tag{6.77c}$$

where Q is the rock quality index as described in Chapter 5. The above relationships are only applicable to $Q > 1$ and generally hard rocks.

Barton (2002) suggested the following general relation for estimating the deformation modulus of rock masses:

$$E_m = 10 \left(Q \frac{\sigma_c}{100} \right)^{1/3} \quad \text{(GPa)} \tag{6.78}$$

which is similar to Eq. (6.63) in that it considers the effect of the unconfined compressive strength of intact rock σ_c.

6.4.1.4   Methods Relating Deformation Modulus With RMi

Palmström and Singh (2001) suggested the following correlations for estimating the rock mass deformation modulus E_m from RMi:

$$E_m = 5.6 \text{RMi}^{0.375} \quad \text{(GPa)} \quad (0.1 < \text{RMi} < 1) \tag{6.79a}$$

$$E_m = 7 \text{RMi}^{0.4} \quad \text{(GPa)} \quad (1 < \text{RMi} < 30) \tag{6.79b}$$

where RMi is the rock mass index as described in Chapter 5.

6.4.1.5   Methods Relating Deformation Modulus With Seismic P-Wave Velocity

Barton (2002) presented the following correlation for estimating the rock mass deformation modulus E_m from the seismic P-wave velocity:

$$E_m = 10 \times 10^{(v_p - 3.5)/3} \quad \text{(GPa)} \tag{6.80}$$

where v_p is the seismic P-wave velocity of the rock mass in km/s.

6.4.1.6   Methods Relating Deformation Modulus With Unconfined Compressive Strength

Rowe and Armitage (1984) correlated the rock mass deformation modulus deduced from a large number of field tests of drilled shafts under axial loading with the average unconfined compressive strength σ_c of weak rock deposits in which the drilled shafts were founded as follows:

$$E_m = 0.215\sqrt{\sigma_c} \quad (\text{GPa}) \tag{6.81}$$

where σ_c is in MPa.

Radhakrishnan and Leung (1989) found good agreement between the rock mass deformation modulus values obtained from back analysis of load-settlement relationship of large diameter drilled shafts in weathered sedimentary rocks and those obtained from Eq. (6.81).

Palmström and Singh (2001) also proposed a simple relation to estimate E_m from σ_c:

$$E_m = 0.2\sigma_c \quad (\text{GPa}) \tag{6.82}$$

where σ_c is in MPa.

6.4.1.7 Comments

Although the empirical methods are most widely used in practice to estimate rock mass deformation modulus, there are limitations for them:

1. The anisotropy of the rock mass caused by discontinuities is not considered. If the index properties in a single direction are used, the estimated deformation modulus may not be representative of the deformation modulus in other directions.
2. Different empirical relations often give very different deformation modulus values even for rock masses at the same site, as can be clearly seen from Fig. 6.19, which shows the predicted E_m values of 13 rock masses listed in Table 6.12. So it is important that the estimation of rock mass deformation modulus should not rely only on a single empirical relation. Instead, various empirical relations should be used to get an idea on the possible range of the rock mass deformation modulus.

6.4.2   Equivalent Continuum Approach for Estimating Rock Mass Deformation Modulus

The equivalent continuum approach treats jointed rock mass as an equivalent anisotropic continuum with deformability that reflects the deformation properties of the intact rock and those of the discontinuities.

6.4.2.1 Rock Mass Containing Persistent Discontinuities

For rock masses containing persistent discontinuities, analytical expressions for their deformation properties have been derived by a number of researchers, including Singh (1973), Kulhawy (1978), Gerrard (1982a,b, 1991), Amadei (1983), Oda et al. (1984), Fossum (1985), Yoshinaka and Yambe (1986), Oda (1988), Amadei and Savage (1993), and Zhang (2010). The basic idea used

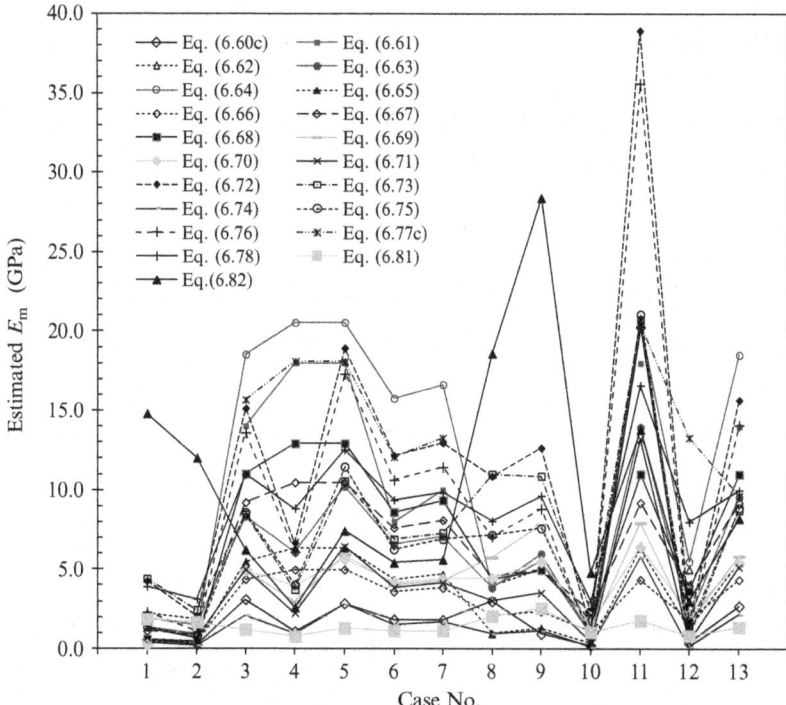

FIG. 6.19 Estimated rock mass deformation modulus values from empirical methods based on RQD, RMR, GSI, and/or Q.

by different researchers to derive the expressions for deformation properties is essentially the same, ie, the average stresses are assumed to distribute throughout the rock mass and the overall average strains of the rock mass are contributed by both the intact rock and the discontinuities. The only difference is the method for determining the additional deformation due to the discontinuities. Some of the typical results are presented in the following.

The 3D equivalent continuum model presented by Kulhawy (1978) for a rock mass containing three orthogonal discontinuity sets is shown in Fig. 6.20. The intact rock material is defined by Young's modulus E_r and Poisson's ratio v_r, while the discontinuities are described by normal stiffness k_n, shear stiffness k_s, and mean discontinuity spacing s. The properties of the equivalent orthotropic elastic mass are given as:

$$E_{mi} = \left(\frac{1}{E_r} + \frac{1}{s_i k_{ni}}\right)^{-1}$$ (6.83)

$$G_{mij} = \left(\frac{1}{G_r} + \frac{1}{s_i k_{si}} + \frac{1}{s_j k_{sj}}\right)^{-1}$$ (6.84)

TABLE 6.12 Summary of Rock Properties of 13 Rock Masses

#	Rock	E_r (GPa)	σ_c (MPa)	RQD (%)	RMR	Q	GSI	Reference
1	Granite	31.5[a]	74	8.5	24	0.08	19	Ozsan et al. (2007)
2	Diorite	19.5[a]	60	1.5	21	0.05	16	
3	Limestone (L1)	24.8[a]	31	54	57	4.23	52	El-Naqa and Kuisi (2002)
4	Limestone (L2)	10.4[a]	13	50	59	5.29	54	
5	Limestone (R1)	29.6[a]	37	48	59	5.29	54	
6	Limestone (R2)	21.6[a]	27	45	54	3.04	59	
7	Marly Limestone	22.4[a]	28	44	55	3.39	50	
8	Andesite	41.9	93	41	34	0.56	41	Ozsan and Akin (2002)
9	Basalt	40.0	142	15	38	0.63	42.5	
10	Tuff	11.6	24	10	21	0.11	31	
11	Basalt (d1)	60.9	69	77	59	6.6	52	Justo et al. (2006)
12	Basalt (d2)	5.3[a]	15	42.5	38	3.4	39	
13	Limestone	25.7	41	50	57	2.4[b]	52	Alber and Heiland (2001)

[a]*Estimated using modulus ratio (MR $= E_r/\sigma_c$) values from Table 6.2.*
[b]*Estimated from GSI using correlation GSI $= 9 \ln Q + 44$ [Eq. (5.30)].*

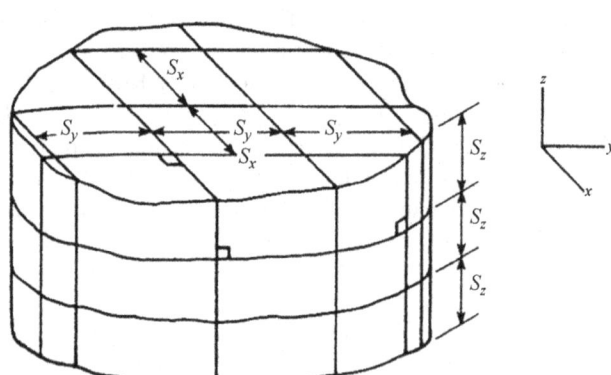

FIG. 6.20   Rock mass model of Kulhawy (1978).

$$\nu_{mij} = \nu_{mik} = \nu_r \frac{E_{mi}}{E_r} \qquad (6.85)$$

for $i = x, y, z$ with $j = y, z, x$ and $k = z, x, y$. These equations describe the rock mass elastic properties completely. The single discontinuity model is a special case of the foregoing in which $s_x = s_y = \infty$. Singh (1973), Amadei (1983), Chen (1989), and Amadei and Savage (1993) obtained the same expressions as above for the deformation properties of rock masses containing three orthogonal discontinuity sets.

For engineering convenience, it is useful to define a modulus reduction factor, α_E, which represents the ratio of the rock mass deformation modulus to the intact rock deformation modulus. This factor can be obtained by re-writing Eq. (6.83) as:

$$\alpha_E = \frac{E_{mi}}{E_r} = \left(1 + \frac{E_r}{s_i k_{ni}}\right)^{-1} \qquad (6.86)$$

The relationship is plotted in Fig. 6.21. It shows smaller values of α_E in rock masses with softer discontinuities (larger E_r/k_n values).

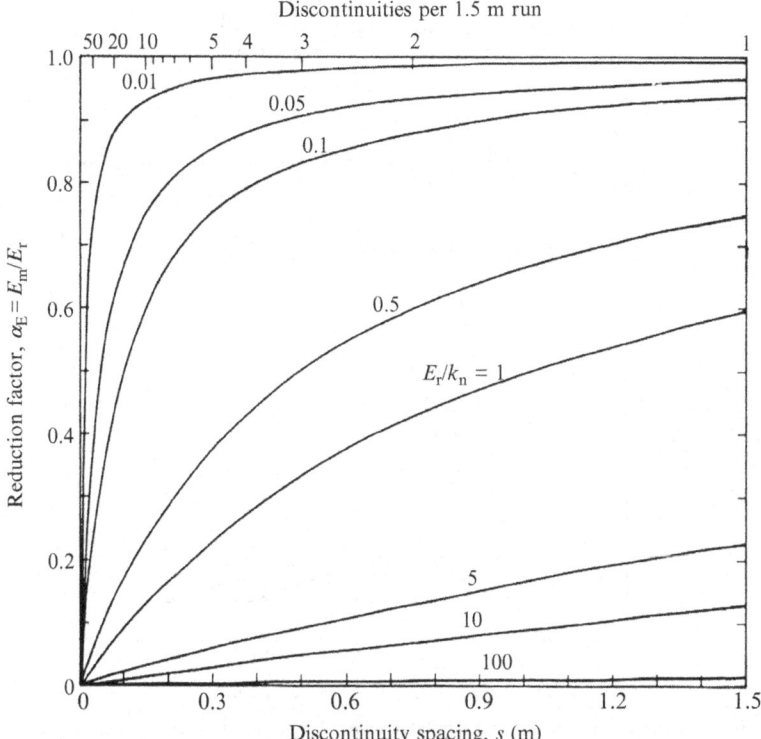

FIG. 6.21 Modulus reduction factor versus discontinuity spacing. *(From Kulhawy, 1978.)*

Unfortunately, the mean discontinuity spacing is not easy to obtain directly and, in normal practice, RQD values are determined instead. Using a physical model, RQD can be correlated with the number of discontinuities per 1.5 m (5 ft) core run, a common measure in practice, as shown in Fig. 6.22. Combining Figs. 6.21 and 6.22 yields Fig. 6.23, which relates α_E and RQD with E_r/k_n as an additional parameter.

Consider a jointed rock mass under uniaxial loading as shown in Fig. 6.24. The constitutive relation in the n, s, t coordinate system can be defined from the single discontinuity model of Kulhawy (1978). In the global coordinate system x, y, z, the constitutive relation can be determined using the second tensor coordinate transformation rules. In matrix form, it gives (Amadei and Savage, 1993):

$$(\varepsilon)_{xyz} = (A)_{xyz}(\sigma)_{xyz} \tag{6.87}$$

where $(\varepsilon)^t_{xyz} = (\varepsilon_x, \varepsilon_y, \varepsilon_z, \gamma_{xy}, \gamma_{yz}, \gamma_{zx})$; and $(\sigma)^t_{xyz} = (\sigma_x, \sigma_y, \sigma_z, \tau_{xy}, \tau_{yz}, \tau_{zx})$. The components $a_{ij} = a_{ji}$ $(i, j = 1\text{--}6)$ of the compliance matrix $(A)_{xyz}$ depend on the dip angle θ as follows:

$$a_{11} = \frac{1}{E_r} + \frac{\sin^4\theta}{k_n s} + \frac{\sin^2\theta}{4k_s s} \tag{6.88a}$$

$$a_{12} = -\frac{\nu_r}{E_r} + \frac{\sin^2 2\theta}{4}\left(\frac{1}{k_n s} - \frac{1}{k_s s}\right) \tag{6.88b}$$

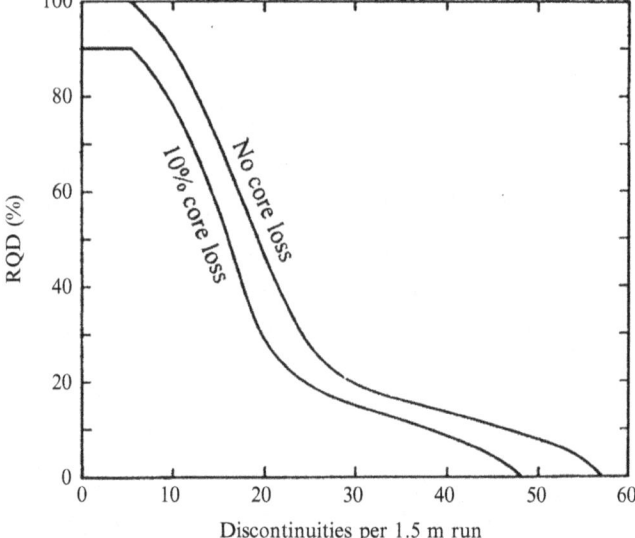

FIG. 6.22 RQD versus number of discontinuities per 1.5 m run. *(From Kulhawy, 1978.)*

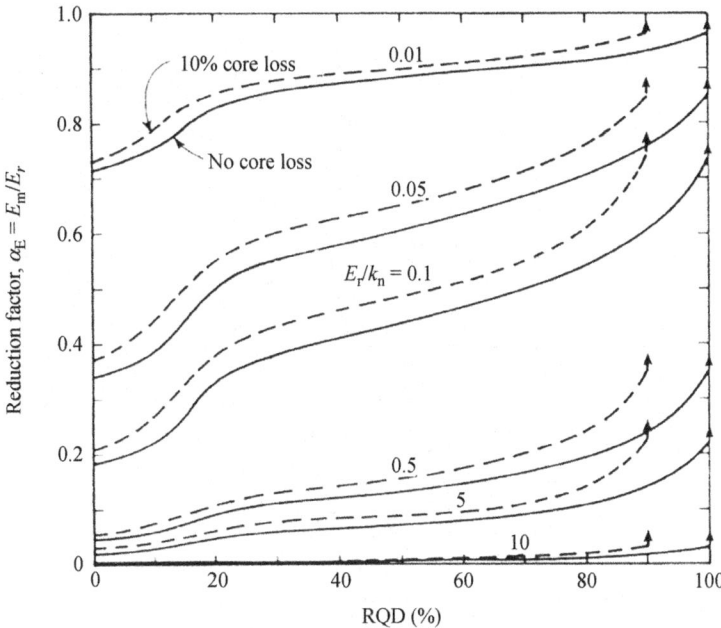

FIG. 6.23   Modulus reduction factor versus RQD. *(From Kulhawy, 1978.)*

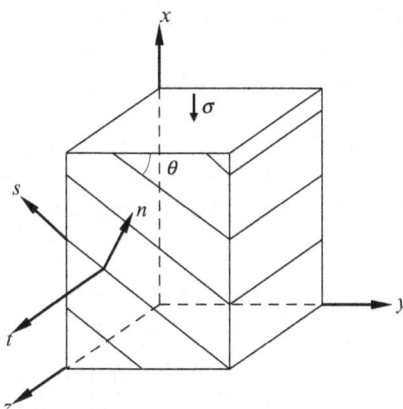

FIG. 6.24   Jointed rock mass under uniaxial loading. *(Based on Amadei, B., Savage, W.Z., 1993. Effect of joints on rock mass strength and deformability. In: Hudson, J.A. (Ed.), Comprehensive Rock Engineering – Principle, Practice and Projects, vol. 1. Pergamon, Oxford, UK, pp. 331–365.)*

$$a_{13} = a_{23} = -\frac{\nu_r}{E_r} \tag{6.88c}$$

$$a_{22} = \frac{1}{E_r} + \frac{\cos^4\theta}{k_n s} + \frac{\sin^2\theta}{4k_s s} \tag{6.88d}$$

$$a_{33} = \frac{1}{E_r} \tag{6.88e}$$

$$a_{14} = \frac{\sin 2\theta \cos 2\theta}{2k_s s} + \frac{\sin^2\theta \sin 2\theta}{k_n s} \tag{6.88f}$$

$$a_{24} = -\frac{\sin 2\theta \cos 2\theta}{2k_s s} + \frac{\cos^2\theta \sin 2\theta}{k_n s} \tag{6.88g}$$

$$a_{44} = \frac{1}{G_r} + \frac{\sin^2 2\theta}{k_n s} + \frac{\cos^2 2\theta}{k_s s} \tag{6.88h}$$

$$a_{55} = \frac{1}{G_r} + \frac{\cos^2\theta}{k_s s} \tag{6.88i}$$

$$a_{56} = \frac{\sin 2\theta}{2k_s s} \tag{6.88j}$$

$$a_{66} = \frac{1}{G_r} + \frac{\sin^2\theta}{k_s s} \tag{6.88k}$$

All other components a_{ij} vanish. Note that for the orientation of the discontinuities considered here, the jointed rock mass has a plane of elastic symmetry normal to the z-axis. If the discontinuity set is inclined with respect to x and z axes or if the rock sample under consideration has two or three orthogonal discontinuity sets, then new expressions must be derived.

Fossum (1985) derived a constitutive model for a rock mass that contains randomly oriented discontinuities with constant normal stiffness k_n and shear stiffness k_s. He assumed that if the discontinuities are randomly oriented, the mean discontinuity spacing would be the same in all directions taken through a representative sample of the rock mass. Arguing that the mechanical properties of the jointed rock mass would be isotropic, Fossum derived the following expressions for the bulk modulus K_m and shear modulus G_m of the equivalent elastic continuum:

$$K_m = \frac{E_r}{9}\left[\frac{3(1+\nu_r)sk_n + 2E_r}{(1+\nu_r)(1-2\nu_r)sk_n + (1-\nu_r)E_r}\right] \tag{6.89}$$

$$G_m = \frac{E_r}{30(1+\nu_r)}\left[\frac{9(1+\nu_r)(1-2\nu_r)sk_n + (7-5\nu_r)E_r}{(1+\nu_r)(1-2\nu_r)sk_n + (1-\nu_r)E_r}\right] + \frac{2}{5}\left[\frac{E_r sk_s}{2(1+\nu_r)sk_s + E_r}\right] \tag{6.90}$$

The equivalent Young's modulus and Poisson's ratio can then be obtained from:

$$E_m = \frac{9K_m G_m}{3K_m + G_m} \tag{6.91}$$

$$\nu_m = \frac{3K_m - 2G_m}{2(3K_m + G_m)} \tag{6.92}$$

At large values of mean discontinuity spacing the equivalent modulus E_m and Poisson's ratio ν_m approach the values E_r and ν_r for the intact rock material, respectively. At very small values of mean discontinuity spacing the equivalent modulus E_m and Poisson's ratio ν_m are given by the following expressions:

$$E_m \rightarrow \frac{2E_r(7 - 5\nu_r)}{3(1 - \nu_r)(9 + 5\nu_r)} \quad \text{as } s \rightarrow 0 \tag{6.93}$$

$$\nu_m \rightarrow \frac{(1 + 5\nu_r)}{(9 + 5\nu_r)} \quad \text{as } s \rightarrow 0 \tag{6.94}$$

Zhang (2010) derived simple expressions for estimating the deformation properties of heavily jointed rock masses which can be reasonably considered an isotropic continuum, using a geometric averaging method. The method assumes that all discontinuity sets in the rock mass have the same average discontinuity spacing s, the same elastic normal stiffness k_n, and the same elastic shear stiffness k_s. With this assumption, the contribution of each discontinuity set to the overall deformability of the rock mass depends only on its orientation (denoted by α and β). Therefore, the average contribution of all discontinuity sets to the overall isotropic rock mass deformability can be estimated by considering the averaging process over the range of α and β. Using this averaging approach, E_m and G_m of the equivalent isotropic rock mass can be obtained as:

$$E_m = \frac{1}{\dfrac{1}{E_r} + \dfrac{1}{3}\left[\dfrac{21}{32k_n s} + \dfrac{11}{32k_s s}\right]} \tag{6.95}$$

$$G_m = \frac{1}{\dfrac{1}{G_r} + \dfrac{1}{3}\left[\dfrac{11}{32k_n s} + \dfrac{21}{32k_s s}\right]} \tag{6.96}$$

where s, k_n, and k_s are, respectively, the average discontinuity spacing, the normal stiffness and the shear stiffness for all of the discontinuity sets.

Fig. 6.25 shows a comparison of the E_m/E_r values obtained from the expressions by Kulhawy (1978) (Eq. 6.83), Fossum (1985) (Eqs. 6.89–6.91), and Zhang (2010) (Eq. 6.95), respectively. The same properties of the rock mass as used by Fossum (1985) are used here: $E_r = 35$ GPa, $\nu_r = 0.25$, $k_s = 10$ GPa/m, and $k_n = 20$ GPa/m. It can be seen that the E_m/E_r values from the

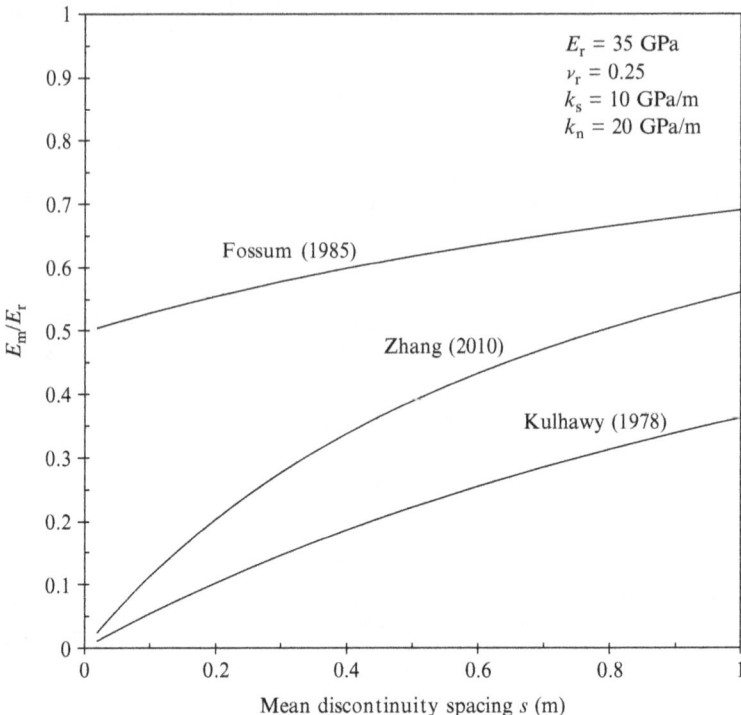

FIG. 6.25 Comparison of E_m/E_r, respectively, from the relations by Kulhawy (1978), Fossum (1985), and Zhang (2010).

expression by Zhang (2010) are bounded by those from the relations by Kulhawy (1978) and Fossum (1985). The relation by Kulhawy (1978) gives the smallest E_m/E_r value because it represents the deformability in the most deformable direction (ie, the direction perpendicular to a discontinuity set). The relation by Fossum (1985) gives the highest E_m/E_r value because it considers rock masses containing randomly distributed discontinuities and thus the derived deformability represents the average of the lowest to the highest deformability related to individual discontinuities in all directions. Unlike Fossum (1985), Zhang (2010) considers a rock mass containing many discontinuity sets with discontinuities in each set being parallel. As Fossum (1985), however, Zhang (2010) uses the average contribution of all of the discontinuity sets to the deformability of the rock mass.

Considering the fact that the available methods did not consider the statistical nature of jointed rock masses, Dershowitz et al. (1979) presented a statistically based analytical model to examine the rock mass deformability. The statistical model is shown in Fig. 6.26. The rock mass is taken as a 3D circular cylinder. Deformation is assumed to accrue both from the elasticity of intact rock and from the displacement along discontinuities. The displacements along

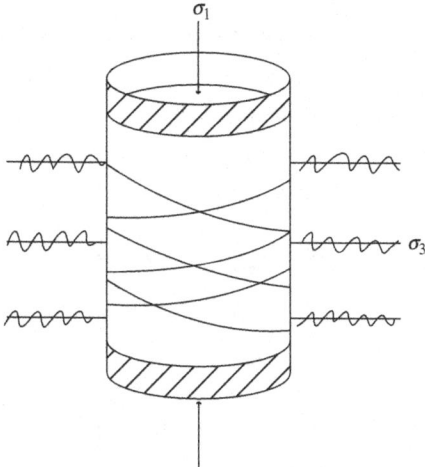

FIG. 6.26 Statistical model for jointed rock. *(Based on Dershowitz, W.S., Baecher, G.B., Einstein, H.H., 1979. Prediction of rock mass deformability. Proc. 4th Int. Cong. on Rock Mech., Montreal, Canada, 1, 605–611.)*

intersecting discontinuities are assumed to be independent. In this model, the compatibility of lateral displacements across jointed blocks is approximated by constraining springs. Inputs to the model include stiffness and deformation modulus, stress state, and discontinuity geometry. The intact rock deformability is expressed by Young's modulus E_r, set at 200,000 kg/cm^2, a typical value. The discontinuity stiffness is represented by normal stiffness k_n set at 1,000,000 kg/cm^3, and shear stiffness k_s set at 200,000 kg/cm^3. The stress state is described by a vertical major principal stress σ_1, and a horizontal "confining" stress σ_3. The confining stress σ_3 is determined from initial stress σ_{30} and a spring constant k_g as follows:

$$\sigma_3 = \sigma_{30} + k_g \delta_y \tag{6.97}$$

where δ_y is the calculated horizontal displacement; σ_{30} is set to 50 kg/cm^2; and k_g is set at 2500 kg/cm^3, a value chosen to maximize the increase of stress with lateral strain without causing rotation of principal planes.

The discontinuity geometry is described by three parameters: the mean spacing s_m, the mean orientation θ_m and the dispersion according to the Fisher model k. The discontinuity spacing is assumed to follow a negative exponential distribution and orientation a Fisher distribution (Table 6.13).

Some of the results are shown in Figs. 6.27–6.30. The results show that the proposed model is consistent with the data of E_m/E_r versus RQD (see Figs. 6.14 and 6.15), to the extent that the relationships between deformation modulus and RQD are of similar form.

TABLE 6.13 Distribution Assumptions for Deformation Model

Discontinuity Property	Distribution Form
Spacing	Negative exponential: $\lambda e^{-\lambda s}$, $\lambda = (\text{mean spacing})^{-1}$
Size (persistence)	Completely persistent
Orientation	Fisher: $\dfrac{\kappa e^{-\kappa \cos\alpha}}{4\pi \sinh\kappa}$, $\kappa = $ dispersion; $\alpha = $ angle from mean pole
Normal stiffness	Deterministic
Shear stiffness	Deterministic

Based on Dershowitz, W.S., Baecher, G.B., Einstein, H.H., 1979. Prediction of rock mass deformability. Proc. 4th Int. Cong. on Rock Mech., Montreal, Canada, 1, 605–611.

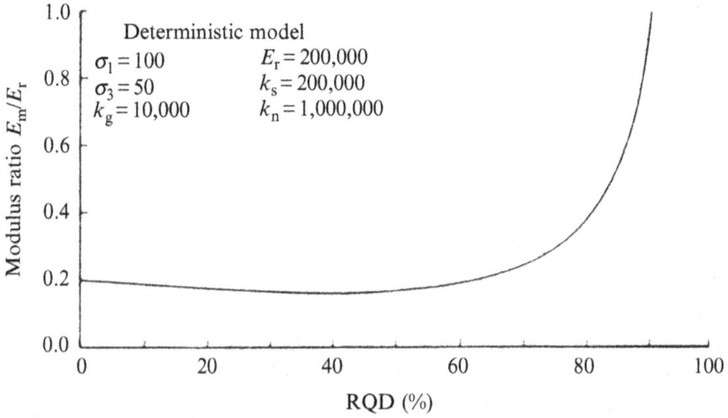

FIG. 6.27 Relationship between E_m/E_r and RQD, parallel discontinuities. *(From Dershowitz, W.S., Baecher, G.B., Einstein, H.H., 1979. Prediction of rock mass deformability. Proc. 4th Int. Cong. on Rock Mech., Montreal, Canada, 1, 605–611.)*

The model proposed by Dershowitz et al. (1979) has the following limitations:

1) The analysis applies only to "hard" rock. Shears and weathering can only be accommodated through changes in discontinuity stiffnesses, which is inadequate.
2) The analysis is for infinitesimal strains. Finite strains would violate the assumption of independence among discontinuity displacements.
3) The analysis is for a homogeneous deterministic stress field specified extraneous to the discontinuity pattern. Real rock masses may have complex stress distributions strongly influenced by the actual jointing pattern.
4) The boundary conditions are highly idealized.

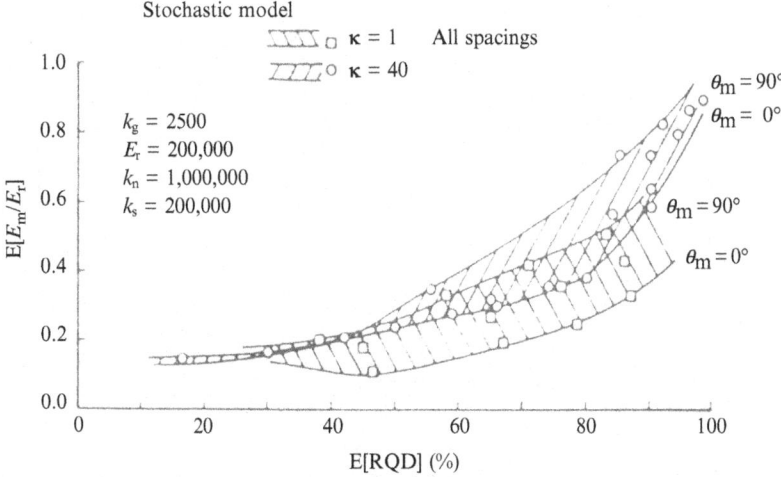

FIG. 6.28 Relationship between E[E_m/E_r] and E[RQD], subparallel discontinuities distributed according to Fisher. *(From Dershowitz, W.S., Baecher, G.B., Einstein, H.H., 1979. Prediction of rock mass deformability. Proc. 4th Int. Cong. on Rock Mech., Montreal, Canada, 1, 605–611.)*

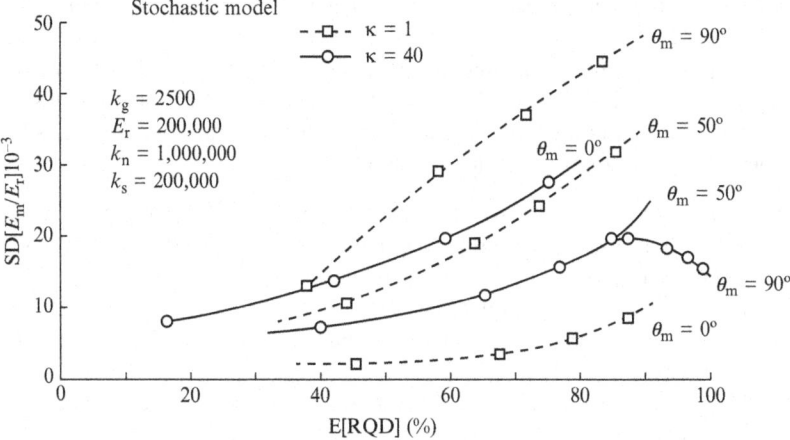

FIG. 6.29 Relationship between SD[E_m/E_r] and E[RQD], subparallel discontinuities distributed according to Fisher. *(From Dershowitz, W.S., Baecher, G.B., Einstein, H.H., 1979. Prediction of rock mass deformability. Proc. 4th Int. Cong. on Rock Mech., Montreal, Canada, 1, 605–611.)*

6.4.2.2   Rock Mass Containing Nonpersistent Discontinuities

For rock masses containing nonpersistent discontinuities, relationships between the deformation properties and the fracture tensor parameters in two and three dimensions have been derived by Kulatilake et al. (1992, 1993) and Wang (1992) based on the discrete element method (DEM) analysis results of

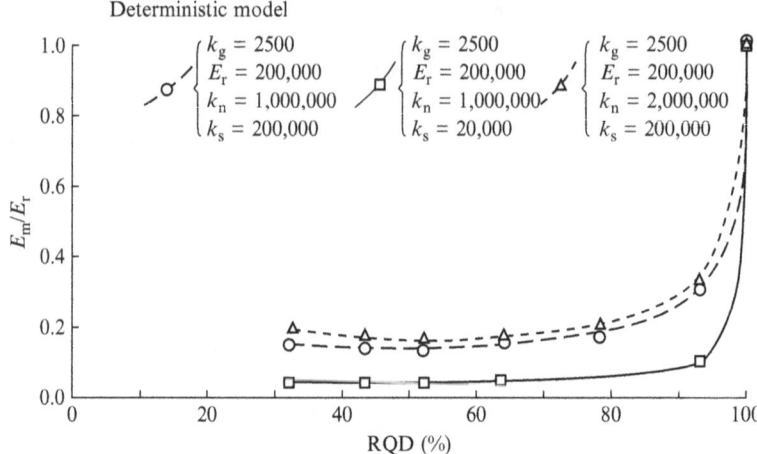

FIG. 6.30   Effect of stiffness values on modulus ratio E_m/E_r, parallel discontinuities. *(From Dershowitz, W.S., Baecher, G.B., Einstein, H.H., 1979. Prediction of rock mass deformability. Proc. 4th Int. Cong. on Rock Mech., Montreal, Canada, 1, 605–611)*

generated rock mass blocks. The procedure used to evaluate the effect of discontinuities and the obtained relationships between the deformation properties and the fracture tensor parameters in three dimensions are outlined in the following.

The procedure for evaluating the effect of discontinuities on the deformability of rock masses is shown in Fig. 6.31. The first step is the generation of nonpersistent discontinuities in 2 m cubical rock blocks. The discontinuities are generated in a systematic fashion as follows:

(1) In each rock block, a certain number of discontinuities having a selected orientation and a selected discontinuity size are placed to represent a discontinuity set.
(2) Discontinuities are considered as 2D circular discs.
(3) Discontinuity center locations are generated according to a uniform distribution.
(4) Either a single discontinuity set or two discontinuity sets are included in each rock block.

The generated discontinuity networks in the rock blocks are given in Table 6.14.

The second step is the generation of fictitious discontinuities according to the actual nonpersistent discontinuity network generated in the rock block. In order to use the DEM for 3D analyses of a generated rock block, the block should be discretized into polyhedra. Since a typical nonpersistent discontinuity network in 3D may not discretize the block into polyhedra, it is necessary to create some type of fictitious discontinuities so that when they are combined with the actual discontinuities, the block was discretized into polyhedra. Before the generation of fictitious discontinuities, the actual disc-shaped

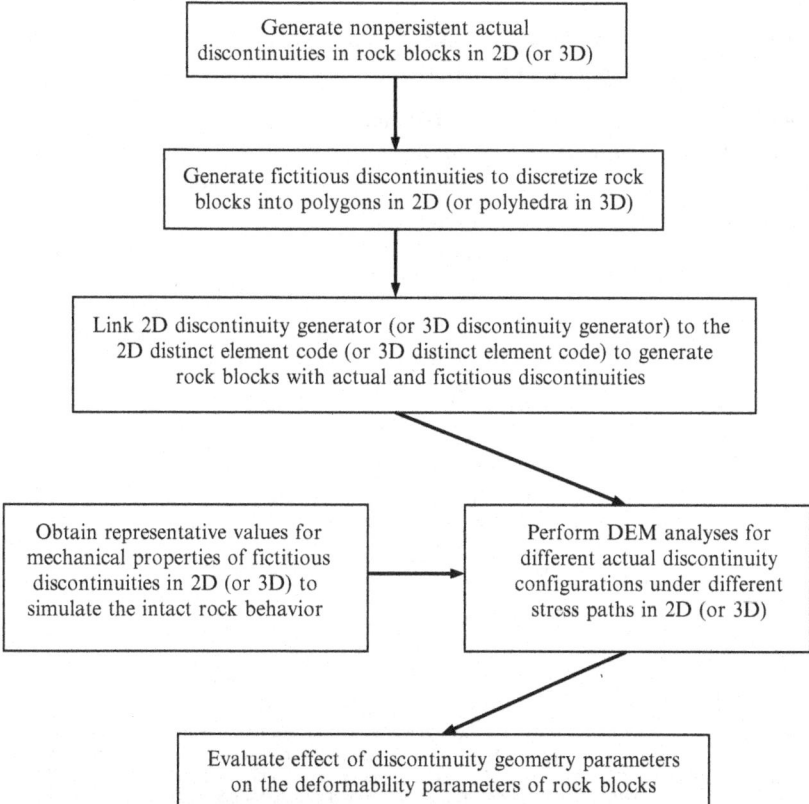

FIG. 6.31 Procedure for evaluating the effect of discontinuity geometry parameters on the deformability properties of jointed rock mass. *(Based on Kulatilake, P.H.S.W., Wang, S., Stephansson, O., 1993. Effect of finite size joints on the deformability of jointed rock in three dimensions. Int. J. Rock Mech. Min. Sci. Geomech. Abstr. 30 (5), 479–501.)*

discontinuities are converted into square-shaped ones having the same area. In order for the fictitious discontinuities to simulate the intact rock behavior, an appropriate constitutive model and associated parameter values for the fictitious discontinuities have to be found. From the investigation performed on 2D rock blocks, Kulatilake et al. (1992) found that by choosing the mechanical properties of the fictitious discontinuities in the way given below, it is possible to make the fictitious discontinuities behave as the intact rock:

(a) The strength parameters of the fictitious discontinuities are the same as those of the intact rock.

(b) $G_r/k_s = 0.008–0.012$.

(c) $k_n/k_s = 2–3$, with the most appropriate value being E_r/G_r.

For the intact rock (granitic gneiss) studied by Kulatilake et al. (1992, 1993) and Wang (1992), the approximate parameters of the fictitious discontinuities are

TABLE 6.14 Generated Discontinuity Networks of Actual Discontinuities in the Rock Block for 3D DEM Analysis

# of Discontinuity Sets	Orientation α/β	Discontinuity Size/Block Size	# of Discontinuities	Discontinuity Location
One set	60°/45°	0.1–0.9 with step 0.1	5, 10, 20, 30	Uniform distribution
	94.42°/ 37.89°	0.3, 0.5, 0.6, 0.7, 0.9	5, 10, 20, 30	
	30°/45°	0.3, 0.5, 0.6, 0.7, 0.8, 0.9	5, 10, 20,	
	90°/45°	0.3, 0.5, 0.6, 0.7, 0.8, 0.9	5, 10, 20	
	68.2°/72.2 °	0.3, 0.6, 0.7, 0.8	5, 10, 20, 30	
	248.9°/79.8°	0.3, 0.6, 0.7, 0.8	5, 10, 20, 30	
Two sets	60°/45°, 240°/60°	0.1, 0.2, 0.3, 0.5, 0.6, 0.7	10, 10	

Based on Kulatilake, P.H.S.W., Ucpirti, H., Wang, S., Radberg, G., Stephansson, O., 1992. Use of the distinct element method to perform stress analysis in rock with non-persistent joints and to study the effect of joint geometry parameters on the strength and deformability of rock. Rock Mech. Rock Eng., 25, 253–274; Kulatilake, P.H.S.W., Wang, S., Stephansson, O., 1993. Effect of finite size joints on the deformability of jointed rock in three dimensions. Int. J. Rock Mech. Min. Sci. Geomech. Abstr. 30 (5), 479–501.

shown in Table 6.15. The mechanical parameters of the actual discontinuities used by them are also shown in the table. The constitutive models used for the intact rock and discontinuities (both actual and fictitious) are shown in Figs. 6.32 and 6.33, respectively.

The third step is the DEM analysis of the rock block (using the 3D distinct element code 3DEC) under different stress paths and the evaluation of the effect of discontinuities on the deformation parameters of the rock mass. In order to estimate different property values of the jointed rock block, Kulatilake et al. (1993) and Wang (1992) used the following stress paths:

(1) The rock block is first subjected to an isotropic compressive stress of 5 MPa in three perpendicular directions (x, y, z); then, for each of the three directions, eg, the z-direction, the compressive stress σ_z is increased, while keeping the confining stresses in the other two directions (σ_x and σ_y) the same, until the failure of the rock occurs (see Fig. 6.34). From the

TABLE 6.15 Values for the Mechanical Parameters of Intact Rock, Actual and Fictitious Discontinuities Used by Kulatilake et al. (1992, 1993) and Wang (1992)

Intact Rock or Discontinuities	Parameter	Assigned Value
Intact rock	Young's modulus E_r	60 GPa
	Poisson's ratio ν_r	0.25
	Cohesion c_r	50 MPa
	Tensile strength t_r	10 MPa
	Friction coefficient $\tan \phi_r$	0.839
Fictitious discontinuities	Normal stiffness k_n	5000 GPa/m
	Shear stiffness k_s	2000 GPa/m
	Cohesion c_j	50 MPa
	Dilation coefficient d_j	0
	Tensile strength t_j	10 MPa
	Friction coefficient $\tan \phi_j$	0.839
Actual discontinuities	Normal stiffness k_n	67.2 GPa/m
	Shear stiffness k_s	2.7 GPa/m
	Cohesion c_j	0.4 MPa
	Tensile strength t_j	0
	Friction coefficient $\tan \phi_j$	0.654

analysis results, it is possible to estimate the deformation modulus of the rock block in each of the three directions and the related Poisson's ratios.

(2) The rock block is first subjected to an isotropic compressive stress of 5 MPa in three perpendicular directions (x, y, z); then, on each of the three perpendicular planes, eg, the x-y plane, the rock is subjected to an increasing shear stress as shown in Fig. 6.35. The analysis results can be used to estimate the shear modulus of the rock block on each of the three perpendicular planes.

In the DEM analysis, during the loading process, displacements are recorded simultaneously on each block face in the direction(s) needed to calculate the required block strains. On each block face, five points are selected to record the displacement. The average value of these five displacements is considered as the mean displacement of this face for block strain calculations. To make it possible to estimate the deformation properties of the rock block from the DEM

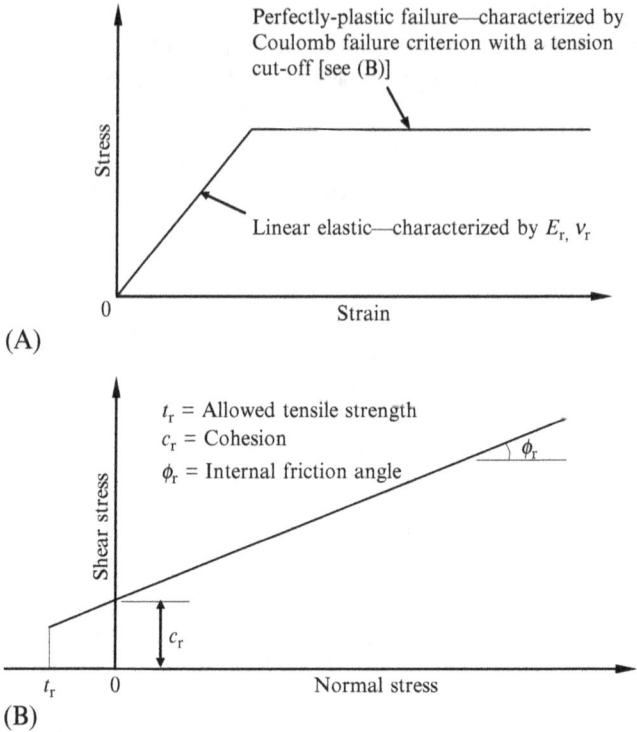

(A)

(B)

FIG. 6.32 Constitutive model assumed for intact rock: (A) stress versus strain; and (B) Coulomb failure criterion with a tension cut-off. *(Based on Kulatilake, P.H.S.W., Wang, S., Stephansson, O., 1993. Effect of finite size joints on the deformability of jointed rock in three dimensions. Int. J. Rock Mech. Min. Sci. Geomech. Abstr. 30 (5), 479–501.)*

analysis results, Kulatilake et al. (1993) and Wang (1992) assumed that the rock block is orthotropic in the x, y, z directions, regardless of the actual orientations of the discontinuities, ie,

$$\begin{Bmatrix} \Delta\varepsilon_x \\ \Delta\varepsilon_y \\ \Delta\varepsilon_z \\ \Delta\varepsilon_{xy} \\ \Delta\varepsilon_{xz} \\ \Delta\varepsilon_{yz} \end{Bmatrix} = \begin{bmatrix} \dfrac{1}{E_x} & \dfrac{-\nu_{yx}}{E_y} & \dfrac{-\nu_{zx}}{E_z} & 0 & 0 & 0 \\ & \dfrac{1}{E_y} & \dfrac{-\nu_{zy}}{E_z} & 0 & 0 & 0 \\ & & \dfrac{1}{E_3} & 0 & 0 & 0 \\ & & & \dfrac{1}{G_{xy}} & 0 & 0 \\ & & & & \dfrac{1}{G_{xz}} & 0 \\ \text{Sym.} & & & & & \dfrac{1}{G_{yz}} \end{bmatrix} \begin{Bmatrix} \Delta\sigma_x \\ \Delta\sigma_y \\ \Delta\sigma_z \\ \Delta\sigma_{xy} \\ \Delta\sigma_{xz} \\ \Delta\sigma_{yz} \end{Bmatrix} \qquad (6.98)$$

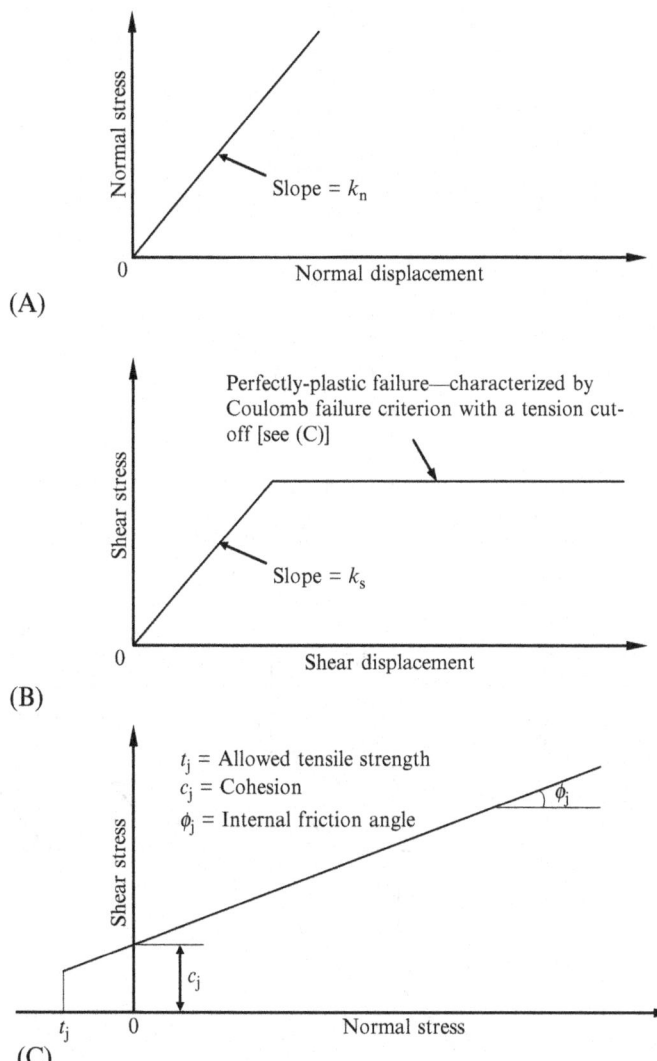

FIG. 6.33 Constitutive model assumed for joints: (A) normal stress versus normal displacement; (B) shear stress versus shear displacement; and (C) Coulomb failure criterion with a tension cut-off. *(Based on Kulatilake, P.H.S.W., Wang, S., Stephansson, O., 1993. Effect of finite size joints on the deformability of jointed rock in three dimensions. Int. J. Rock Mech. Min. Sci. Geomech. Abstr. 30 (5), 479–501.)*

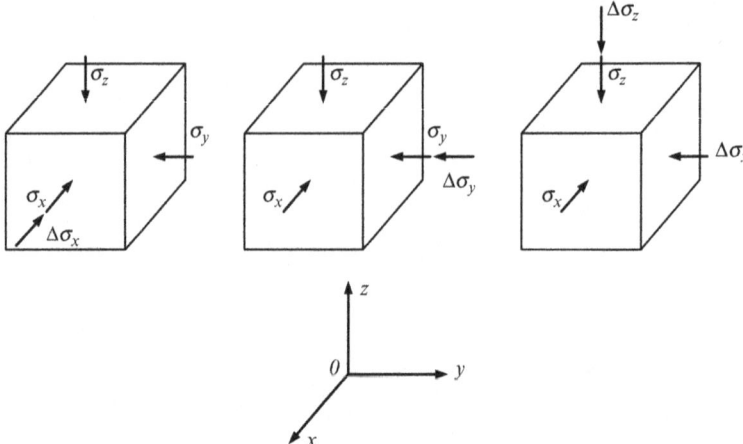

FIG. 6.34 Stress paths of first type used to perform DEM analysis of generated rock blocks. *(Based on Kulatilake, P.H.S.W., Wang, S., Stephansson, O., 1993. Effect of finite size joints on the deformability of jointed rock in three dimensions. Int. J. Rock Mech. Min. Sci. Geomech. Abstr. 30 (5), 479–501.)*

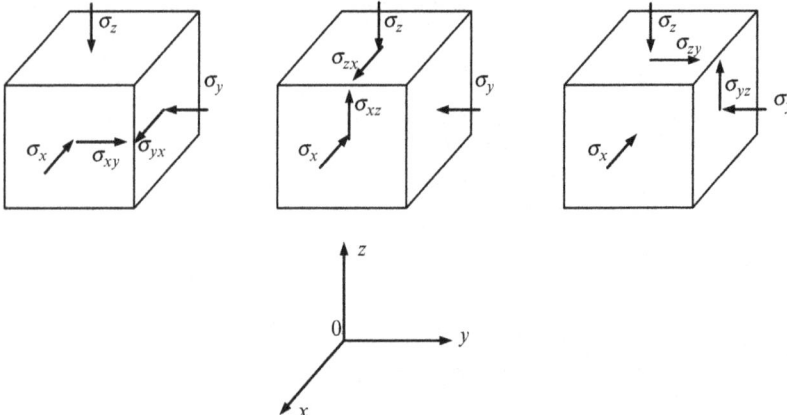

FIG. 6.35 Stress paths of second type used to perform DEM analysis of generated rock blocks. *(Based on Kulatilake, P.H.S.W., Wang, S., Stephansson, O., 1993. Effect of finite size joints on the deformability of jointed rock in three dimensions. Int. J. Rock Mech. Min. Sci. Geomech. Abstr. 30 (5), 479–501.)*

With the above constitutive model, the deformation moduli E_x, E_y, E_z and Poisson's ratios ν_{xy}, ν_{xz}, ν_{yx}, ν_{yz}, ν_{zx}, ν_{zy} can be estimated from the DEM analysis results of rock blocks under stress path 1 (Fig. 6.34), and the shear moduli G_{xy}, G_{xz}, and G_{yz} can be estimated from the DEM analysis results of rock blocks under stress path 2 (Fig. 6.35).

To reflect the effect of discontinuity geometry parameters on the deformation properties, Kulatilake et al. (1993) and Wang (1992) used the fracture tensor defined by Oda (1982) as an overall measure of the discontinuity parameters: discontinuity density, orientation, size, and the number of discontinuity sets. For thin circular discontinuities, the general form of the fracture tensor at the 3D level for the kth discontinuity set can be expressed as (see also Chapter 4 about the discussion of fracture tensors):

$$F_{ij}^{(k)} = 2\pi\rho \int_0^\infty \iint_{\Omega/2} r^3 n_i n_j f(\boldsymbol{n}, r) d\Omega dr \qquad (6.99)$$

where ρ is the average number of discontinuities per unit volume (discontinuity density), r is the radius of the circular discontinuity (discontinuity size), $\boldsymbol{n}$ is the unit vector normal to the discontinuity plane, $f(\boldsymbol{n}, r)$ is the discontinuity probability density function of $\boldsymbol{n}$ and r, $\Omega/2$ is a solid angle corresponding to the surface of a unit hemisphere, and n_i and n_j ($i, j = x, y, z$) are the components of vector $\boldsymbol{n}$ in the rectangular coordinate system considered (see Fig. 6.36). The solid angle $d\Omega$ is also shown in Fig. 6.36. If the distributions of the size and orientation of the discontinuities are independent of each other, Eq. (6.99) can be rewritten as:

$$F_{ij}^{(k)} = 2\pi\rho \int_0^\infty r^3 f(r) dr \iint_{\Omega/2} n_i n_j f(\boldsymbol{n}) d\Omega \qquad (6.100)$$

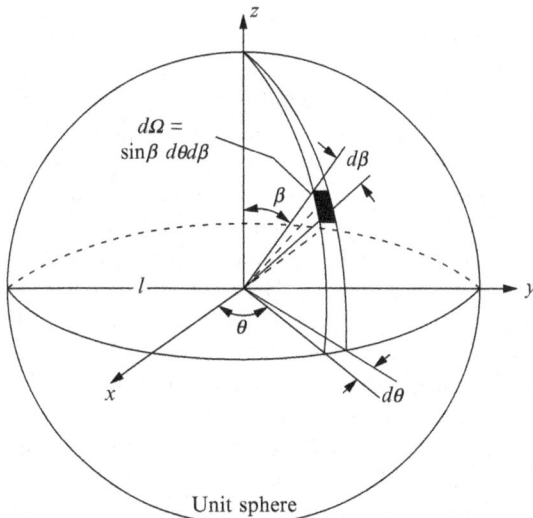

FIG. 6.36 Unit sphere used to define the solid angle $d\Omega$. *(Based on Oda, M., 1982. Fabric tensor for discontinuous geological materials. Soils Found., Tokyo, Japan, 22, 96–108.)*

where $f(\mathbf{n})$ and $f(r)$ are the probability density functions of the unit normal vector $\mathbf{n}$ and size r, respectively. If there are more than one discontinuity sets in the rock mass, the fracture tensor for the rock mass can be obtained by:

$$F_{ij} = \sum_{k=1}^{N} F_{ij}^{(k)} \tag{6.101}$$

where N is the number of discontinuity sets in the rock mass. The fracture tensor F_{ij} can also be written in matrix form as:

$$\mathbf{F}\left(F_{ij}\right) = \begin{bmatrix} F_{xx} & F_{xy} & F_{xz} \\ & F_{yy} & F_{yz} \\ \text{Sym.} & & F_{zz} \end{bmatrix} \tag{6.102}$$

Since the diagonal components of the fracture tensor F_{xx}, F_{yy}, and F_{zz} express the combined effect of discontinuity density and discontinuity size in the x, y and z directions, respectively, Kulatilake et al. (1993) and Wang (1992) showed the obtained deformation properties as in Figs. 6.37 and 6.38. Putting the data in Figs. 6.37A–C and Figs. 6.38A–C together, respectively, Figs. 6.39 and 6.40 are obtained, which show that the deformation properties of jointed rock masses are related to the corresponding components of the fracture tensor. As for the Poisson's ratios of the generated rock blocks, Kulatilake et al. (1993) and Wang (1992) found that they are between 50 and 190% of the intact rock Poisson's ratio.

6.4.2.3 Comments

In the equivalent continuum approach, the elastic properties of the equivalent rock mass are essentially derived by examining the behavior of two rock blocks having the same volume and by using an averaging process. One volume is a representative sample of the rock mass whereas the second volume is cut from the equivalent continuum and is subject to homogeneous (average) stresses and strains. Therefore, the equivalent continuum approach requires that the representative sample of the rock mass be large enough to contain a large number of discontinuities. On the other hand, the corresponding equivalent continuum volume must also be sufficiently small to make negligible stress and strain variations across it. This leads to a dilemma which is typical in modeling continuous or discontinuous composite media.

Numerous researchers have used the equivalent continuum approach to derive the expressions for the equivalent continuum deformation properties. Most of these expressions are based on the assumption that the discontinuities are persistent. This is a conservative assumption since, in reality, most of the discontinuities are nonpersistent with finite size.

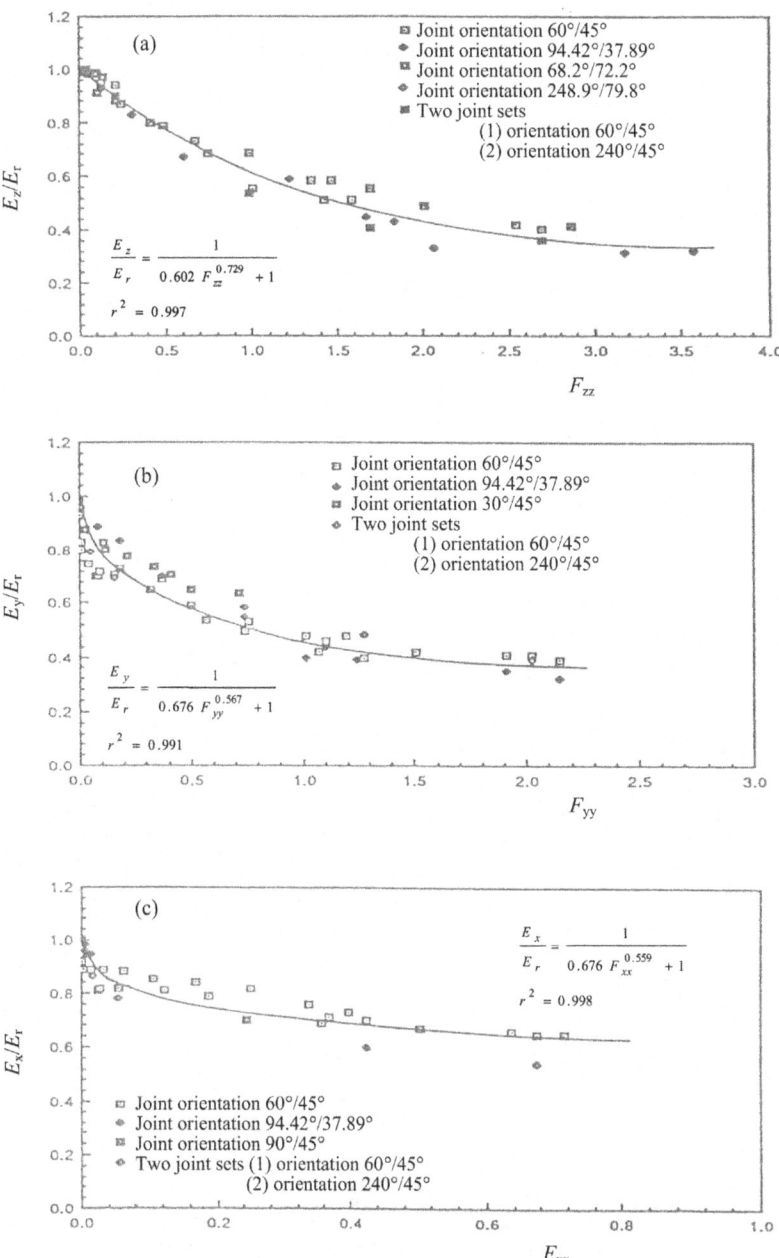

FIG. 6.37 Relations between rock block deformation moduli and fracture tensor components for different discontinuity networks: (A) E_z/E_r vs. F_{zz}; (B) E_y/E_r vs. F_{yy}; and (C) E_x/E_r vs. F_{xx}. *(Based on Kulatilake, P.H.S.W., Wang, S., Stephansson, O., 1993. Effect of finite size joints on the deformability of jointed rock in three dimensions. Int. J. Rock Mech. Min. Sci. Geomech. Abstr. 30 (5), 479–501.)*

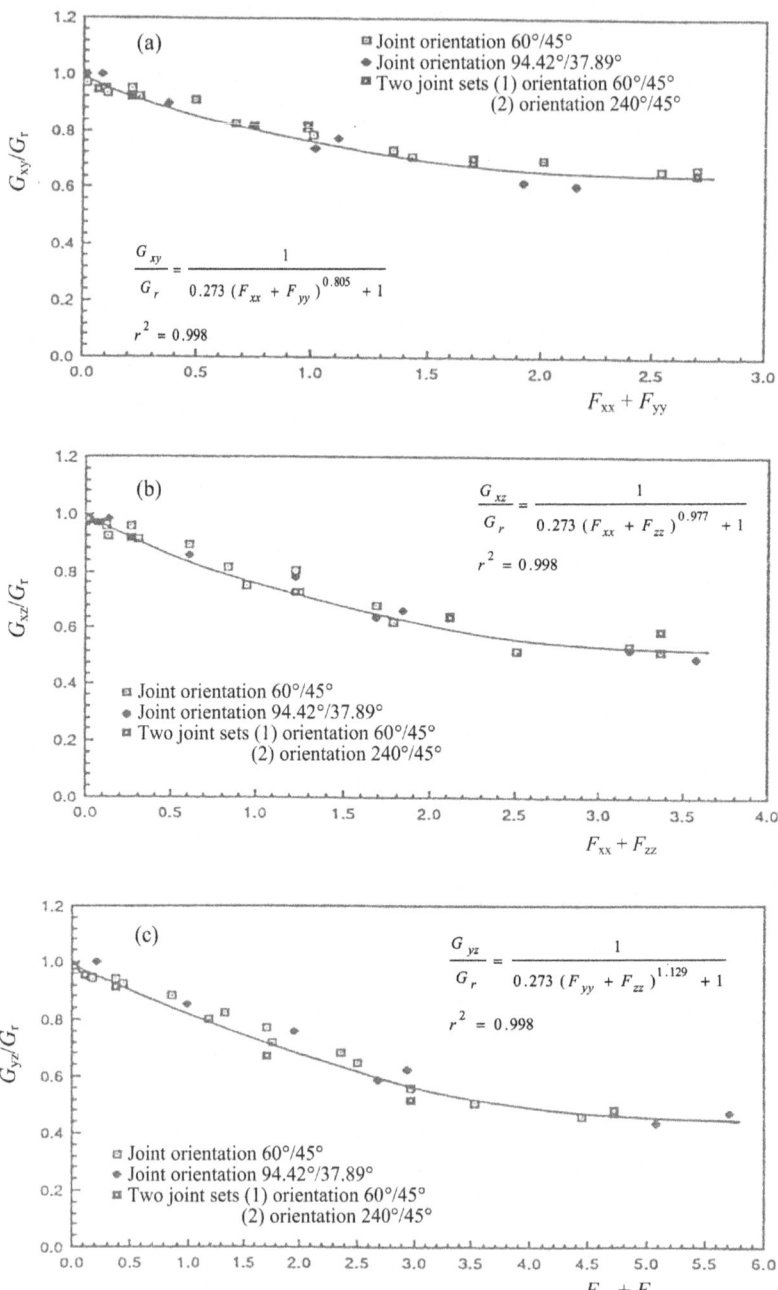

FIG. 6.38 Relations between rock block shear moduli and summation of corresponding fracture tensor components for different discontinuity networks: (A) G_{xy}/G_r vs. $(F_{xx}+F_{yy})$; (B) G_{xz}/G_r vs. $(F_{xx}+F_{zz})$; and (C) G_{yz}/G_r vs. $(F_{yy}+F_{zz})$. *(Based on Kulatilake, P.H.S.W., Wang, S., Stephansson, O., 1993. Effect of finite size joints on the deformability of jointed rock in three dimensions. Int. J. Rock Mech. Min. Sci. Geomech. Abstr. 30 (5), 479–501.)*

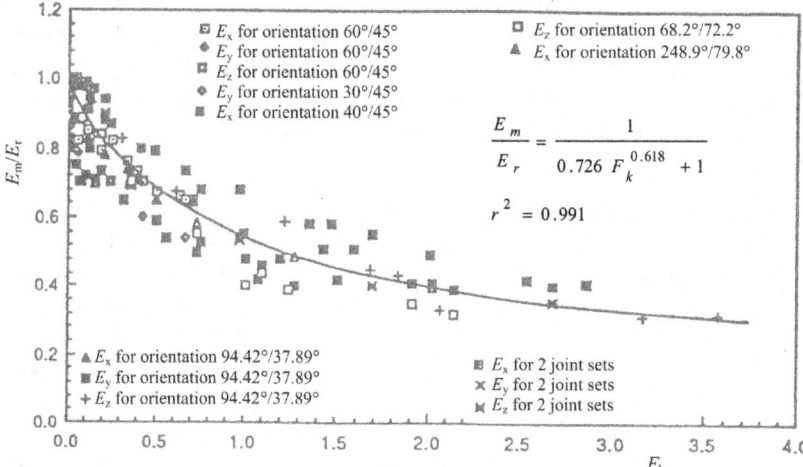

FIG. 6.39 Relations between rock block deformation modulus in any direction E_m and the fracture tensor components in the same direction. *(Based on Kulatilake, P.H.S.W., Wang, S., Stephansson, O., 1993. Effect of finite size joints on the deformability of jointed rock in three dimensions. Int. J. Rock Mech. Min. Sci. Geomech. Abstr. 30 (5), 479–501.)*

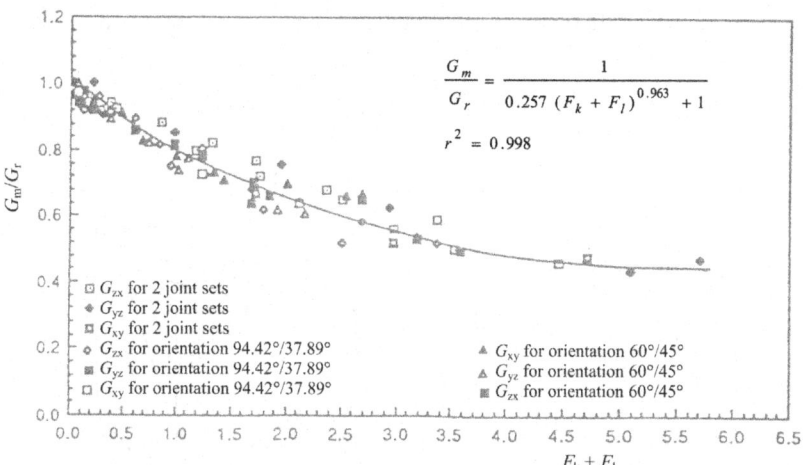

FIG. 6.40 Relations between rock block shear modulus on any plane G_m and the summation of fracture tensor components on that plane. *(Based on Kulatilake, P.H.S.W., Wang, S., Stephansson, O., 1993. Effect of finite size joints on the deformability of jointed rock in three dimensions. Int. J. Rock Mech. Min. Sci. Geomech. Abstr. 30 (5), 479–501.)*

For a rock mass containing nonpersistent discontinuities, Kulatilake et al. (1992, 1993) and Wang (1992) derived relationships between the deformation properties and the fracture tensor parameters based on the DEM analysis results of generated rock mass blocks. However, there exist limitations for the method they used and thus for the relationships they derived:

1. The generated rock mass block is assumed to be orthotropic in the x, y, z directions, regardless of the actual orientations of the discontinuities. The appropriateness of this assumption is questionable. For example, the two blocks shown in Fig. 6.41 have the same fracture tensor F_{ij}, block 1 containing three orthogonal discontinuity sets while block 2 containing only one discontinuity set. It is appropriate to assume that block 1 is orthotropic in the x, y, z directions. However, it is obviously inappropriate to assume that block 2 is orthotropic in the x, y, z directions.

2. To do DEM analysis on the generated rock mass block, fictitious discontinuities are introduced so that when they are combined with actual discontinuities, the block is discretized into polyhedra. To make the fictitious discontinuities behave as the intact rock, appropriate mechanical properties have to be assigned to the fictitious discontinuities. From the investigation performed on 2D rock blocks, Kulatilake et al. (1992) found a relationship between the mechanical properties of the fictitious discontinuities and those of the intact rock. However, even if the mechanical properties of the fictitious discontinuities are chosen from this relationship, the fictitious discontinuities can only approximately behave as the intact rock. So the introduction of fictitious discontinuities brings further errors to the final analysis results.

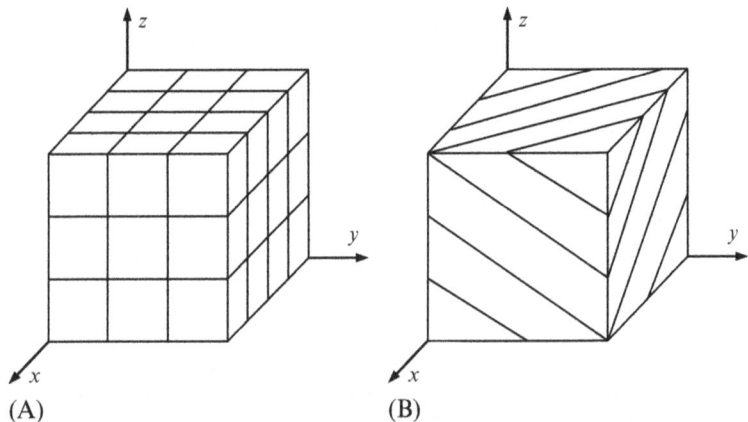

(A)   (B)

FIG. 6.41 Two rock blocks having the same fracture tensor but different discontinuity sets: (A) Rock block with three orthogonal discontinuity sets; and (B) Rock block with one discontinuity set.

3. Discontinuity persistence ratio PR (defined as the ratio of the actual area of a discontinuity to the cross-section area of the discontinuity plane within the rock block) should have a great effect on the deformability of rock masses. However, the relationships derived by Kulatilake et al. (1992, 1993) and Wang (1992) does not show any effect of PR on the deformability of jointed rock masses.

4. The conclusion that E_i/E_r ($i=x, y, z$) is related only to F_{ii} ($i=x, y, z$) is questionable. This can be clearly seen from the two rock blocks shown in Fig. 6.42. The two blocks have the same fracture tensor component F_{zz}. From Fig. 6.39, the two blocks will have the same deformation modulus in the z-direction. However, block 2 is obviously more deformable than block 1 in the z-direction.

6.4.3   Poisson's Ratio of Rock Mass

The presence of discontinuities also influences the Poisson's ratio of rock masses. Although the rock mass deformation modulus E_m can be empirically correlated to the intact rock deformation modulus E_r as shown in the previous subsection, there seems to be no such correlation between the rock mass Poisson's ratio ν_m and the intact rock Poisson's ratio ν_m (Gercek, 2007). By treating a jointed rock mass as an equivalent continuum, its Poisson's ratio can be determined using the method by Kulhawy (1978) (Eq. 6.85), Fossum (1985) (Eq. 6.92) or Zhang (2010) (based on the E_m and G_m from Eqs. (6.95) and (6.96), respectively) Numerical studies such as those by Kulatilake et al. (1992, 1993) and Wang (1992) presented in the previous subsection can also be conducted to predict the value of Poisson's ratio for jointed rock masses. In the majority of

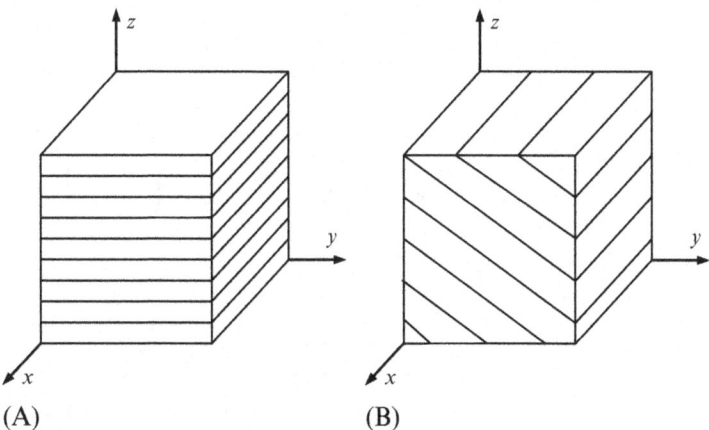

(A)   (B)

FIG. 6.42   Two rock blocks having the same fracture tensor component in z-direction but different discontinuity orientations: (A) Rock block with discontinuity normal parallel to z-axis; and (B) Rock block with discontinuity normal inclined from z-axis.

cases, the values of Poisson's ratio for rock masses are larger than those for intact rocks, and sometimes, unusually high values (>0.5) can be obtained, indicating the anisotropy induced by the discontinuities.

The Poisson's ratio of a rock mass can also be determined by performing in situ tests such as borehole pressure cell test, large flat jack test, plate loading test, and dynamic test in which seismic wave velocities are measured. It needs to be noted that, depending on the specific test method, the volume of rock mass involved can be significantly different. To determine the Poisson's ratio for a very large volume of rock mass, the dynamic test can be conducted; but it needs to be noted that the obtained Poisson's ratio is a dynamic one and needs to be converted to the static one by using an empirical relation such as Eq. (6.23).

6.5   SCALE EFFECT ON ROCK DEFORMABILITY

The scale effect on the deformability of rock masses can be simply seen from the difference between rock mass deformation modulus measured in the field and intact rock modulus measured in the laboratory. Heuze (1980) concluded that the rock mass deformation modulus measured in the field ranges between 20 and 60% of the intact rock modulus measured in the laboratory. The tests on Lac du Bonnet granite specimens of different diameters by Jackson and Lau (1990) showed that the elastic modulus decreases by about 10% when the specimen diameter increases from 45 to 300 mm (Fig. 6.43). Fig. 6.44 shows the variation of measured dynamic modulus (E_{dyn}) with the test volume of rock. The (static) modulus values of the intact rock (E_r) and the rock mass (E_m) are also shown in this figure. One simple and apparent explanation to the

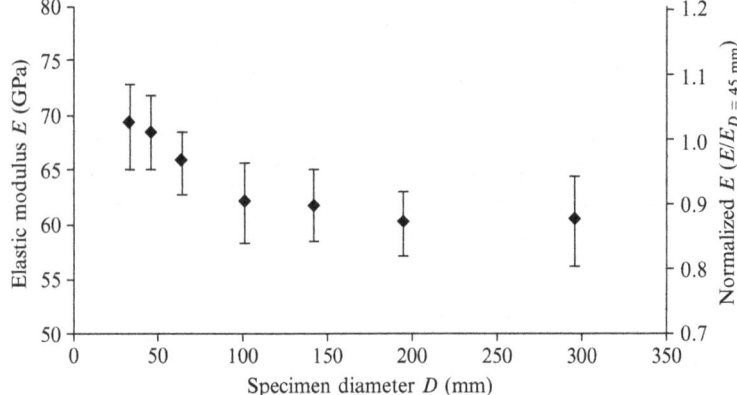

FIG. 6.43   Variation of elastic modulus of Lac du Bonnet granite with specimen diameter (error bars indicate ±1.0 standard deviation from the mean). *(Based on Jackson, R., Lau, J.S.O., 1990. The effect of specimen size on the laboratory mechanical properties of Lac du Bonnet grey granite. In: Cunha, P. (Ed.), Scale Effects in Rock Masses. Balkema, Rotterdam.)*

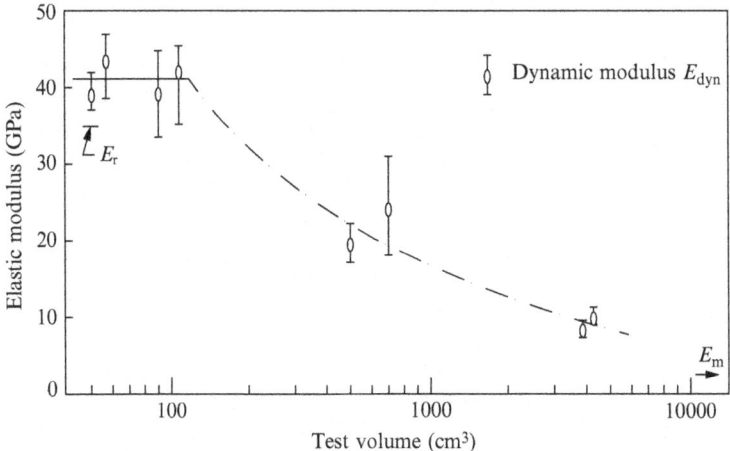

FIG. 6.44   Effect of test volume on the elastic modulus of rock. *(Based on Lo, K.Y., Yung, T.C.B., Lukajic, B., 1987. A field meted for the determination of rock-mass modulus. Can. Geotech. J., Ottawa, Canada, 24, 406–413.)*

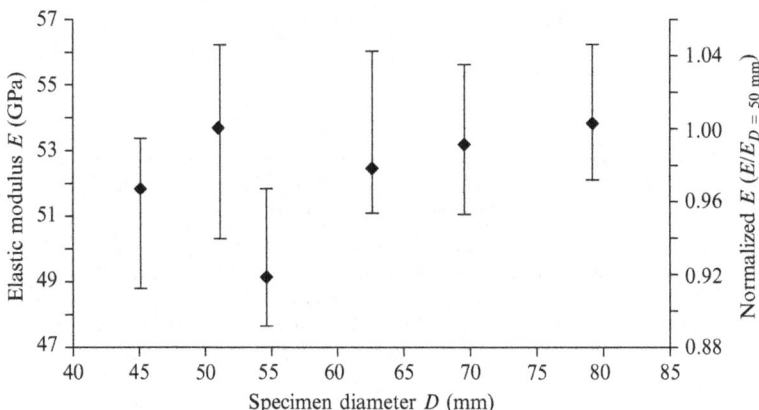

FIG. 6.45   Variation of elastic modulus of limestone with specimen diameter (error bars indicate minimum and maximum values in data set). *(Based on Thuro, K., Plinninger, R.J., Zah, S., Schutz, S., 2001. Scale effects in rock strength properties. Part 1: Unconfined compressive test and Brazilian test. In: Rock Mechanics – A Challenge for Society, ISRM, Espoo, June 3–7, 2001, pp. 169–174.)*

reduction of rock mass deformation modulus with greater test volume is that the effect of discontinuities is included in the rock mass.

It needs to be noted that there are also test results that show no obvious scale effect (Fig. 6.45) or even increase of elastic modulus with larger rock specimen diameter (Fig. 6.46).

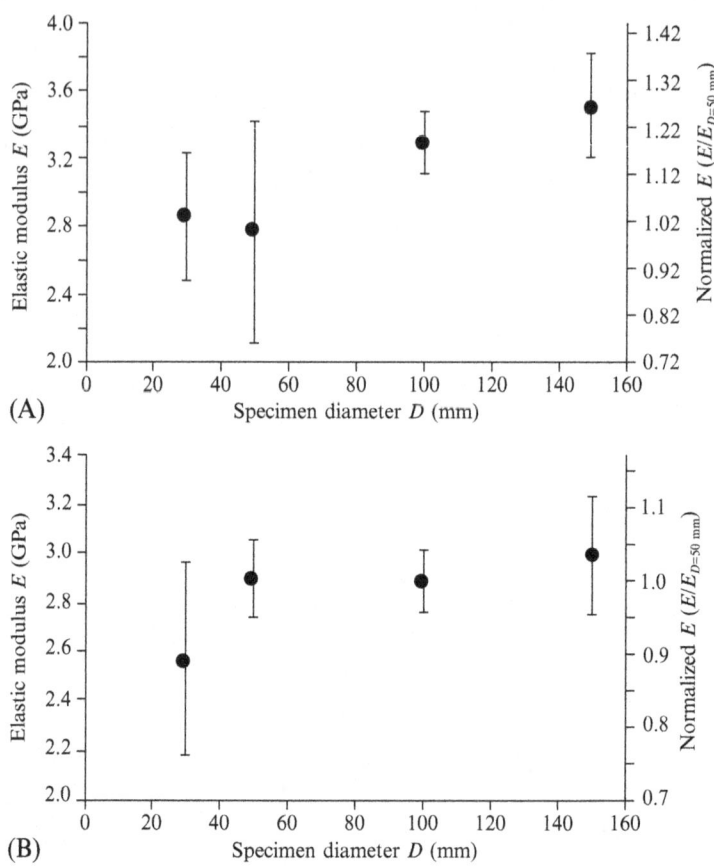

FIG. 6.46 Variation of elastic modulus of Ohya Stone (welded tuff) with specimen diameter (error bars indicate ±1.0 standard deviation from the mean): (A) loaded horizontally to the depositional surface; and (B) loaded vertically to the depositional surface. *(Based on Yuki, N., Aoto, S., Yoshinaka, R., Yoshihiro, O., Terada, M., 1995. The scale and creep effect on the strength of welded tuff. In: Yoshinaka, R., Kikuchi, K. (Eds.), International Workshop on Rock Foundation. Balkema, Tokyo.)*

As for Poisson's ratio, the tests by Jackson and Lau (1990) on Lac du Bonnet granite show a slight decrease with increasing specimen size (Fig. 6.47).

6.6 EFFECT OF CONFINING STRESS ON ROCK DEFORMABILITY

Although the effect of confining stress on rock deformability is not considered in many rock mechanics problems, research results have shown that rock deformation modulus increases significantly with the confining stress

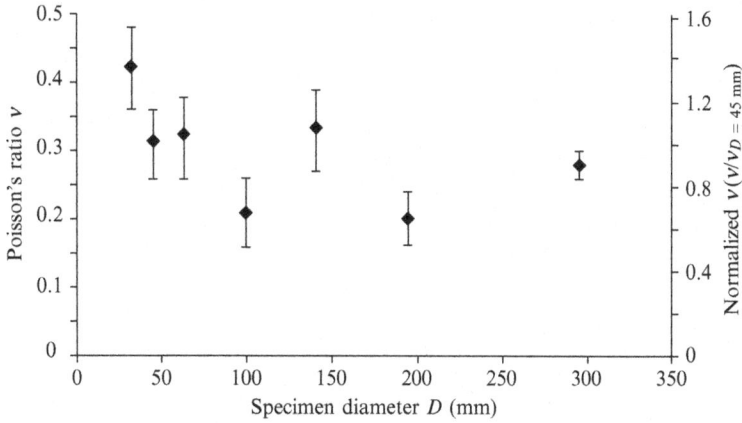

FIG. 6.47 Variation of Poisson's ratio of Lac du Bonnet granite with specimen diameter (error bars indicate ±1.0 standard deviation from the mean). *(Based on Jackson, R., Lau, J.S.O., 1990. The effect of specimen size on the laboratory mechanical properties of Lac du Bonnet grey granite. In: Cunha, P. (Ed.), Scale Effects in Rock Masses. Balkema, Rotterdam.)*

(Gustkiewicz, 1985; Arora, 1987; Zimmerman, 1991; Verman et al., 1997; Asef and Reddish, 2002; Asef and Najibi, 2013). Arora (1987) undertook comprehensive experimental studies on the effect of confining stress on the deformation modulus of jointed rock masses. He conducted triaxial tests on three types of rocks: plaster of Paris, Jamrani sandstone and Agra sandstone at σ_c of 11.3, 55 and 110 MPa, respectively. The test specimens contain clean and rough-broken discontinuities created at various inclinations ranging from 0 to 90 degrees. Using the axial stress versus strain plot, the deformation modulus was calculated at 50% of the maximum stress. Fig. 6.48 shows the normalized deformation modulus against the normalized unconfined strength of the jointed rock mass, leading to the development of the following expression:

$$\frac{E_{m(\sigma_3=0)}}{E_{m(\sigma_3)}} = 1 - \exp\left(-0.1\frac{\sigma_{cm}}{\sigma_3}\right) \tag{6.103}$$

where $E_{m(\sigma_3=0)}$ is the deformation modulus of the jointed rock mass at unconfined stress state; $E_{m(\sigma_3)}$ is the deformation modulus of the jointed rock mass at triaxial stress state with $\sigma_2 = \sigma_3$; and σ_{cm} is the unconfined compressive strength of the jointed rock mass. The equation can also be used for intact rock deformation modulus E_r if σ_{cm} is substituted by σ_c.

Verman et al. (1997) obtained an empirical expression showing the variation of the deformation modulus of rock masses with depth:

$$E_m = 0.4H^\alpha 10^{\frac{RMR-20}{38}} \tag{6.104}$$

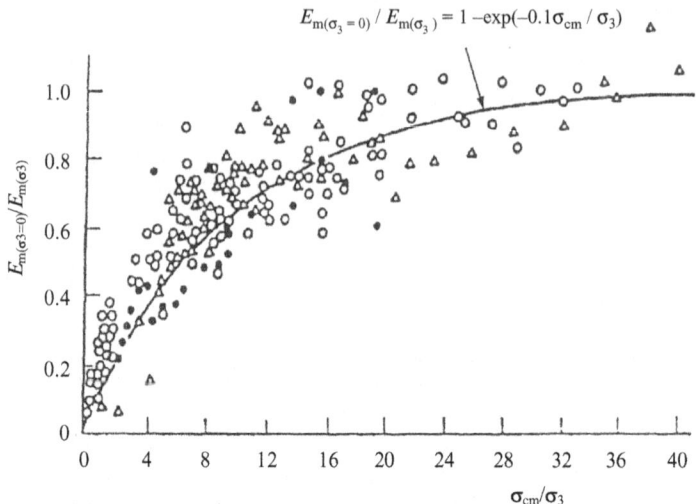

FIG. 6.48 Variation of $E_{m(\sigma3=0)}/E_{m(\sigma3)}$ with σ_{cm}/σ_3. *(From Arora, V.K., 1987. Strength and deformational behavior of jointed rocks. PhD thesis, IIT Delhi, India.)*

where α is a variable depending on RMR ($\alpha=0.3$ and 0.16 respectively at RMR $=68$ and 31); and H is the depth in meters.

Asef and Reddish (2002) showed that Eq. (6.103) significantly overestimates the deformation modulus at a given confining stress when compared with Eq. (6.104). By re-analyzing Arora's original data, Asef and Reddish (2002) derived the following empirical equation:

$$\frac{E_{m(\sigma_3)}}{E_{m(\sigma_3=0)}} = \frac{200\dfrac{\sigma_3}{\sigma_{cm}}+b}{\dfrac{\sigma_3}{\sigma_{cm}}+b} \tag{6.105}$$

where $b=15+\exp(-0.18\sigma_c)$, $E_{m(\sigma3=0)}$ is the deformation modulus of the jointed rock mass at unconfined stress state; $E_{m(\sigma3)}$ is the deformation modulus of the jointed rock mass at triaxial stress state with $\sigma_2=\sigma_3$; σ_{cm} is the unconfined compressive strength of the jointed rock mass; and σ_c is the unconfined compressive strength of the intact rock. The equation can also be used for intact rock deformation modulus E_r if σ_{cm} is substituted by σ_c. Fig. 6.49 is the comparison of Eq. (6.105) with new test data of Asef and Reddish (2002).

The confining stress may also affect the Poisson's ratio of rock as shown in Table 6.8.

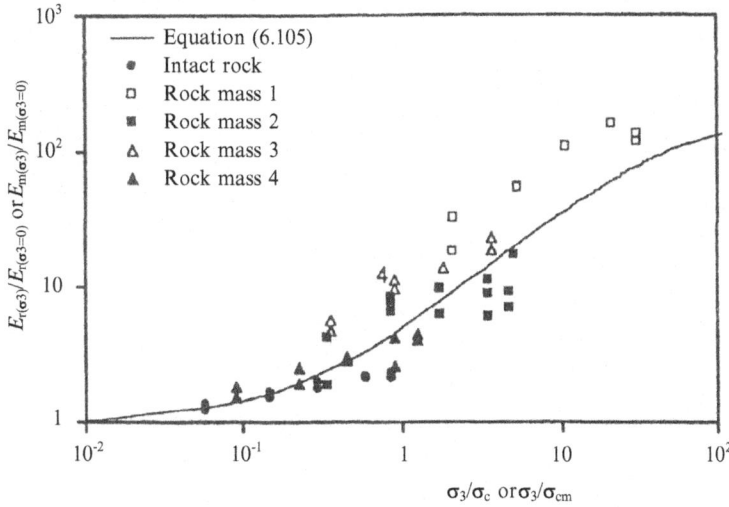

FIG. 6.49 Variation of $E_{r(\sigma3)}/E_{r(\sigma3=0)}$ or $E_{m(\sigma3)}/E_{m(\sigma3=0)}$ with σ_3/σ_c or σ_3/σ_{cm}. *(From Asef, M.R., Reddish, D.J., 2002. The impact of confining stress on the rock mass deformation modulus. Geotechnique, 52, 235–241.)*

6.7 ANISOTROPY OF ROCK DEFORMABILITY

Anisotropy is one of the key aspects of rock properties. Some intact rocks, such as sandstone, shale, limestone, schist, slate, and gneiss belonging to sedimentary and metamorphic groups, show strong deformability anisotropy. Fig. 6.50 shows the anisotropy of elastic modulus for diatomite, siltshale and mudshale under conditions of unconfined compression. The highest and lowest values of the elastic modulus correspond to the directions parallel and perpendicular to the stratification plane, respectively. The tests by Cho et al. (2012) on Asan gneiss, Boryeong shale and Yeoncheon schist also show the anisotropy of elastic modulus in a trend similar to that in Fig. 6.50. The degree of deformability anisotropy can be quantified by the deformability anisotropy ratio R_E defined as

$$R_E = \frac{E_{max}}{E_{min}} \tag{6.106}$$

where E_{max} and E_{min} are the maximum and minimum elastic modulus values, respectively. Table 6.16 lists the values of R_E for different rocks.

Rock masses containing discontinuities also display deformability anisotropy. The equivalent continuum models presented in Section 6.4.2 clearly show the anisotropy of rock mass deformability due to the presence of discontinuities.

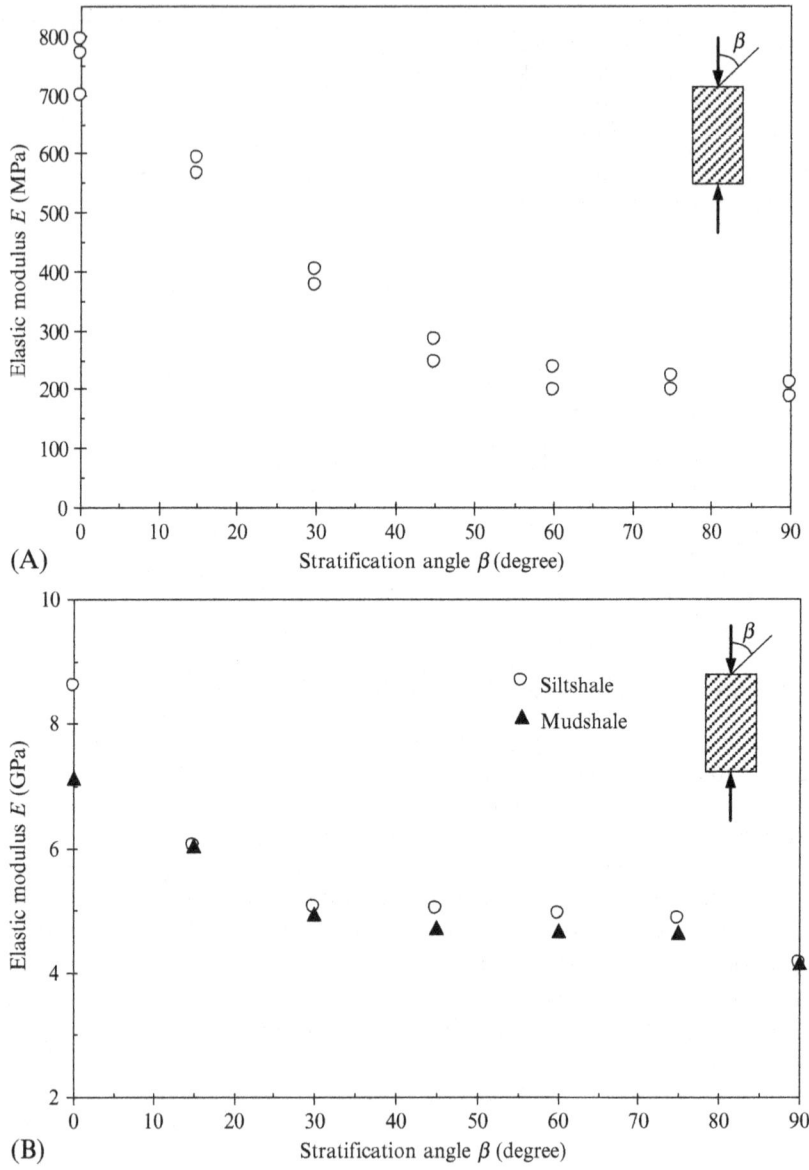

FIG. 6.50 Anisotropy of elastic modulus of intact rocks: (A) Montagne d'Andance diatomite (Data from Allirot and Boehler, 1979); and (B) Siltshale and mudshale (Data from Ajalloeian and Lashkaripour, 2000).

TABLE 6.16 Deformability Anisotropy Ratio R_E for Different Rocks

Rock	Anisotropy Ratio R_E	Reference
Cedillo slate	7.1	Peres Rodrigues (1979)
Rothbach sandstone	5.31	Louis et al. (2005)
Yeoncheon schist	4.0	Cho et al. (2012)
Diatomite	3.81	Allirot and Boehler (1979)
Boryeong shale	2.7	Cho et al. (2012)
Mudstone (dry)	2.06	Angabini (2003)
Siltshale	2.06	Ajalloeian and Lashkaripour (2000)
Mudshale	1.72	Ajalloeian and Lashkaripour (2000)
Asan gneiss	1.7	Cho et al. (2012)
Carboniferous mudstone	1.52	King et al. (1994)
Marble	1.5	Lepper (1949)
Bentheim sandstone	1.45	Louis et al. (2005)
Hast Schist	1.44	Read et al. (1987)
Hornfel	1.31	Chang and Haimson (2005)
Metapelite	1.29	Chang and Haimson (2005)
Adamswiller sandstone	1.26	Gatelier et al. (2002)
Sandstone	1.23	Müller (1930)

REFERENCES

AASHTO, 1989. Standard Specifications for Highway Bridges, 14th ed. American association of state highway and transportation officials, Washington DC.

Ajalloeian, R., Lashkaripour, G.R., 2000. Strength anisotropies in mudrocks. Bull. Eng. Geol. Environ. 59, 195–199.

Alber, M., Heiland, J., 2001. Investigation of a limestone pillar failure, Part 1: geology, laboratory testing and numerical modeling. Rock Mech. Rock. Eng. 34, 167–186.

Alejano, L.R., González, J., Muralha, J., 2012. Comparison of different techniques of tilt testing and basic friction angle variability assessment. Rock Mech. Rock. Eng. 45, 1023–1035.

Allirot, D., Boehler, J.-P., 1979. Evolution of mechanical properties of a stratified rock under confining pressure. In: Proceedings of 4th International Congress Rock Mechanics, vol. 1, Montreux. Balkema, Rotterdam, pp. 15–22 (in French).

Amadei, B., 1983. Brebbia, C.A., Orszag, S.A. (Eds.), Rock Anisotropy and the Theory of Stress Measurements, Lecture Notes in Engineering. Springer-Verlag, Berlin.

Amadei, B., Savage, W.Z., 1989. Anisotropic nature of jointed rock mass strength. J. Geotech. Eng. ASCE 115, 525–542.

Amadei, B., Savage, W.Z., 1993. Effect of joints on rock mass strength and deformability. In: Hudson, J.A. (Ed.), Comprehensive Rock Engineering—Principle, Practice and Projects, vol. 1. Pergamon, Oxford, UK, pp. 331–365.

Angabini, A., 2003. Anisotropy of rock elasticity behavior and of gas migration in a Variscan Carboniferous rock mass in the South Limburg, the Netherlands. Eng. Geol. 67, 353–372.

Arora, V.K., 1987. Strength and deformational behavior of jointed rocks. Ph.D. Thesis, IIT Delhi, India.

ASCE, 1996. Rock Foundations: Technical Engineering and Design Guides as Adapted from the US Army Corps of Engineers. No. 16, ASCE Press, New York, NY.

Asef, M.R., Najibi, A.R., 2013. The effect of confining pressure on elastic wave velocities and dynamic to static Young's modulus ratio. Geophysics 78 (3), D135–D142.

Asef, M.R., Reddish, D.J., 2002. The impact of confining stress on the rock mass deformation modulus. Geotechnique 52, 235–241.

ASTM, 2004. Annual book of ASTM standards. In: Soil and Rock.American Society for Testing and Materials, vol. 4.08. West Conshohocken, Philadelphia, PA.

Aufmuth, R.E., 1973. A systematic determination of engineering criteria for rocks. Bull. Assoc. Eng. Geol. 11, 235–245.

Aydan, Ö., Sato, A., Yagi, M., 2014. The inference of geo-mechanical properties of soft rocks and their degradation from needle penetration tests. Rock Mech. Rock. Eng. 47, 1867–1890.

Bandis, S.C., Lumsden, A.C., Barton, N.R., 1981. Experimental studies of scale effects on the shear behavior of rock joints. Int. J. Rock Mech. Min. Sci. Geomech. Abstr. 18, 1–21.

Bandis, S.C., Lumsden, A.C., Barton, N.R., 1983. Fundamentals of rock joint deformation. Int. J. Rock Mech. Min. Sci. Geomech. Abstr. 20, 249–268.

Barton, N., 1987. Predicting the behavior of underground openings in rock. In: Manuel Rocha Memorial Lecture. Norwegian Geotech. Inst., Lisbon, Oslo.

Barton, N., 2002. Some new Q value correlations to assist in site characterization and tunnel design. Int. J. Rock Mech. Min. Sci. Geomech. Abstr. 39, 185–216.

Barton, N., Bandis, S.C., 1990. Review of predictive capabilities of JRC-JCS model in engineering practice. In: Barton, N., Stephansson, O. (Eds.), Proceedings of the International Symposium on Rock Joints. Loen, Norway. Balkema, Rotterdam, pp. 603–610.

Barton, N., Choubey, V., 1977. The shear strength of rock joints in theory and practice. Rock Mech. 10, 1–54.

Barton, N., Loset, F., Lien, R., Lunde, J., 1980. Application of the Q-system in design decisions. In: Bergman, M. (Ed.), Subsurface Space. vol. 2, pp. 553–561.

Beiki, M., Bashari, A., Majdi, A., 2010. Genetic programming approach for estimating the deformation modulus of rock mass using sensitivity analysis by neural network. Int. J. Rock Mech. Min. Sci. 47, 1091–1103.

Belikov, B.P., Alexandrov, K.S., Rysova, T.W., 1970. Uprugie Svoistva Porodo-Obrasujscich Mineralov I Gornich Porod, Izdat. Nauka, Moskva.

Beverly, B.E., Schoenwolf, D.A., Brierly, G.S., 1979. Correlations of Rock Index Values With Engineering Properties and the Classification of Intact Rock. FHWA, Washington, DC.

Bieniawski, Z.T., 1978. Determining rock mass deformability: experience from case histories. Int. J. Rock Mech. Min. Sci. Geomech. Abstr. 15, 237–248.

Chang, C., Haimson, B., 2005. Non-dilatant deformation and failure mechanism in two Long Valley Caldera rocks under true triaxial compression. Int. J. Rock Mech. Min. Sci. 42, 402–414.

Chen, E.P., 1989. A constitutive model for jointed rock mass with orthogonal sets of joints. J. Appl. Mech. ASME 56, 25–32.

Chen, Y.-L., Ni, J., Shao, W., Azzam, R., 2012. Experimental study on the influence of temperature on the mechanical properties of granite under uni-axial compression and fatigue loading. Int. J. Rock Mech. Min. Sci. 56, 62–66.

Cho, J.-W., Kim, H., Jeon, S., Min, K.-B., 2012. Deformation and strength anisotropy of Asan gneiss, Boryeong shale, and Yeoncheon schist. Int. J. Rock Mech. Min. Sci. 50, 158–169.

Coon, R.F., Merritt, A.H., 1970. Predicting in situ modulus of deformation using rock quality indices. In: Determination of the In Situ Modulus of Deformation of Rock, ASTM STP 477, pp. 154–173.

Deere, D.U., Miller, R.P., 1966. Engineering classification and index properties for intact rock. Technical Report No. AFWL-TR-65-116, Air Force Weapons Lab, Kirtland Air Base, New Mexico.

Deere, D.U., Hendron, A.J., Patton, F.D., Cording, E.J., 1967. Design of surface and near surface construction in rock. In: Fairhurst, C. (Ed.), Failure and Breakage of Rock, Proceedings of 8th U.S. Symposium on Rock Mechanics, pp. 237–302.

Dershowitz, W.S., Baecher, G.B., Einstein, H.H., 1979. Prediction of rock mass deformability. In: Proceedings 4th International Congress on Rock Mechanics, Montreal, Canada, vol. 1, pp. 605–611.

Dincer, I., Acar, A., Cobanoglu, I., Uras, Y., 2004. Correlation between Schmidt hardness, uniaxial compressive strength and Young's modulus for andesites, basalts and tuffs. Bull. Eng. Geol. Environ. 63, 141–148.

Duncan, J.M., Chang, C.Y., 1970. Non-linear analysis of stress and strain in soils. J. Soil Mech. Found. Div. Am. Soc. Civ. Eng. 96, 1629–1655.

Dwivedi, R.D., Goel, R.K., Prasad, V.R.R., Sinha, A., 2008. Thermo-mechanical properties of Indian and other granites. Int. J. Rock Mech. Min. Sci. 45, 303–315.

Ebisu, S., Aydan, O., Komura, S., Kawamoto, T., 1992. Comparative study on various rock mass characterization methods for surface structures. In: Hudson, J.A. (Ed.), Rock Characterization: ISRM Symposium, Eurock '92, Chester, UK, 14–17 September 1992, pp. 203–208.

Eissa, E.A., Kazi, A., 1988. Relation between static and dynamic Young's moduli of rocks. Int. J. Rock Mech. Min. Sci. Geomech. Abstr. 29, 479–482.

El-Naqa, A., Kuisi, M.A., 2002. Engineering geological characterisation of the rock masses at Tannur Dam site. S. Jordan. Environ. Geol. 42, 817–826.

Fossum, A.F., 1985. Effective elastic properties for a randomly jointed rock mass. Int. J. Rock Mech. Min. Sci. Geomech. Abstr. 22, 467–470.

Gardner, W.S., 1987. Design of drilled piers in the Atlantic Piedmont. In: Smith, R.E. (Ed.), Foundations and Excavations in Decomposed Rock of the Piedmont Province, GSP No. 9, pp. 62–86, ASCE.

Gatelier, N., Pellet, F., Loret, B., 2002. Mechanical damage of an anisotropic porous rock in cyclic triaxial tests. Int. J. Rock Mech. Min. Sci. 39, 249–257.

Gercek, H., 2007. Poisson's ratio values for rocks. Int. J. Rock Mech. Min. Sci. 44, 1–13.

Gerrard, C.M., 1982a. Elastic models of rock masses having one, two and three sets of joints. Int. J. Rock Mech. Min. Sci. Geomech. Abstr. 19, 15–23.

Gerrard, C.M., 1982b. Joint compliances as a basis for rock mass properties and the design of supports. Int. J. Rock Mech. Min. Sci. Geomech. Abstr. 19, 285–305.

Gerrard, C.M., 1991. The equivalent elastic properties of stratified and jointed rock masses. In: Beer, G., Brooker, J.R., Carter, J.P. (Eds.), Proceedings of International Conference on Computer Methods and Advances in Geomechanics, Cairns. Balkema, Rotterdam, pp. 333–337.

Gokceoglu, C., Sonmez, H., Kayabasi, A., 2003. Predicting the deformation moduli of rock masses. Int. J. Rock Mech. Min. Sci. 40, 701–710.

Goodman, R.E., Taylor, R.L., Brekke, T.L., 1968. A model for the mechanics of jointed rock. J. Soil Mech. Found. Div. Am. Soc. Civ. Eng. 96, 637–659.

Gustkiewicz, J., 1985. Deformation and failure of the Rowa Ruda sandstone in a three-axial state of stress with gas under pressure in the pores. Archiwum Gornictwa 30, 401–424.

Harrison, J.P., 1999. Selection of the threshold value in RQD assessments. Int. J. Rock Mech. Min. Sci. 36, 673–685.

Heuze, F.E., 1980. Scale effects in the determination of rock mass strength and deformability. Rock Mech. 12, 167–192.

Heuze, F.E., 1983. High-temperature mechanical, physical and thermal properties of granitic rocks – a review. Int. J. Rock Mech. Min. Sci. Geomech. Abstr. 20, 3–10.

Hoek, E., 2004. Personal communication.

Hoek, E., Brown, E.T., 1997. Practical estimates of rock mass strength. Int. J. Rock Mech. Min. Sci. 34, 1165–1186.

Hoek, E., Diederichs, M.S., 2006. Empirical estimation of rock mass modulus. Int. J. Rock Mech. Min. Sci. 36, 203–215.

Homand-Etienne, H., Houpert, R., 1989. Thermally induced micro-cracking in granites: characterization and analysis. Int. J. Rock Mech. Min. Sci. Geomech. Abstr. 26, 125–134.

ISRM, 1978. Suggested methods for the quantitative description of discontinuities in rock masses. International Society for Rock Mechanics, Commission on Standardization of Laboratory and Field Tests. Int. J. Rock Mech. Min. Sci. Geomech. Abstr. 15, 319–368.

ISRM, 1979. Suggested methods for determining the uniaxial compressive strength and deformability of rock materials. International Society for Rock Mechanics, Commission on Standardization of Laboratory and Field Tests. Int. J. Rock Mech. Min. Sci. Geomech. Abstr. 16, 135–140.

Jackson, R., Lau, J.S.O., 1990. The effect of specimen size on the laboratory mechanical properties of Lac du Bonnet grey granite. In: Cunha, P. (Ed.), Scale Effects in Rock Masses. Balkema, Rotterdam.

Johnston, I.W., Donald, I.B., Bennett, A.G., Edwards, J., 1980. The testing of large diameter pile rock sockets with a retrievable test rig. In: Proceedings, 3rd Australia-New Zealand Conference on mechanics, Wellington, 1, pp. 105–108.

Justo, J.L., Justo, E., Durand, P., Azanon, J.M., 2006. The foundation of a 40-storey tower in jointed basalt. Int. J. Rock Mech. Min. Sci. 43, 267–281.

Karakul, H., Ulusay, R., 2013. Empirical correlations for predicting strength properties of rocks from P-wave velocity under different degrees of saturation. Rock Mech. Rock. Eng. 46, 981–999.

Katz, O., Reches, Z., Roegiers, J.-C., 2000. Evaluation of mechanical rock properties using a Schmidt hammer. Int. J. Rock Mech. Min. Sci. 37, 723–728.

Kayabasi, A., Gokceoglu, C., Ercanoglu, M., 2003. Estimating the deformation modulus of rock masses: a comparative study. Int. J. Rock Mech. Min. Sci. 40, 55–63.

King, M.S., 1983. Static and dynamic elastic properties of rock from the Canadian shield. Int. J. Rock Mech. Min. Sci. Geomech. Abstr. 20, 237–241.

King, M.S., Andrea, M.O., Shams, K.M., 1994. Velocity anisotropy of carboniferous mudstones. Int. J. Rock Mech. Min. Sci. Geomech. Abstr. 31, 261–263.

Kulatilake, P.H.S.W., Ucpirti, H., Wang, S., Radberg, G., Stephansson, O., 1992. Use of the distinct element method to perform stress analysis in rock with non-persistent joints and to study the effect of joint geometry parameters on the strength and deformability of rock. Rock Mech. Rock. Eng. 25, 253–274.

Kulatilake, P.H.S.W., Wang, S., Stephansson, O., 1993. Effect of finite size joints on the deformability of jointed rock in three dimensions. Int. J. Rock Mech. Min. Sci. Geomech. Abstr. 30 (5), 479–501.

Kulhawy, F.H., 1978. Geomechanical model for rock foundation settlement. J. Geotech. Eng. 104 (2), 211–227.

Kwasniewski, M., Oitaben, P.R., 2009. Effect of water on the deformability of rocks under uniaxial compression. In: ISRM Regional Symposium EUROCK2009: Rock Engineering in Difficult Ground Conditions-Soft Rock and Karst, Dubrovnik, Cavtat, Croatia, pp. 271–276.

Lashkaripour, G.R., 2002. Predicting mechanical properties of mudrock from index parameters. Bull. Eng. Geol. Environ. 61, 73–77.

Lee, Y.H., Carr, J.R., Barr, D.J., Haas, C.J., 1990. The fractal dimension as measure of the roughness of rock discontinuity profiles. Int. J. Rock Mech. Min. Sci. Geomech. Abstr. 27, 453–464.

Leite, M.H., Ferland, F., 2001. Determination of unconfined compressive strength and Young's modulus of porous materials by indentation tests. Eng. Geol. 59, 267–280.

Lepper Jr., H.A., 1949. Compression tests on oriented specimens of Yule marble. Am. J. Sci. 247, 570–575.

Louis, L., David, C., Metz, V., Robion, P., Menéndez, B., Kissel, C., 2005. Microstructural control on the anisotropy of elastic and transport properties in undeformed sandstones. Int. J. Rock Mech. Min. Sci. 42, 911–923.

McCann, D.M., Entwisle, D.C., 1992. Determination of Young's modulus of the rock mass from geophysical well logs. In: Hurst, A., Griffiths, C.M., Worthington, P.F. (Eds.), Geological Applications of Wireline Logs II. Geological Society Special Publication No. 65, pp. 317–325.

McLaren, J.R., Titchel, I., 1981. Physical properties of granite relevant to near field conditions in a nuclear waste depository. In: AERE, Harwell, Report AERE-R-10046.

McWilliams, P.C., Kerkering, J.C., Miller, S.M., 1990. Fractal characterization of rock fracture roughness for estimating shear strength. In: Rossmanith, H.P. (Ed.). Proceedings of International Conference on Mechanics of Jointed and Faulted Rock, Vienna, Aurtria. Balkema, Rotterdam, pp. 331–336.

Mitri, H.S., Edrissi, R., Henning, J., 1994. Finite element modeling of cable-bolted stopes in hard-rock underground mines. In: SME Annual Meeting, Albuquerque, New Mexico, pp. 94–116.

Müller, O., 1930. Untersuchungen an Karbongesteinen zur Klärung von Gebirgsdruckfragen. Glückauf 66, 1601–1612.

Nicholson, G.A., Bieniawski, Z.T., 1990. A nonlinear deformation modulus based on rock mass classification. Int. J. Min. Geol. Eng. 8, 181–202.

Oda, M., 1982. Fabric tensor for discontinuous geological materials. Soils Found., Tokyo, Japan, vol. 22, pp. 96–108.

Oda, M., 1988. An experimental study of the elasticity of mylonite rock with random cracks. Int. J. Rock Mech. Min. Sci. Geomech. Abstr. 25, 59–69.

Oda, M., Suzuki, K., Maeshibu, T., 1984. Elastic compliance for rock-like materials with random cracks. Soils Found., Tokyo, Japan, vol. 24, pp. 27–40.

Odling, N.E., 1994. Natural fracture profiles, fractal dimension and joint roughness coefficients. Rock Mech. Rock. Eng. 27, 135–153.

Ohen, H.A., 2003. Calibrated wireline mechanical rock properties method for predicting and preventing wellbore collapse and sanding. SPE 82236.

Ozsan, A., Akin, M., 2002. Engineering geological assessment of the proposed Urus Dam, Turkey. Eng. Geol. 66, 271–281.

Ozsan, A., Ocal, A., Akin, M., Bassarir, H., 2007. Engineering geological appraisal of the Sulakyurt dam site, Turkey. Bull. Eng. Geol. Environ. 66, 483–492.

Palchik, V., 2011. On the ratios between elastic modulus and uniaxial compressive strength of heterogeneous carbonate rocks. Rock Mech. Rock. Eng. 44, 121–128.

Palchik, V., Hatzor, Y.H., 2002. Crack damage stress as a composite function of porosity and elastic matrix stiffness in dolomites and limestones. Eng. Geol. 63, 233–245.

Palmström, A., Singh, R., 2001. The deformation modulus of rock masses—comparisons between in situ tests and indirect estimates tunnelling. Tunn. Undergr. Space Technol. 16, 115–131.

Pappalardo, G., 2015. Correlation between P-wave velocity and physical–mechanical properties of intensely jointed dolostones, Peloritani Mounts, NE Sicily. Rock Mech. Rock. Eng. 48, 1711–1721.

Peres Rodrigues, F., 1979. The anisotropy of the moduli of elasticity and of the ultimate stresses in rocks. In: Proceedings of 4th International Congress Rock Mechanics, vol. 2, Montreux. Balkema, Rotterdam, pp. 517–523.

Poon, C.Y., Sayles, R.S., Jones, T.A., 1992. Surface measurement and fractal characterization of naturally fractured rocks. J. Phys. D: Appl. Phys. 25, 1269–1275.

Priest, S.D., 1993. Discontinuity Analysis for Rock Engineering. Chapman & Hall, London.

Rabbani, E., Sharif, F., KoolivandSalooki, M., Moradzadeh, A., 2012. Application of neural network technique for prediction of uniaxial compressive strength using reservoir formation properties. Int. J. Rock Mech. Min. Sci. 56, 100–111.

Radhakrishnan, R., Leung, C.F., 1989. Load transfer behavior of rock-socketed piles. J. Geotech. Eng. ASCE 115 (6), 755–768.

Ramamurthy, T., 2004. A geo-engineering classification for rocks and rock masses. Int. J. Rock Mech. Min. Sci. 41, 89–101.

Read, S.A.L., Perrin, N.D., Brown, I.R., 1987. Measurement and analysis of laboratory strength and deformability characteristics of schistose rocks. In: Proceedings of 6th International Conference on Rock Mechanics, vol. 1, Montreal, Canada, pp. 233–238.

Read, S.A.L., Richards, L.R., Perrin, N.D., 1999. Applicability of the Hoek–Brown failure criterion to New Zealand greywacke rocks. In: Vouille, G., Berest, P. (Eds.), Proceedings of the 9th International Congress on Rock Mechanics, vol. 2 Paris, pp. 655–660.

Roberds, W.J., Iwano, M., Einstein, H.H., 1990. Probabilistic mapping of rock joint surfaces. In: Barton, N., Stephanson, O. (Eds.). Proceedings International Symposium on Rock Joints, Loen, Norway. Balkema, Rotterdam, pp. 681–691.

Rowe, R.K., Armitage, H.H., 1984. The Design of Piles Socketed into Weak Rock. Faculty of Engineering Science, the University of Western Ontario, London, Ont., Research Report GEOT-11-84.

Sachpazis, C.I., 1990. Correlating Schmidt hardness with compressive strength and Young's modulus of carbonate rocks. Bull. Int. Assoc. Eng. Geol. 42, 75–84.

Schultz, R.A., 1996. Relative scale and the strength and deformability of rock masses. J. Struct. Geol. 18, 1139–1149.

Serafim, J.L., Pereira, J.P., 1983. Considerations of the geomechanics classification of Bieniawski. In: Proceedings International Symposium Engineering Geology and Underground Construction, vol. 1, Lisbon, II33–II42.

Shalabi, F.I., Cording, E.J., Al-Hattamleh, O.H., 2007. Estimation of rock engineering properties using hardness tests. Eng. Geol. 90, 138–147.

Singh, B., 1973. Continuum characterization of jointed rock masses. Part I—the constitutive equations. Int. J. Rock Mech. Min. Sci. Geomech. Abstr. 10, 311–335.

Sonmez, H., Gokceoglu, C., Ulusay, R., 2004. Indirect determination of the modulus of deformation of rock masses based on the GSI system. Int. J. Rock Mech. Min. Sci. 41, 849–857.

Sonmez, H., Gokceoglu, C., Nefeslioglu, H.A., Kayabasi, A., 2006. Estimation of rock modulus. Int. J. Rock Mech. Min. Sci. 43, 224–235.

Stacey, T.R., van Veerden, W.L., Vogler, U.W., 1987. Properties of intact rock. In: Bell, F.G. (Ed.), Ground Engineer's Reference Book. Butterworths, London, UK.

Stephans, R.E., Banks, D.C., 1989. Moduli of deformation studies of the foundation and abutments of the Portugues Dam—Puerto Rico. In: Rock Mechanics as a Guide for Efficient Utilization of Natural Resources: Proceedings of 30th U.S. Symposium on Rock Mechanics, Morgantown, WV, pp. 31–38.

Stimpson, B., 1981. A suggested technique for determining the basic friction angle of rock surfaces using core. Int. J. Rock Mech. Min. Sci. Geomech. Abstr. 18, 63–65.

Tse, R., Cruden, D.M., 1979. Estimating joint roughness coefficients. Int. J. Rock Mech. Min. Sci. Geomech. Abstr. 16, 303–307.

Vasarhelyi, B., 2003. Some observations regarding the strength and deformability of sandstones in case of dry and saturated conditions. Bull. Eng. Geol. Environ. 62, 245–249.

Vasarhelyi, B., 2005. Statistical analysis of the influence of water content on the strength of the Miocene limestone. Rock Mech. Rock. Eng. 38, 69–76.

Verman, M., Singh, B., Viladkar, M.N., Jethwa, J.L., 1997. Effect of tunnel depth on modulus of deformation of rock mass. Rock Mech. Rock. Eng. 30, 121–127.

Voss, R., 1988. Fractals in nature. In: Peitgen, H., Saupe, D. (Eds.), The Science of Fractal Images. Springer, New York, pp. 21–69.

Wang, S., 1992. Fundamental studies of the deformability and strength of jointed rock masses at three dimensional level. Ph.D. Dissertation, University of Arizona, Tucson.

Wyllie, D.C., 1999. Foundations on Rock, second ed. E & FN Spon, London/New York.

Xu, S., Grasso, P., Mahtab, A., 1990. Use of Schmidt hammer for estimating mechanical properties of weak rock. In: 6th International IAEG Congress. Balkema, Rotterdam, pp. 511–519.

Yasar, E., Erdoğan, Y., 2004. Correlating sound velocity with the density, compressive strength and Young's modulus of carbonate rocks. Int. J. Rock Mech. Min. Sci. 41, 871–875.

Yilmaz, I., 2010. Influence of water content on the strength and deformability of gypsum. Int. J. Rock Mech. Min. Sci. 47, 342–347.

Yilmaz, I., Sendir, H., 2002. Correlation of Schmidt hardness with unconfined compressive strength and Young's modulus in gypsum from Sivas (Turkey). Eng. Geol. 66, 211–219.

Yilmaz, I., Yuksek, G., 2009. Prediction of the strength and elasticity modulus of gypsum using multiple regression, ANN, and ANFIS models. Int. J. Rock Mech. Min. Sci. 46, 803–810.

Yoshinaka, R., Yambe, T., 1986. Joint stiffness and the deformation behavior of discontinuous rock. Int. J. Rock Mech. Min. Sci. Geomech. Abstr. 23, 19–28.

Yu, X., Vayssade, B., 1990. Joint profiles and their roughness parameters. In: Barton, N., Stephanson, O. (Eds.), Proceedings of International Symposium on Rock Joints, Loen, Norway. Balkema, Rotterdam, pp. 781–785.

Zhang, L., 2010. Method for estimating the deformability of heavily jointed rock masses. J. Geotech. Geoenviron. 136 (9), 1242–1250.

Zhang, L., Einstein, H.H., 2004. Estimating the deformation modulus of rock masses. Int. J. Rock Mech. Min. Sci. 41, 337–341.

Zhang, L., Mao, X., Liu, R., Guo, X., Ma, D., 2014. The mechanical properties of mudstone at high temperatures: an experimental study. Rock Mech. Rock. Eng. 47, 1479–1484.

Zimmer, M., 2003. Seismic velocities in unconsolidated sands: measurements of pressure, sorting and compaction effects. Ph.D. Thesis, Stanford University.

Zimmerman, R.R., 1991. Deformability of sandstones. Developments in Petroleum Science. Elsevier, Amsterdam.

Chapter 7

Strength

7.1 INTRODUCTION

Determining the strength of rock masses is one of the major problems confronting designers of engineering structures on or in rock. The strength of a rock mass depends not only on the rock material (intact rock), but also on the discontinuities that separate the intact rock blocks. Because of the discontinuities, a rock mass almost always has significantly lower strength than the corresponding intact rock.

Unconfined compressive strength is the most commonly used measurement of rock strength. Tensile strength is also used as a measurement of rock strength in many cases. The different correlations for estimating the unconfined compressive and tensile strengths of intact rocks and rock masses are presented in Sections 7.2 and 7.4, respectively. Since rocks are seldom loaded in only one direction, the strength criteria considering the effect of minor principal stress (ie, the two-dimensional (2D) strength criteria) for intact rocks and rock masses are also presented in the two sections. In Section 7.5, the strength criteria considering the effect of both the minor and medium principal stresses (ie, the three-dimensional (3D) strength criteria) are discussed.

In many cases, the behavior of a rock mass is controlled by sliding along discontinuities. To analyze the stability of a rock mass, it is necessary to know the shear strength of discontinuities. So several shear strength models for rock discontinuities are presented in Section 7.3.

Rock masses usually exhibit a strain-softening postpeak behavior and it is important to determine the residual strength of rock masses in order to properly design engineering structures in or on rock. Section 7.6 discusses the evaluation of residual strength of rock masses.

Rocks show strong scale-dependent and anisotropic strength properties. The scale effect on and anisotropy of rock strength are briefly discussed in Sections 7.7 and 7.8, respectively.

7.2 STRENGTH OF INTACT ROCK

7.2.1 Unconfined Compressive Strength of Intact Rock

The typical ranges of the unconfined compressive strength of different rocks are listed in Table 7.1. The procedure for measuring the unconfined compressive

Engineering Properties of Rocks. http://dx.doi.org/10.1016/B978-0-12-802833-9.00007-9

TABLE 7.1 Typical Range of Unconfined Compressive Strength of Intact Rocks

Rock Category	General Description	Rock	Unconfined Compressive Strength, $\sigma_c{}^a$ (MPa)
A	Carbonate rocks with well-developed crystal cleavage	Dolostone	33–310
		Limestone	24–290
		Carbonatite	38–69
		Marble	38–241
		Tactite-Skarn	131–338
B	Lithified argillaceous rock	Argillite	29–145
		Claystone	1–8
		Marlstone	52–193
		Phyllite	24–241
		Siltstone	10–117
		Shale[b]	7–35
		Slate	145–207
C	Arenaceous rocks with strong crystals and poor cleavage	Conglomerate	33–221
		Sandstone	67–172
		Quartzite	62–379
D	Fine-grained igneous crystalline rock	Andesite	97–179
		Diabase	21–572
E	Coarse-grained igneous and metamorphic crystalline rock	Amphibolite	117–276
		Gabbro	124–310
		Gneiss	24–310
		Granite	14–338
		Quartz diorite	10–97
		Quartz monozonite	131–159
		Schist	10–145
		Syenite	179–427

[a]Range of unconfined compressive strength reported by various investigators.
[b]Not including oil shale.
From AASHTO, 1996. Standard Specifications for Highway Bridges, 16th ed. American Association of State Highway and Transportation Officials, Washington, DC.

strength has been standardized by both the American Society for Testing and Materials (ASTM) and the International Society for Rock Mechanics (ISRM). Although the method is relatively simple, it is time consuming and expensive; also, it requires well-prepared rock cores, which is often difficult for weak rocks and especially for shales. Therefore, indirect tests such as point load, Schmidt hammer, needle penetration, cone indenter and seismic wave velocity tests are often carried out to estimate the unconfined compressive strength based on empirical correlations.

7.2.1.1   Unconfined compressive strength versus porosity

Porosity has a great effect on the strength of intact rocks. The unconfined compressive strength of intact rocks decreases with increasing porosity. Rshewski and Novik (1978) recommended the following relationship between them:

$$\sigma_c = \alpha(1 - \beta n)^2 \tag{7.1}$$

where n is the porosity in the unit of %; and α and β are constants which can be obtained by fitting analysis of experimental data. For limestone, they found $\alpha = 277$ MPa and β between 0.02 and 0.05. The same type of relationship was also used by Nabaei et al. (2010) for sandstones with $\alpha = 254$ MPa and $\beta = 0.027$, and carbonates with $\alpha = 276$ MPa and $\beta = 0.03$.

Tuğrul and Zarif (1999) derived the following simple linear relation between unconfined compressive strength σ_c and porosity n for granitic rocks from Turkey:

$$\sigma_c = 183 - 16.55n \quad (r^2 = 0.69) \tag{7.2}$$

where σ_c is in MPa; n is in %; and r^2 is the determination coefficient. Pappalardo (2015) also derived similar linear relations between unconfined compressive strength σ_c and porosity n for dolostones at two different sites:

$$\sigma_c = 170 - 17.57n \quad (r^2 = 0.69) \quad \text{(Taormina site)} \tag{7.3}$$

$$\sigma_c = 158 - 16.10n \quad (r^2 = 0.84) \quad \text{(Castelmola site)} \tag{7.4}$$

where σ_c, n, and r^2 are as defined earlier.

Lashkaripour (2002) found that the relationship between unconfined compressive strength σ_c and porosity n for mudrocks can be described by the following hyperbolic function:

$$\sigma_c = 210.1n^{-0.821} \quad (r^2 = 0.67) \tag{7.5}$$

where σ_c is in MPa; n is in %; and r^2 is the determination coefficient. In oil industry, the following hyperbolic empirical relations have been used for estimating the unconfined compressive strength σ_c of shales from porosity n (Rabbani et al., 2012):

$$\sigma_c = 1.001n^{-1.143} \quad \text{for low porosity } (< 0.1) \text{ and high strength } (\approx 79 \text{ MPa}) \text{ shale}$$
$$\text{(7.6a)}$$

$$\sigma_c = 2.92n^{-0.96} \quad \text{for mostly high porosity Tertiary shale} \quad \text{(7.6b)}$$

where σ_c is in MPa and n has no unit.

According to Palchik and Hatzor (2004), the relationship between unconfined compressive strength σ_c and porosity n can be described by the following negative exponential function:

$$\sigma_c = ae^{-bn} \tag{7.7}$$

where σ_c is in MPa; n is in %; and a and b are constants which can be obtained by fitting analysis of experimental results. Table 7.2 shows the values of a and b for several types of rocks.

Using the experimental data of gypsum, Yilmaz and Yuksek (2009) derived the following logarithmic relation between unconfined compressive strength σ_c and porosity n:

$$\sigma_c = -28.4 \ln(n) + 78.99 \quad (r^2 = 0.80) \tag{7.8}$$

TABLE 7.2 Values of a and b in Eq. (7.7) for Several Types of Rocks

a	b	r^2	Rock Type	Reference
74.4	0.048	0.79	Sandstone	Palchik (1999)
210.1	0.821	0.67	Mudrocks: claystone, clay shale, mudstone, mud shale, siltstone, and silt shale	Lashkaripour (2002)
273.1	0.076	0.87	Chalk	Palchik and Hatzor (2004)
195.0	0.210	0.79	Sandstone, limestone, basalt, and granodiorite	Tuğrul (2004)
135.9	0.048		Sandstones	Rabbani et al. (2012)
143.8	0.0695		Carbonates and limestones ($5 < n < 20\%$, $30 < \sigma_c < 150$ MPa)	
135.9	0.048		Carbonates and limestones ($0 < n < 20\%$, $10 < \sigma_c < 300$ MPa)	

Notes: For the values of a and b listed in the table, the unconfined compressive strength σ_c is in the unit of MPa and the porosity n is in %. r^2 is the determination coefficient.

where σ_c is in MPa; n is in %; and r^2 is the determination coefficient.

Fig. 7.1 shows the variation of unconfined compressive strength with porosity for various geomaterials from polluted sludge to hard rock (Adachi and Yoshida, 2002).

7.2.1.2 Unconfined compressive strength versus density

Since density is closely related to the degree of porosity, it also affects the strength of intact rocks. The unconfined compressive strength of intact rocks increases with higher density. Based on an extensive study of different types of rocks (basalt, diabase, dolomite, gneiss, granite, limestone, marble, quartzite, rock salt, sandstone, schist, siltstone, and tuff), Deere and Miller (1966) derived the following simple linear relation between unconfined compressive strength σ_c and dry density ρ_d:

$$\sigma_c = 198.4\,\rho_d - 362.7 \quad \left(r^2 = 0.36\right) \tag{7.9}$$

where σ_c is in MPa; ρ_d is in kg/m^3; and r^2 is the determination coefficient.

Tuğrul and Zarif (1999) also derived a similar linear relation between unconfined compressive strength σ_c and dry density ρ_d for granitic rocks from Turkey:

$$\sigma_c = 566.2\,\rho_d - 1347 \quad \left(r^2 = 0.67\right) \tag{7.10}$$

where σ_c, ρ_d, and r^2 are as defined earlier.

According to Vasarhelyi (2005) and Del Potro and Hürlimann (2009), the relationship between unconfined compressive strength σ_c and density ρ can also be described by an exponential function:

$$\sigma_c = ae^{b\rho} \tag{7.11}$$

where a and b are constants which can be obtained by fitting analysis of experimental results. For example, Del Potro and Hürlimann (2009) obtained the values of 0.626 and 1.746 for a and b, respectively, with a determination coefficient r^2 of 0.93, based on the experimental data of volcanic rocks, where σ_c is in MPa and ρ is in kg/m^3.

Fig. 7.2 shows the data and trend line of unconfined compressive strength σ_c versus dry density ρ_d for chalks from different locations (Bowden et al., 2002).

7.2.1.3 Unconfined compressive strength versus seismic wave velocity

Seismic wave velocity has been used to estimate the unconfined compressive strength of rocks by different researchers. Table 7.3 lists a number of empirical correlations between the unconfined compressive strength σ_c and the P-wave velocity v_p. It is noted that different correlations may give very different

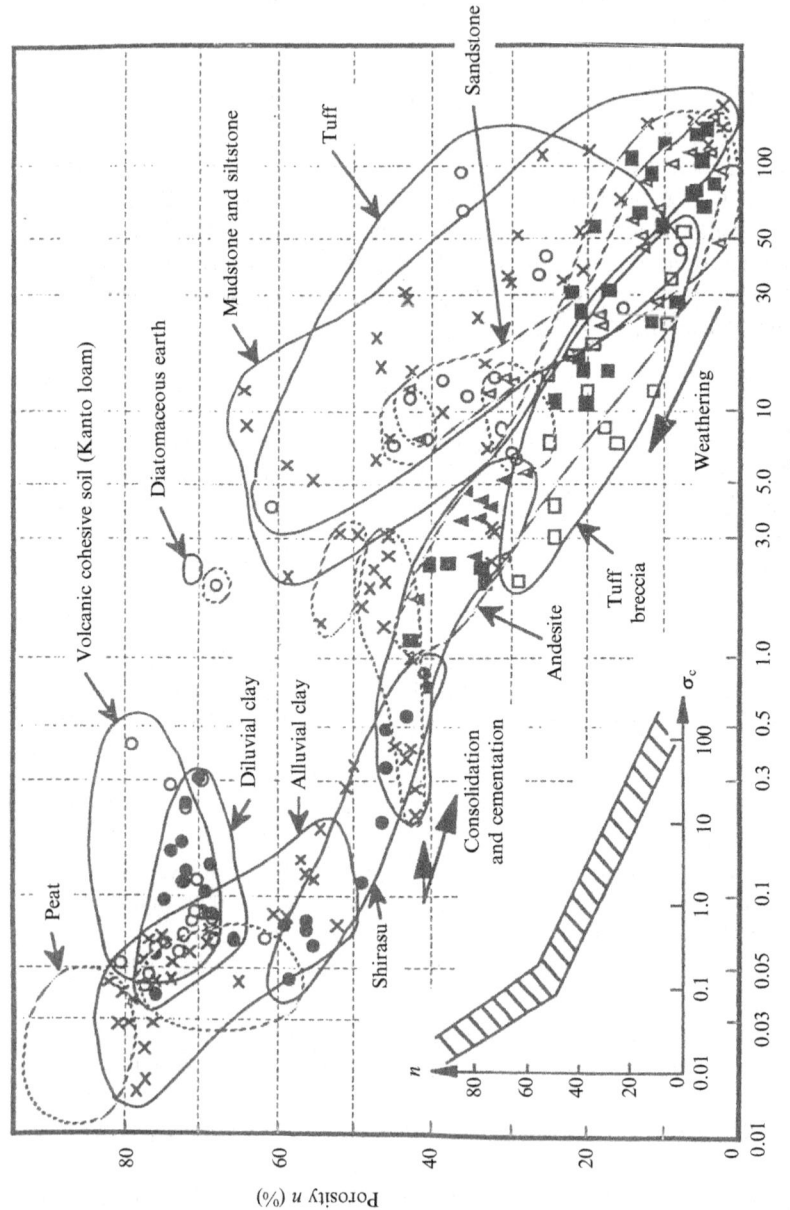

FIG. 7.1 Variation of unconfined compressive strength σ_c with porosity n for various geomaterials. (Based on Adachi, T., Yoshida, N., 2002. In situ investigation on mechanical characteristics of soft rocks. In: Sharma, V.M., Saxena, K.R. (Eds.), In-Situ Characterization of Rocks. Balkema, Lisse, pp. 131–186.)

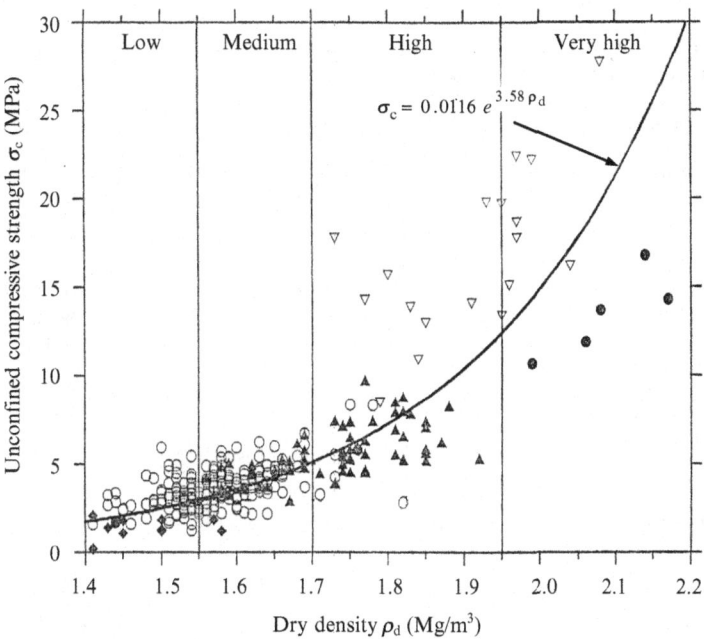

FIG. 7.2 Unconfined compressive strength σ_c versus dry density ρ_d for chalks. *(Based on Bowden, A.J., Spink, T.W., Mortmore, R.N., 2002. Engineering description of chalk: its strength, hardness and density. Q. J. Eng. Geol. Hydrogeol. 35, 355–361.)*

TABLE 7.3 Correlations Between Unconfined Compressive Strength σ_c and P-Wave Velocity v_p

Correlation	r^2	Rock Type	Reference
$\sigma_c = 35.0 v_p - 31.5$		Sandstone	Freyburg (1972)
$\sigma_c = 2.45 v_p^{1.82}$		Limestone	Militzer and Stoll (1973)
$\log \sigma_c = 0.358 v_p + 0.283$		Limestone	Golubev and Rabinovich (1976)
$\log \sigma_c = 0.444 v_p + 0.003$		Schist	

Continued

TABLE 7.3 Correlations Between Unconfined Compressive Strength σ_c and P-Wave Velocity v_p—cont'd

Correlation	r^2	Rock Type	Reference
$\sigma_c = -0.98v_p + 0.68v_p^2 + 0.98$		Sandy and shaly rocks	Gorjainov and Ljachovickij (1979)
$\sigma_c = k\rho v_p^2 + A$		Soft rocks	Inoue and Ohomi (1981)
$\sigma_c = 1277e^{(-11.2/v_p)}$		Sandstone	McNally (1987)
$\sigma_c = 36.0v_p - 31.2$		Coal measure rocks	Göktan (1988)
$\sigma_c = 35.54v_p - 55$	0.64	Granitic rocks	Tuğrul and Zarif (1999)
$\sigma_c = 9.95v_p^{1.21}$	0.69	Dolomite, sandstone, limestone, marl, diabase, serpentine, hematite	Kahraman (2001b)
$\sigma_c = 31.5v_p - 63.7$	0.80	Dolomite, marble, and limestone	Yasar and Erdoğan (2004b)
$\sigma_c = 22.03v_p^{1.247}$	0.72	Granites	Sousa et al. (2005)
$\sigma_c = 64.35e^{1.4v_p}$		Granite, andesite	Cha et al. (2006)
$\sigma_c = 3.9348e^{0.6129v_p}$	0.82	Gypsum	Yilmaz and Yuksek (2009)
$\sigma_c = av_p^b$ where		9 Sedimentary, 3 volcanic, and 2 volcano-sedimentary rocks	Karakul and Ulusay (2013)
$a = 6.987e^{-0.64S_r}$	0.99		
$b = 1.608e^{0.064S_r}$	0.96		
$\sigma_c = 37.5v_p - 116$	0.82	Dolostone at Taormina	Pappalardo (2015)
$\sigma_c = 34.6v_p - 108$	0.88	Dolostone at Castelmola	

Notes: σ_c is the unconfined compressive strength in MPa; ρ is the rock density in g/cm^2; v_p is the P-wave velocity in km/s; S_r is the degree of saturation from 0 to 1; and r^2 is the determination coefficient.

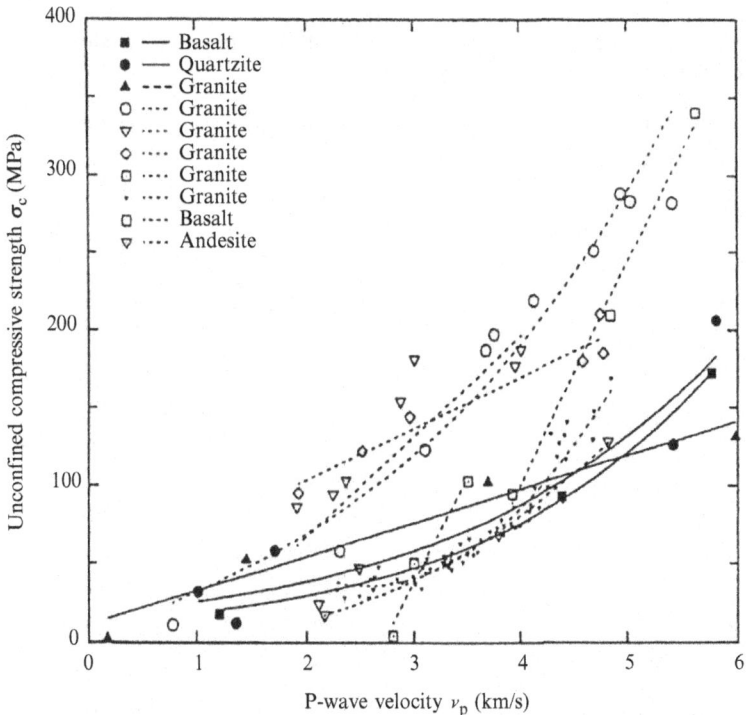

FIG. 7.3 Variation of unconfined compressive strength σ_c with P-wave velocity v_p for fresh and weathered crystalline rocks. *(From Gupta, A.S., Rao, K.S., 1998. Index properties of weathered rocks: inter-relationships and applicability. Bull. Eng. Geol. Env. 57, 161–172.)*

unconfined compressive strength values. Fig. 7.3 clearly shows the wide range of the unconfined compressive strength values for different rocks at the same P-wave velocity.

7.2.1.4 Unconfined compressive strength versus point load index

The point load index is an indirect measure of rock strength (see Chapter 3 about the point load test). There exist a large number of empirical correlations between unconfined compressive strength σ_c and point load index $I_{s(50)}$. Table 7.4 lists some of them. The ratio of σ_c to $I_{s(50)}$ varies widely. To obtain reliable results for a specific site, a series of unconfined compression tests should be carried out to calibrate the point load tests.

Palchik and Hatzor (2004) investigated the influence of porosity on the relation between σ_c and $I_{s(50)}$ for porous chalks. They showed that the ratio $\sigma_c/I_{s(50)}$ is not constant, but is porosity dependent. An increase in porosity from 18% to 40% leads to a decrease of $\sigma_c/I_{s(50)}$ from 18 to 8. Kahraman et al. (2005) also investigated the influence of porosity on the relation between σ_c and $I_{s(50)}$ for different rock types (igneous, sedimentary, and metamorphic). There is a

TABLE 7.4 Correlations Between Unconfined Compressive Strength σ_c and Point Load Index $I_{s(50)}$

Correlation	r^2	Rock Type	Reference
$\sigma_c = 15.3 I_{s(50)} + 16.3$			D'Andrea et al. (1965)
$\sigma_c = 20.7 I_{s(50)} + 29.6$			Deere and Miller (1966)
$\sigma_c = 24 I_{s(50)}$		Various rocks	Broch and Franklin (1972)
$\sigma_c = 23 I_{s(50)}$		Sandstones	Bieniawski (1975)
$\sigma_c = 30 I_{s(50)}$		SW England granites	Irfan and Dearman (1978)
$\sigma_c = (10 \text{ to } 29) I_{s(50)}$			Al-Jassar and Hawkins (1979)
$\sigma_c = 29 I_{s(50)}$		Sedimentary rocks	Hassani et al. (1980)
$\sigma_c = 16 I_{s(50)}$		Sedimentary rocks	Read et al. (1980)
$\sigma_c = 20 I_{s(50)}$		Basalts	
$\sigma_c = 18.7 I_{s(50)} - 13.2$			Singh (1981)
$\sigma_c = 14.5 I_{s(50)}$			Forster (1983)
$\sigma_c = 16.5 I_{s(50)} + 51.0$			Gunsallus and Kulhawy (1984)
$\sigma_c = (20 \text{ to } 25) I_{s(50)}$			ISRM (1985)
$\sigma_c = 14.7 I_{s(50)}$		Siltstone	Das (1985)
$\sigma_c = 18 I_{s(50)}$		Sandstone	
$\sigma_c = 12.6 I_{s(50)}$		Shale	
$\sigma_c = 26.5 I_{s(50)}$		Limestone	Hawkins and Olver (1986)
$\sigma_c = 24.8 I_{s(50)}$		Sandstone	
$\sigma_c = (8 \text{ to } 54) I_{s(50)}$			Norbury (1986)
$\sigma_c = 30 I_{s(50)}$		Sedimentary rocks	O'Rourke (1988)
$\sigma_c = (8.6 \text{ to } 16) I_{s(50)}$		Sandstone, shale	Vallejo et al. (1989)
$\sigma_c = 23 I_{s(50)} + 13$			Cargill and Shakoor (1990)
$\sigma_c = (14 \text{ to } 82) I_{s(50)}$			Tsidzi (1991)

Continued

TABLE 7.4 Correlations Between Unconfined Compressive Strength σ_c and Point Load Index $I_{s(50)}$—cont'd

Correlation	r^2	Rock Type	Reference
$\sigma_c = 16 I_{s(50)}$			Ghosh and Srivastava (1991)
$\sigma_c = 9.3 I_{s(50)} + 20.0$			Grasso et al. (1992)
$\sigma_c = 25.67 I_{s(50)}^{0.57}$			
$\sigma_c = 23.37 I_{s(50)}$	0.96	Quartzite rocks	Singh and Singh (1993)
$\sigma_c = 19 I_{s(50)} + 12.7$	0.81	Sandstones	Ulusay et al. (1994)
$\sigma_c = 12.5 I_{s(50)}$	0.53	Granite, tuff	Chau and Wong (1996)
$\sigma_c = 24 I_{s(50)}$		Sandstone, limestone	Smith (1997)
$\sigma_c = 12.6 I_{s(50)}$		Shale	
$\sigma_c = 15.25 I_{s(50)}$	0.96	Granitic rocks	Tuğrul and Zarif (1999)
$\sigma_c = (14.5 \text{ to } 27) I_{s(50)}$		Limestones	Romana (1999)
$\sigma_c = (12 \text{ to } 24) I_{s(50)}$		Sandstones	
$\sigma_c = (10 \text{ to } 15) I_{s(50)}$		Siltstones, mudstones	
$\sigma_c = (5 \text{ to } 10) I_{s(50)}$		Chalk, porous limestones	
$\sigma_c = 21.8 I_{s(50)}$		Shale	Rusnak and Mark (1999)
$\sigma_c = 20.2 I_{s(50)}$		Siltstone	
$\sigma_c = 20.6 I_{s(50)}$		Sandstone	
$\sigma_c = 21.9 I_{s(50)}$		Limestone	
$\sigma_c = 23.6 I_{s(50)} - 2.7$		Coal measure rocks	Kahraman (2001b)
$\sigma_c = 8.4 I_{s(50)} + 9.5$		22 Different rocks	
$\sigma_c = 21.4 I_{s(50)}$	0.85	Mudrocks	Lashkaripour (2002)
$\sigma_c = 24.4 I_{s(50)}$		Strong rocks	Quane and Russel (2003)
$\sigma_c = 3.86 I_{s(50)}^2 + 5.65 I_{s(50)}$		Weak rocks	

Continued

TABLE 7.4 Correlations Between Unconfined Compressive Strength σ_c and Point Load Index $I_{s(50)}$ — cont'd

Correlation	r^2	Rock Type	Reference
$\sigma_c = 10.3 I_{s(50)} + 28.1$	0.76	Sandstones	Zorlu et al. (2004)
$\sigma_c = (8 \text{ to } 18) I_{s(50)}$			Palchik and Hatzor (2004)
$\sigma_c = 23 I_{s(50)}$	0.75	Limestones, marlstones, and sandstones	Tsiambaos and Sabatakakis (2004)
$\sigma_c = 7.3 I_{s(50)}^{1.71}$	0.82		
$\sigma_c = 10.9 I_{s(50)} + 27.4$	0.61	All rocks	Kahraman et al. (2005)
$\sigma_c = 24.8 I_{s(50)} - 39.6$	0.72	Rocks with $n < 1\%$	
$\sigma_c = 10.2 I_{s(50)} + 23.4$	0.75	Rocks with $n > 1\%$	
$\sigma_c = 9.08 I_{s(50)} + 39.3$	0.72	9 Different rocks	Fener et al. (2005)
$\sigma_c = 21 I_{s(50)}$	0.93	Hong Kong granites	Basu and Aydin (2006)
$\sigma_c = 13.3 I_{s(50)} + 7.43$	0.64	6 Different rocks	Yilmaz (2009)
$\sigma_c = 10.5 I_{s(50)} - 3.97$	0.57	Gypsum	Yilmaz and Yuksek (2009)
$\sigma_c = 11.1 I_{s(50)} + 37.7$	0.74	Schistose rocks	Basu and Kamran (2010)
$\sigma_c = 22.8 I_{s(50)}$	0.99	Quartzite A	Singh et al. (2012)
$\sigma_c = 15.8 I_{s(50)}$	0.91	Khondalite	
$\sigma_c = 22.2 I_{s(50)}$	0.78	Quartzite B	
$\sigma_c = 21.9 I_{s(50)}$	0.89	Sandstone	
$\sigma_c = 16.1 I_{s(50)}$	0.71	Rock salt	
$\sigma_c = 14.4 I_{s(50)}$	0.82	Shale	
$\sigma_c = 23.3 I_{s(50)}$	0.97	Gabbro	
$\sigma_c = 23.5 I_{s(50)}$	0.98	Amphibolite	
$\sigma_c = 21.0 I_{s(50)}$	0.96	Epidorite	
$\sigma_c = 22.3 I_{s(50)}$	0.68	Limestone	
$\sigma_c = 22.7 I_{s(50)}$	0.82	Dolomite	
$\sigma_c = 21.5 I_{s(50)}$		All of the above 11	

Continued

TABLE 7.4 Correlations Between Unconfined Compressive Strength σ_c and Point Load Index $I_{s(50)}$ — cont'd

Correlation	r^2	Rock Type	Reference
$\sigma_c = 10.9 I_{s(50)} + 49.0$	0.8	Granite	Mishra and Basu (2012)
$\sigma_c = 11.2 I_{s(50)} + 4.01$	0.84	Schist	
$\sigma_c = 13.0 I_{s(50)} - 5.19$	0.84	Sandstone	
$\sigma_c = 14.6 I_{s(50)}$	0.88	All of the above 3	
$\sigma_c = 20 I_{s(50)}$		Metasiltstone	Li and Wong (2013)
$\sigma_c = 21 I_{s(50)}$		Metasandstone	
$\sigma_c = 20.1 I_{s(50)} - 17.1$	0.8	Travertine, marble	Palassi and Emami (2014)

Notes: Both σ_c and $I_{s(50)}$ are in MPa and r^2 is the determination coefficient.

significant correlation between σ_c and $I_{s(50)}$ for all rock types, but it is not strong. When the rocks are divided into two groups according to porosity values ($n < 1\%$ and $n > 1\%$), stronger correlations are obtained. The slope of the regression line of the rocks having porosity values lower than 1% is much greater than that of the rocks having porosity values higher than 1% (Table 7.4).

7.2.1.5 Unconfined compressive strength versus Schmidt hammer rebound number

The Schmidt hammer rebound test has been briefly described in Chapter 3. Various empirical correlations have been proposed for estimating the unconfined compressive strength of rocks from the Schmidt hammer rebound number. Fig. 7.4 shows a series of empirically determined curves relating the L-type Schmidt hammer rebound number at different orientations to the unconfined compressive strength.

Table 7.5 lists a number of closed-form empirical correlations for estimating the unconfined compressive strength from the Schmidt hammer rebound number. It is noted that different correlations may give very different unconfined compressive strength values. To obtain reliable results for a specific site, a series of unconfined compression tests need to be carried out to calibrate the Schmidt hammer rebound tests. It is also important to specify the hammer type (L or N) so that the right correlation(s) are used.

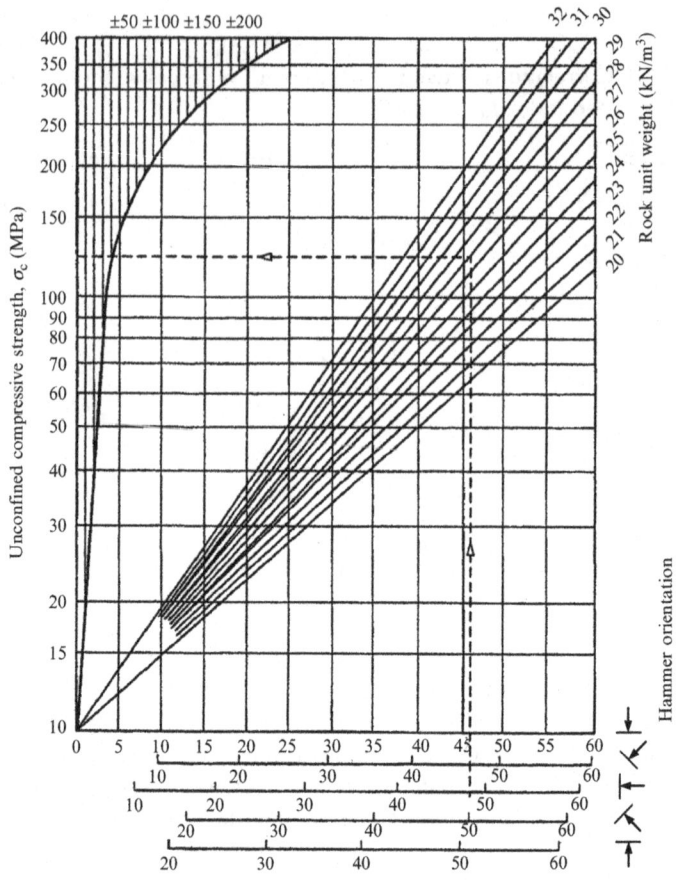

FIG. 7.4   L-type Schmidt hammer rebound number $R_{n(L)}$ versus unconfined compressive strength σ_c. *(Based on Deere, D.U., Miller, R.P., 1966. Engineering classification and index properties for intact rock. Tech. Rep. No. AFWL-TR-65-116, Air Force Weapons Lab, Kirtland Air Force Base, Albuquerque, NM.)*

7.2.1.6   Unconfined compressive strength versus needle penetration index (NPI)

Table 7.6 lists several empirical relations which can be used to estimate the unconfined compressive strength σ_c from the NPI. Note the specific units used for each empirical relation.

7.2.1.7   Unconfined compressive strength versus Shore Sclerscope hardness

Various empirical correlations have been proposed for calculating the unconfined compressive strength of rocks from the Shore Sclerscope hardness. Table 7.7 lists some of them.

TABLE 7.5 Correlations Between Unconfined Compressive Strength σ_c and Schmidt Hammer Rebound Number R_n

Correlation	r^2	Rock Type	Reference
$\sigma_c = 8.59 R_{n(L)} - 240.6$	0.77	28 Lithological units, 3 base rock types	Deere and Miller (1966)
$\sigma_c = 6.9 \times 10^{\left[0.0087 \rho_d R_{n(L)} + 0.16\right]}$	0.89		
$\sigma_c = 6.9 \times 10^{\left[1.348 \log\left(\rho R_{n(L)}\right) - 1.325\right]}$		25 Lithological units	Aufmuth (1973)
$\sigma_c = 12.74 e^{0.185 \rho R_{n(L)}}$		20 Lithological units	Beverly et al. (1979)
$\sigma_c = 0.447 e^{\left(0.045 R_{n(L)} + \rho\right)}$		Different rock types in Northern Silesia	Kidybinski (1980)
$\sigma_c = 0.994 R_{n(L)} - 0.383$	0.70	10 Lithological units	Haramy and DeMarco (1985)
$\ln(\sigma_c) = 0.043 \rho_d R_{n(L)} + 1.2$	0.86	Sandstones	Cargill and Shakoor (1990)
$\ln(\sigma_c) = 0.018 \rho_d R_{n(L)} + 2.9$	0.96	Carbonates	
$\sigma_c = 4.29 R_{n(L)} - 67.5$	0.92	Marble, limestone, dolomite	Sachpazis (1990)
$\sigma_c = e^{a R_{n(L)} + b}$ a and b are coefficients depending on rock type	0.83–0.90	Mica-schist, prasinite, serpentinite, gabbro, mudstone	Xu et al. (1990)
$\sigma_c = 8.36 R_{n(L)} - 416$	0.76	Granitic rocks	Tuğrul and Zarif (1999)
$\sigma_c = 2.208 e^{0.067 R_{n(N)}}$	0.96	Chalk, limestone, sandstone, marble, syenite, granite	Katz et al. (2000)
$\sigma_c = 6.97 e^{0.014 \rho R_{n(N)}}$	0.78	Dolomite, sandstone, limestone, marl, diabase, serpentine, hematite	Kahraman (2001b)
$\sigma_c = e^{0.059 R_{n(L)} + 0.818}$	0.96	Gypsum	Yilmaz and Sendir (2002)

Continued

TABLE 7.5 Correlations Between Unconfined Compressive Strength σ_c and Schmidt Hammer Rebound Number R_n—cont'd

Correlation	r^2	Rock Type	Reference
$\sigma_c = e^{0.053R_{n(L)}+1.332}$	0.88	Limestone, claystone, siltstone, sandstone, marl, marlstone, basalt, calcarenite, ophite	Morales et al. (2004)
$\sigma_c = 4 \times 10^{-6} R_{n(L)}^{4.2917}$	0.80	Limestone, marble, sandstone, basalt	Yasar and Erdoğan (2004a)
$\sigma_c = 2.75R_{n(L)} - 35.83$	0.95	Basalt, andesite, tuff	Dincer et al. (2004)
$\sigma_c = 4.24e^{0.059R_{n(L)}}$	0.66	9 Different rocks	Fener et al. (2005)
$\sigma_c = 3.201R_{n(L)} - 46.59$	0.58	Dolomite, limestone	Shalabi et al. (2007)
$\sigma_c = 4.01e^{0.038R_{n(L)}}$	0.78	9 Different rocks	Yilmaz and Yuksek (2009)
$\sigma_c = e^{-4.04 + 2.28 \ln[R_{n(L)}]}$	0.98	Carbonate rocks	Bruno et al. (2013)

Notes: σ_c is in the unit of MPa; ρ is the rock density in g/cm^3; $R_{n(L)}$ and $R_{n(N)}$ are, respectively, the L- and N-type Schmidt hammer rebound numbers (see Chapter 3 for detailed description of Schmidt hammer rebound test); and r^2 is the determination coefficient.

7.2.1.8   Unconfined compressive strength versus cone indenter number

The unconfined compressive strength σ_c can be estimated from the cone indenter number as follows (Szlavin, 1974; Brook, 1993):

$$\sigma_c = 24.8I_s \tag{7.12a}$$

$$\sigma_c = 35.8I_m \tag{7.12b}$$

$$\sigma_c = 16.5I_w \tag{7.12c}$$

TABLE 7.6 Relations Between Unconfined Compressive Strength σ_c and Needle Penetration Index (NPI)

Correlation	r^2	Rock Type	Units		Reference
			σ_c	NPI	
$\log\sigma_c = 0.978\log NPI + 1.599$	0.84	Artificial cement based samples, mudstone	kgf/cm^2	kgf/mm	Okada et al. (1985)
$\log\sigma_c = 0.982\log NPI - 0.209$	0.76	Pyroclastic flow and fall deposits	kgf/cm^2	kgf/mm	Yamaguchi et al. (1997)
$\sigma_c = 1.5395 NPI^{0.9896}$	0.81	Sandstone, mudstone, conglomerate, greywacke, tuff	MPa	N/mm	Takahashi et al. (1998)
$\sigma_c = 27.3 NPI + 132$	0.70	Sandstone	kN/m^2	N/cm	Naoto et al. (2004)
$\sigma_c = 41.8 NPI - 4$	0.81	Ariaka clay	kN/m^2	N/cm	
$\sigma_c = 0.51 NPI^{0.8575}$	0.77	Marl, siltstone, mudstone, tuff	MPa	N/mm	Erguler and Ulusay (2007, 2009)
$\sigma_c = 0.8244 NPI^{0.6975}$	0.67	All the above rocks	MPa	N/mm	Ulusay and Erguler (2012)
$\sigma_c = 0.402 NPI^{0.929}$	0.79	All the above rocks except conglomerate	MPa	N/mm	
$\sigma_c = 0.2 NPI$	0.62	Marl, siltstone, mudstone, tuff	MPa	N/mm	Aydan et al. (2014)

Note: r^2 is the determination coefficient.

TABLE 7.7 Correlations Between Unconfined Compressive Strength σ_c and Shore Sclerscope Hardness H

Correlation		r^2	Rock Type	Reference
$\sigma_c = 2.1H$	Lower limit			Wuerker (1953)
$\sigma_c = 2.8H$	Average			
$\sigma_c = 3.4H$	Upper limit			
$\sigma_c = 3.54H - 42.8$		0.80	28 Lithological units, 3 base rock types	Deere and Miller (1966)
$\sigma_c = 6.9 \times 10^{[0.0041\rho_d H + 0.62]}$		0.85		
$\sigma_c = 3.54(H - 12)$		0.32		Atkinson (1993); Brook (1993)
$\sigma_c = 0.895H + 41.98$		0.32	Shale	Koncagül and Santi (1999)
$\sigma_c = H^{5.555} \times 10^{-8}$		0.83	Limestone, marble, basalt, sandstone	Yasar and Erdoğan (2004a)
$\sigma_c = 3.326H - 79.76$		0.64	Dolomite, limestone	Shalabi et al. (2007)
$\sigma_c = 1.581H - 62.2$		0.72	Shale	

Notes: σ_c is in the unit of MPa; ρ_d is the dry rock density in g/cm³; and r^2 is the determination coefficient.

where σ_c is in MPa; and I_s, I_m, and I_w are the cone indenter numbers obtained using 40, 110, and 12 N force indenters, respectively (see Chapter 3 for more details).

There are also other forms of empirical relations between unconfined compressive strength σ_c and cone indenter number I_s, such as Eqs. (7.13a) (Stimpson and Acott, 1983) and (7.13b) (Ghose and Chakraborti, 1986):

$$\sigma_c = 45.3I_s - 15.9 \text{ for sedimentary rocks (Stimpson and Acott, 1983)}$$
$$(7.13a)$$

$$\sigma_c = 22.1I_s - 8.45 \text{ for Indian coals (Ghose and Chakraborti, 1986) (7.13b)}$$

where σ_c and I_s are as defined earlier.

7.2.1.9   Effect of water content on unconfined compressive strength

Many researchers have studied the effect of water content on the strength of intact rocks. The unconfined compressive strength of intact rocks decreases as the water content increases and their relationship can be described by the following negative exponential function (Hawkins and McConnel, 1992; Lashkaripour, 2002; Yilmaz, 2010):

$$\sigma_c = ae^{-bw} + c \qquad (7.14)$$

where w is the water content in %; and a, b, and c are constants. Table 7.8 lists the values of a, b, and c for several types of rocks.

There are also other types of closed-form relations between unconfined compressive strength σ_c and water content w, such as the one below derived by Yilmaz and Yuksek (2009) based on the experimental data of gypsum:

$$\sigma_c = -10.16 \ln(w) + 30.58 \quad (r^2 = 0.82) \qquad (7.15)$$

where σ_c is in MPa; w is in %; and r^2 is the determination coefficient.

Fig. 7.5 shows the variation of unconfined compressive strength σ_c with water content w for porous chalks. It is noticed that σ_c in the direction parallel to bedding is larger than that in the direction perpendicular to bedding.

Table 7.9 lists the values of the ratio of unconfined compressive strength at saturated condition $\sigma_{c(saturated)}$ to that at dry condition $\sigma_{c(dry)}$ for different rocks. It can be seen that $\sigma_{c(saturated)}/\sigma_{c(dry)}$ covers a wide range from about 20% to 90%.

TABLE 7.8 Values of a, b, and c in Eq. (7.14) for Several Types of Rocks

a	b	c	r^2	Rock Type	Reference
4.16–84.01	0.0752–6.147	2.97–231		15 British sandstones	Hawkins and McConnel (1992)
83.59	0.4433	0	0.96	Coal measures mudrock (clayshale, mudstone, and mudshale)	Lashkaripour (2002)
14.68	0.8193	24.0	0.93	Gypsum	Yilmaz (2010)

Notes: For the values of a, b, and c listed in the table, the unconfined compressive strength σ_c is in MPa and the water content w is in %. r^2 is the determination coefficient.

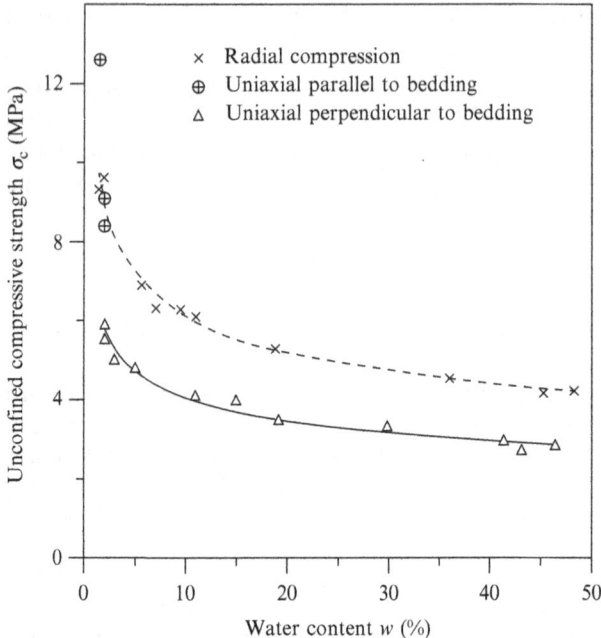

FIG. 7.5   Variation of unconfined compressive strength σ_c with water content w for porous chalks. *(From Talesnick, M.L., Hatzor, Y.H., Tsearsky, M., 2001. The elastic deformability and strength of high porosity anisotropic chalk. Int. J. Rock Mech. Min. Sci. 38, 543–555.)*

7.2.1.10   Effect of temperature on unconfined compressive strength

Temperature also affects the unconfined compressive strength of intact rock. Fig. 7.6 shows the normalized unconfined compressive strength (σ_c/σ_{c0}) values at various temperatures for several granites and one mudstone, where σ_{c0} is the unconfined compressive strength at the lowest test temperature which is 30°C for the British granite, Salisbury granite, Remiremont granite, Senones granite, and Indian granite; 20°C for the Ningbo granite; and 25°C for the mudstone. For the granites, when the temperature is below 200°C, the unconfined compressive strength may slightly increase, stay about the same or decrease with higher temperature. After the temperature goes above 200°C, the unconfined compressive strength decreases with higher temperature for all of the granites. For the mudstone, however, the unconfined compressive strength increases substantially (more than doubled) when the temperature is raised up to 400°C. After the temperature is above 400°C, the unconfined compressive strength decreases when the temperature is higher.

TABLE 7.9 Ratio of Unconfined Compressive Strength at Saturated Condition $\sigma_{c(saturated)}$ to That at Dry Condition $\sigma_{c(dry)}$ for Different Rocks

$\sigma_{c(saturated)}/\sigma_{c(dry)}$	Rock	Reference
0.50	Shale and Quartzitic Sandstone	Colback and Wild (1965)
0.76	Penrith Sandstone	Dyke and Dobereiner (1991)
0.75	Bunter sandstone	
0.66	Waterstone	
0.22–0.92	35 British sandstones	Hawkins and McConnel (1992)
0.97	Oolitic limestone	Lashkaripour and Ghafoori (2002)
0.62	Sandstone and sandy limestone	
0.81	Oolitic limestone and limy sandstone	
0.52	Shale	
0.76	British sandstone	Vasarhelyi (2003)
0.66	Miocene limestone	Vasarhelyi (2005)
0.59	Jastrzębie sandstone	Kwaśniewski and Oitaben (2009)
0.49	Anna mudstone	
0.35	Gypsum	Yilmaz (2010)
0.36–0.69	Limestone	Rajabzadeh et al. (2012)
0.29–0.85	Dolomitic limestone	
0.33–0.64	Marble	

7.2.2   Tensile Strength of Intact Rock

There is a strong correlation between the tensile strength and the unconfined compressive strength. The following simple correlation is usually used as a first estimate of the tensile strength σ_t from the unconfined compressive strength σ_c:

$$\sigma_t = -\frac{\sigma_c}{10} \tag{7.16}$$

Lade (1993) presented the following general relation between σ_t and σ_c for all rock types:

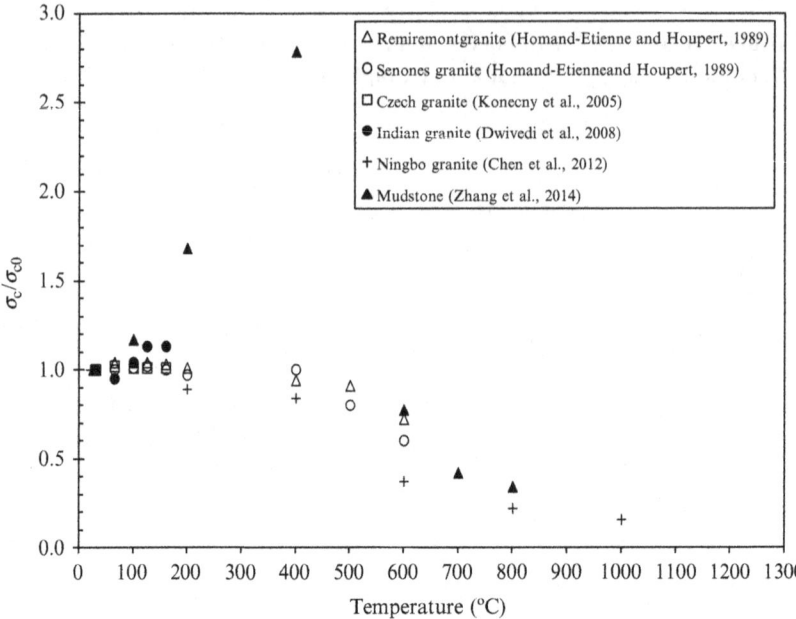

FIG. 7.6 Variation of normalized unconfined compressive strength (σ_c/σ_{c0}) with temperature (σ_{c0} is the unconfined compressive strength of rock at the lowest test temperature for each data set).

$$\sigma_t = T p_a \left(\frac{\sigma_c}{p_a} \right)^t \tag{7.17}$$

where p_a is the atmospheric pressure in the same units as those of σ_t and σ_c; and T and t are dimensionless numbers which vary with rock types. Figs. 7.7A–C show the data of σ_t and σ_c collected by Lade (1993) for igneous, metamorphic, and sedimentary rocks, respectively. Based on fitting analysis of the data, Lade (1993) obtained the values of T and t for these three types of rocks as follows:

Igneous rocks	$T=-0.435$	$t=0.740$
Metamorphic rocks	$T=-0.0518$	$t=1.017$
Sedimentary rocks	$T=-0.316$	$t=0.770$
All rocks	$T=-0.219$	$t=0.825$

In addition to the best fitting lines for each type of rocks, the lines of $|\sigma_t/\sigma_c| = 1/5$, 1/10, 1/20, and 1/50 are also drawn in the figures. It can be seen that the data are widely scattered and $\sigma_t/\sigma_c = -1/10$ (Eq. 7.16) is approximately an average of the whole data.

Because of the strong correlation between the tensile strength and the unconfined compressive strength, the methods for estimating the unconfined

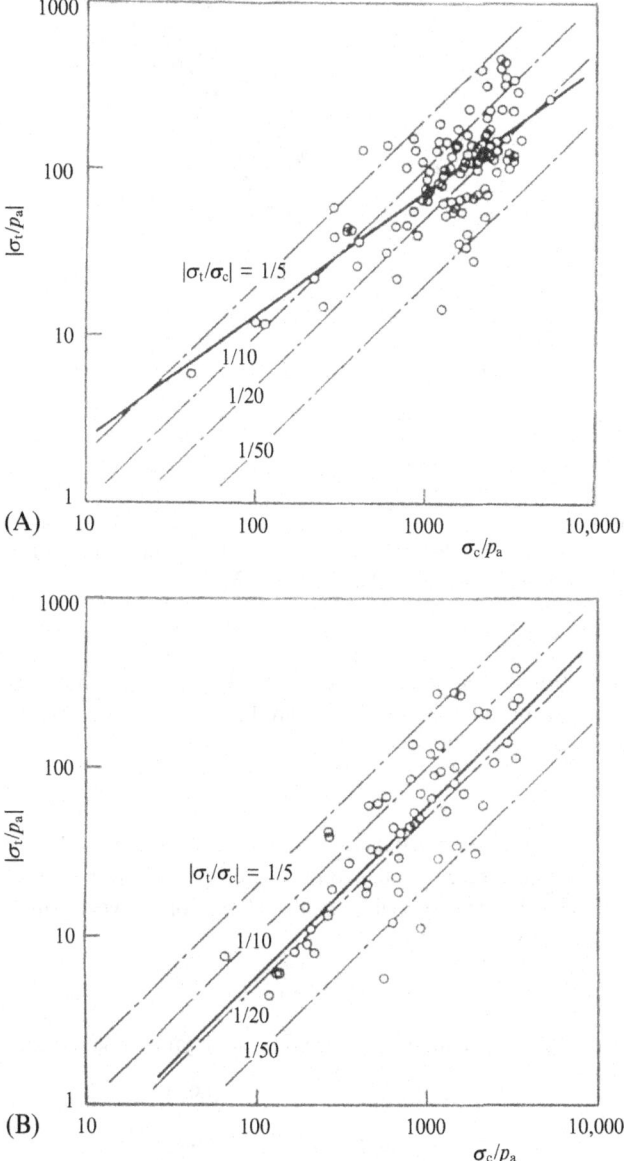

FIG. 7.7 Relation between tensile strength and unconfined compressive strength for (A) igneous rocks; (B) metamorphic rocks; and

(Continued)

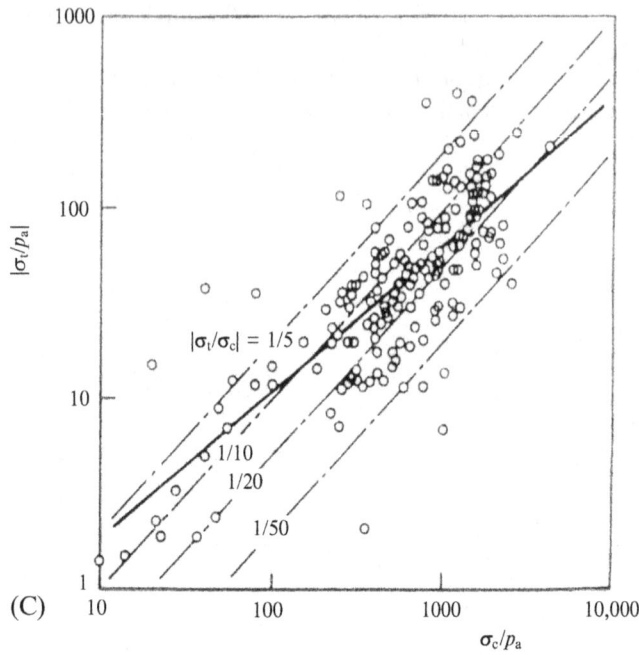

(C)

FIG.7.7, CONT'D   (C) sedimentary rocks. *(Based on Lade, P.V., 1993. Roc strength criteria: the theories and the evidence. In: Hudson, J.A. (Ed.), Comprehensive Rock Engineering—Principle, Practice & Projects, vol. 1. Pergamon, Oxford, pp. 255–284.)*

compressive strength can also be used to estimate the tensile strength. For example, the point load index $I_{s(50)}$ can be used to estimate the tensile strength σ_t:

$$\sigma_t = -1.5 I_{s(50)} \qquad (7.18)$$

Based on the experimental data of nine sedimentary, three volcanic, and two volcano-sedimentary rocks at different degrees of saturation, Karakul and Ulusay (2013) derived the following empirical relation between tensile strength σ_t and P-wave velocity v_t:

$$\sigma_t = -a v_p^b \qquad (7.19)$$

where a and b are coefficients varying with the degree of saturation S_r:

$$a = 1.039 e^{-0.58 S_r} \quad \left(r^2 = 0.99\right) \qquad (7.20a)$$

$$b = 1.395 e^{0.055 S_r} \quad \left(r^2 = 0.96\right) \qquad (7.20b)$$

where S_r is from 0 to 1; and r^2 is the determination coefficient.

The strength criteria which will be described in Section 7.2.3 can also be used to determine the tensile strength of an intact rock. For example, with the Hoek-Brown strength criterion for intact rock, the tensile strength can be estimated from

$$\sigma_t = 0.5\sigma_c \left[m_i - \left(m_i^2 + 4 \right)^{0.5} \right] \qquad (7.21)$$

where m_i is a material constant for the intact rock, which depends on the rock type (texture and mineralogy) as tabulated in Table 7.10. This equation shows that the ratio of σ_t/σ_c varies with rock types represented by the different m_i values. For the possible range of m_i from 4 to 33 (Table 7.10),

$$\sigma_t = -(0.03 \text{ to } 0.24)\sigma_c \qquad (7.22)$$

The tensile strength of intact rock also decreases when water content is higher. According to Vasarhelyi (2005), the reduction percentage for the tensile strength of a rock from dry condition to saturated condition is about the same as that for the unconfined compressive strength of the same rock. Fig. 7.8 shows the variation of tensile strength σ_t with water content w for porous chalks.

Temperature also affects the tensile strength of intact rock. Fig. 7.9 shows the normalized tensile strength (σ_t/σ_{t0}) values at various temperatures for several granites, where σ_{t0} is the tensile strength at the lowest test temperature 30°C. It can be clearly seen that the tensile strength decreases with higher temperature. After the temperature is above 1000°C, the tensile strength decreases to almost zero.

7.2.3   Empirical Strength Criteria of Intact Rock

Various empirical strength criteria of intact rock have been developed by different researchers (eg, Bieniawski, 1974; Hoek and Brown, 1980; Johnston, 1985; Lade, 1993; Chang and Haimson, 2005; Al-Ajmi and Zimmerman, 2005; Kwaśniewski, 2011; Peng et al., 2014). Four of them are described as follows.

7.2.3.1   Hoek-Brown criterion

For intact rock, the Hoek-Brown criterion may be expressed in the following form (Hoek and Brown, 1980):

$$\sigma_1' = \sigma_3' + \sigma_c \left(m_i \frac{\sigma_3'}{\sigma_c} + 1 \right)^{0.5} \qquad (7.23)$$

where σ_c is the unconfined compressive strength of the intact rock; σ_1' and σ_3' are the major and minor effective principal stresses, respectively; and m_i is a material constant for the intact rock as tabulated in Table 7.10.

TABLE 7.10 Values of Parameter m_i for Different Rocks

Rock Type	Class	Group	Texture			
			Coarse	Medium	Fine	Very Fine
Sedimentary	Clastic		Conglomerate (21 ± 3)[a]	Sandstone 17 ± 4	Siltstone 7 ± 2	Claystone 4 ± 2
			Breccia (19 ± 5)		Greywacke (18 ± 3)	Shale (6 ± 2)
						Marl (7 ± 2)
	Non clastic	Carbonate	Crystalline limestone (12 ± 3)	Sparitic limestone (10 ± 2)	Micritic limestone (9 ± 2)	Dolomite (9 ± 3)
		Evaporite		Gypsum 8 ± 2	Anhydrite 12 ± 2	
		Organic				Chalk 7 ± 2
Metamorphic	Nonfoliated		Marble 9 ± 3	Hornfels (19 ± 4)	Quartzite 20 ± 3	
				Metasandstone (19 ± 3)		
	Slightly foliated		Migmatite (29 ± 3)	Amphibolite 26 ± 6	Gneiss 28 ± 5	
	Foliated[b]			Schist 12 ± 3	Phyllite (7 ± 3)	Slate 7 ± 4

Continued

TABLE 7.10 Values of Parameter m_i for Different Rocks—cont'd

Rock Type	Class	Group	Coarse	Medium	Fine	Very Fine
Igneous	Plutonic	Light	Granite 32±3 Granodiorite (29±3)	Diorite 25±5		
		Dark	Gabbro 27±3 Norite 20±5	Dolerite (16±5)		
	Hypabyssal		Porphyrie (20±5)		Diabase (15±5)	Peridotite (25±5)
	Volcanic	Lava		Rhyolite (25±5) Andesite 25±5	Dacite (25±3) Basalt (25±5)	Obsidian (19±3)
		Pyroclastic	Agglomerate (19±3)	Breccia (19±5)	Tuff (13±5)	

[a]Values in parenthesis are estimates.
[b]These values are for intact rock specimen tests normal to bedding or foliation. The value of m_i will be significantly different if failure occurs along a weakness plane.
Based on Hoek, E., Brown, E.T., 1997. Practical estimates of rock mass strength. Int. J. Rock Mech. Min. Sci. 34, 1165–1186; Marinos, P., Hoek, E., 2001. Estimating the geotechnical properties of heterogeneous rock masses such as flysch. Bull. Eng. Geol. Env. 60, 85–92.

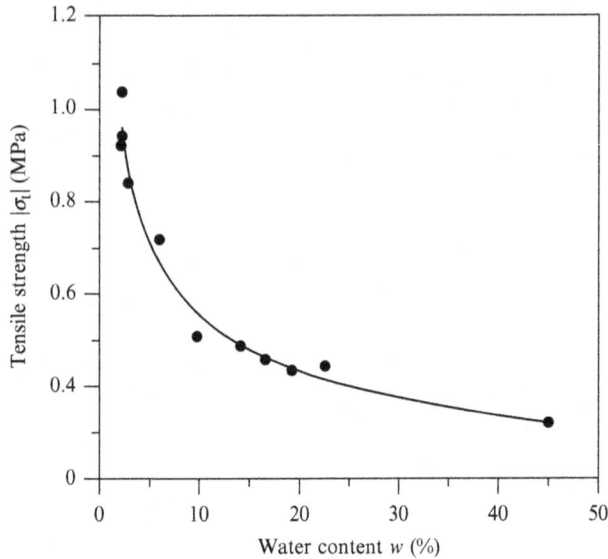

FIG. 7.8   Variation of tensile strength σ_t with water content w for porous chalks. *(From Talesnick, M.L., Hatzor, Y.H., Tsearsky, M., 2001. The elastic deformability and strength of high porosity anisotropic chalk. Int. J. Rock Mech. Min. Sci. 38, 543–555.)*

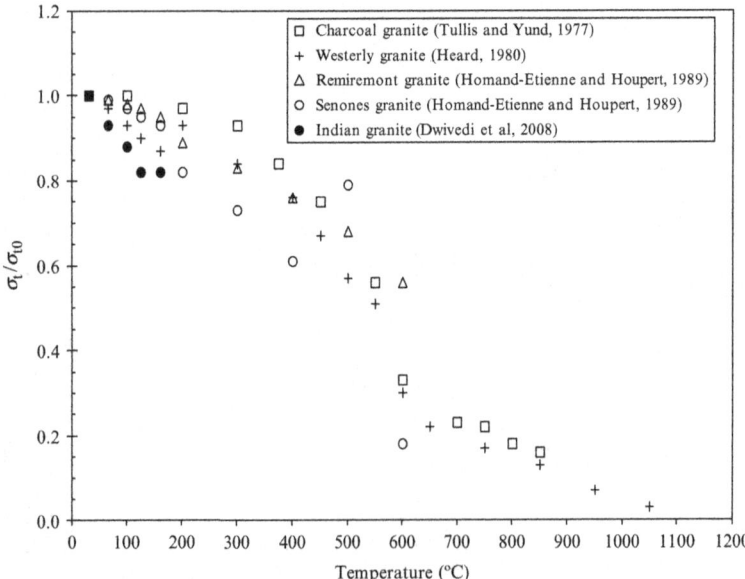

FIG. 7.9   Variation of normalized tensile strength (σ_t/σ_{t0}) with temperature (σ_{t0} is the tensile strength of rock at the lowest test temperature for each data set).

TABLE 7.11 Parameter b in the Bieniawski-Yudhbir Criterion

Rock Type	b
Tuff, shale, limestone	2
Siltstone, mudstone	3
Quartzite, sandstone, dolerite	4
Norite, granite, quartz diorite, chert	5

From Yudhbir, F., Lemanza, W., Prinzl, F., 1983. An empirical failure criterion for rock masses. In: Proc. 5th Int. Congress on Rock Mechanics. vol. 1. ISRM, Melbourne, pp. B1–B8.

7.2.3.2 Bieniawski-Yudhbir criterion

Bieniawski (1974) proposed a strength criterion for intact rock as follows:

$$\frac{\sigma_1'}{\sigma_c} = 1 + b \left(\frac{\sigma_3'}{\sigma_c} \right)^{0.65} \tag{7.24}$$

where b is a parameter which can be determined from Table 7.11.

7.2.3.3 Johnston criterion

Based on experimental data of a wide range of geotechnical material, from lightly overconsolidated clays through hard rocks, Johnston (1985) proposed the following strength criterion for intact rock:

$$\sigma_{1n}' = \left(\frac{M}{B} \sigma_{3n}' + 1 \right)^B \tag{7.25}$$

where σ_{1n}' and σ_{3n}' are the normalized effective principal stresses at failure, obtained by dividing the effective principal stresses, σ_1' and σ_3', by the relevant unconfined compressive strength, σ_c; and B and M are intact material constants. By placing $\sigma_{3n}' = 0$, the unconfined compressive strength is correctly modeled with the right-hand side of Eq. (7.25) becoming one.

When $B = 1$, the criterion simplifies to

$$\sigma_{1n}' = M\sigma_{3n}' + 1 \tag{7.26}$$

which for

$$M = \frac{1 + \sin\phi'}{1 - \sin\phi'} \tag{7.27}$$

is identical to the normalized Mohr-Coulomb criterion.

The parameter B, which describes the nonlinearity of a failure envelope, is essentially independent of the material type, and is a function of the unconfined compressive strength:

$$B = 1 - 0.0172(\log\sigma_c)^2 \tag{7.28}$$

TABLE 7.12 A Range of Rock Types

Type	General Rock Type	Examples
A	Carbonate rocks with well developed crystal cleavage	Dolomite, limestone, marble
B	Lithified argillaceous rocks	Mudstone, siltstone, shale, slate
C	Arenaceous rocks with strong crystals and poorly-developed crystal cleavage	Sandstone, quartzite
D	Fine grained polymineralic igneous crystalline rocks	Andesite, dolerite, diabase, rhyolite
E	Coarse grained polymineralic igneous and metamorphic crystalline rocks	Amphibolite, gabbro, gneiss, granite, norite, quartz diorite

From Hoek, E., Brown, E.T., 1980. Empirical strength criterion for rock masses. J. Geotech. Eng. ASCE 106, 1013–1035.

The parameter M, which describes the slope of a failure envelope at $\sigma'_{3n} = 0$, is found to be a function of both the unconfined compressive strength and the material type. For the material types shown in Table 7.12, M can be estimated by (no result is obtained for Type D material because of lack of data):

$$\text{Type A,} \quad M = 2.065 + 0.170(\log \sigma_c)^2 \tag{7.29a}$$

$$\text{Type B,} \quad M = 2.065 + 0.231(\log \sigma_c)^2 \tag{7.29b}$$

$$\text{Type C,} \quad M = 2.065 + 0.270(\log \sigma_c)^2 \tag{7.29c}$$

$$\text{Type E,} \quad M = 2.065 + 0.659(\log \sigma_c)^2 \tag{7.29d}$$

7.2.3.4 Ramamurthy criterion

Ramamurthy and his coworkers (Ramamurthy et al., 1985; Ramamurthy, 1986, 1993) modified the Coulomb theory to represent the nonlinear shear strength behavior of rocks. For intact rock, the strength criterion is in the following form:

$$\frac{\sigma'_1 - \sigma'_3}{\sigma'_3} = B_r \left(\frac{\sigma_c}{\sigma'_3} \right)^{\alpha_r} \tag{7.30}$$

where σ_1' and σ_3' are the major and minor principal effective stresses; σ_c is the unconfined compressive strength; α_r is the slope of the curve between $(\sigma'_1 - \sigma'_3)/\sigma'_3$ and σ_c/σ_3', with a mean value of 0.8 for most intact rocks; and B_r is a material constant of intact rock, equal to $(\sigma'_1 - \sigma'_3)/\sigma'_3$ when $\sigma_c/\sigma'_3 = 1$. The values of B_r vary from 1.8 to 3.0 depending on the type of rock (Table 7.13).

TABLE 7.13 Mean Values of Parameter B_r for Different Rocks

Rock Type	Metamorphic and Sedimentary Rocks				Igneous Rocks
	Argillaceous	Arenaceous	Chemical		
			Limestone	Marble	Andesite
	Shales				Diorite
	Slates				Norite
	Mudstone		Anhydrite		Liparite Granite
Siltstone	Claystone	Sandstone	Quartzite	Rock salt	Dolomite	Basalt Charnockite
Clays						
Tuffs						
Loess						
B_r	1.8	2.2	2.2	2.6	2.4	2.8	2.6 3.0
Mean value	2.0		2.4		2.6		2.8

Based on Ramamurthy, T., 1993. Strength and modulus responses of anisotropic rocks. In: Hudson, J.A. (Ed.), Comprehensive Rock Engineering—Principle, Practice and Projects, vol. 1. Pergamon, Oxford, pp. 313–329.

The values of α_r and B_r can be estimated by conducting a minimum of two triaxial tests at confining pressures greater than 5% of σ_c for the rock. The above expression is applicable in the ductile region and in most of the brittle region. It underestimates the strength when σ_3' is less than 5% of σ_c and also ignores the tensile strength of the rock. To account for the tensile strength, the following expression can be used:

$$\frac{\sigma_1' - \sigma_3'}{\sigma_3' + \sigma_t} = B \left(\frac{\sigma_c}{\sigma_3' + \sigma_t} \right)^\alpha \tag{7.31}$$

where σ_t is the tensile strength of rock preferably obtained from Brazilian tests; $\alpha = 0.67$ for most rocks; and B is a material constant. The values of α and B in Eq. (7.31) can be obtained by two triaxial tests conducted at convenient confining pressures greater than 5% of σ_c for the rock. In the absence of these tests, the value of B can be estimated as $1.3(\sigma_c/\sigma_t)^{1/3}$.

7.2.4   Mohr-Coulomb Parameters of Intact Rock

Since the Mohr-Coulomb failure criterion is often used in the analyses of rock mechanics problems, it is necessary to estimate the cohesion and friction parameters of the intact rock:

$$\tau_f = c_i + \sigma_n' \tan \phi_i \tag{7.32}$$

where τ_f is the shear strength of the intact rock; c_i and ϕ_i are the cohesion and internal friction angle of the intact rock, respectively; and σ_n' is the effective normal stress on the sliding plane. It needs to be noted that the "primes" for c_i and ϕ_i have been omitted for brevity although they are for the effective stress conditions.

Table 7.14 lists the representative peak values of c_i and ϕ_i for different rocks. Robertson (1970), while recognizing that there is considerable variation, suggested that the peak cohesion be about 16% of the unconfined compressive strength. If the Mohr-Coulomb criterion is used to represent the residual strength (the minimum strength reached by the rock subjected to deformation beyond the peak), the subscript "r" may be used with the cohesion and friction angle terms in Eq. (7.32). The residual cohesion will approach zero and the residual internal friction angle will lie between zero and the peak internal friction angle.

7.3   STRENGTH OF ROCK DISCONTINUITIES

Discontinuities usually have negligible tensile strength and a shear strength that is, under most circumstances, significantly smaller than that of the surrounding intact rock material. The following describes several shear strength models for rock discontinuities.

TABLE 7.14 Typical Peak Cohesion c_i and Internal Friction Angle ϕ_i for Different Rocks

Rock	Porosity (%)	c_i (MPa)	ϕ_i (degree)	Range of Confining Pressure (MPa)
Berea sandstone	18.2	27.2	27.8	0–200
Bartlesville sandstone		8.0	37.2	0–203
Pottsville sandstone	14.0	14.9	45.2	0–68.9
Repetto siltstone	5.6	34.7	32.1	0–200
Muddy shale	4.7	38.4	14.4	0–200
Stockton shale		0.34	22.0	0.8–4.1
Edmonton bentonitic shale (water content 30%)	44.0	0.3	7.5	0.1–3.1
Sioux quartzite		70.6	48.0	0–203
Texas slate; loaded				
30 degree to cleavage		26.2	21.0	34.5–276
90 degree to cleavage		70.3	26.9	34.5–276
Georgia marble	0.3	21.2	25.3	5.6–68.9
Wolf Camp limestone		23.6	34.8	0–203
Indiana limestone	19.4	6.7	42.0	0–9.6
Hasmark dolomite	3.5	22.8	35.5	0.8–5.9
Chalk	40.0	0	31.5	10–90
Blaine anhydrite		43.4	29.4	0–203
Inada biotite granite	0.4	55.2	47.7	0.1–98
Stone mountain granite	0.2	55.1	51.0	0–68.9
Nevada test site basalt	4.6	66.2	31.0	3.4–34.5
Schistose gneiss				
30 degree to schistocity	0.5	46.9	28.0	0–69
90 degree to schistocity	1.9	14.8	27.6	0–69

From Goodman, R.E., 1989. Introduction to Rock Mechanics. Wiley, New York, NY.

7.3.1 Mohr-Coulomb Model

The simplest shear strength model of discontinuities is the Mohr-Coulomb failure criterion, which can be expressed by

$$\tau_f = c_j + \sigma'_n \tan\phi_j \tag{7.33}$$

where τ_f is the shear strength of the discontinuity; c_j and ϕ_j are the cohesion and internal friction angle of the discontinuity, respectively; and σ_n' is the effective normal stress on the discontinuity plane. It needs to be noted that the "primes" for c_j and ϕ_j have been omitted for brevity although they are for the effective stress conditions. The Mohr-Coulomb model can be used for planar, clean (no filling) discontinuities.

7.3.2 Bilinear Shear Strength Model

Natural discontinuities contain undulations and asperities and their shear strength-normal stress relation is usually nonlinear. Patton (1966) addressed this problem by formulating the bilinear model as shown in Fig. 7.10. At normal stresses less than or equal to σ_0' the shear strength is given by

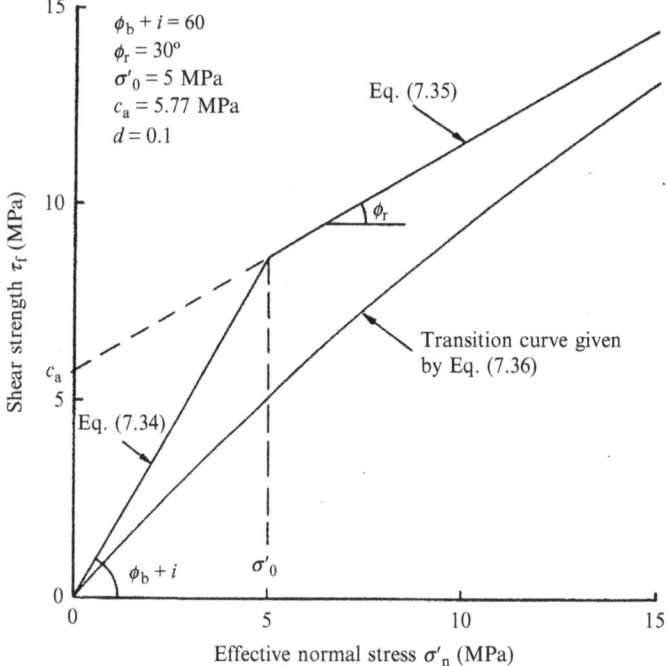

FIG. 7.10 Bilinear shear strength model (Eqs. 7.34, 7.35) with empirical transition curve (Eq. 7.36).

$$\tau_f = \sigma'_n \tan(\phi_b + i) \tag{7.34}$$

where ϕ_b is the basic friction angle for an apparently smooth surface of the rock material; and i is the effective roughness angle. Table 6.7 in Chapter 6 lists the typical values of ϕ_b for different rocks.

At normal stresses $\geq \sigma_0'$ the shear strength is given by

$$\tau_f = c_a + \sigma'_n \tan \phi_r \tag{7.35}$$

where c_a is the apparent cohesion derived from the asperities and ϕ_r is the residual friction angle of the rock material forming the asperities.

Jaeger (1971) proposed the following shear strength model to provide a curved transition between the two straight lines of the Patton (1966) model (Fig. 7.10):

$$\tau_f = c_a\left(1 - e^{-d\sigma'_n}\right) + \sigma'_n \tan \phi_r \tag{7.36}$$

where d is an experimentally determined empirical parameter which controls the shape of the transition curve.

7.3.3 Barton Model

A direct, practical approach to predicting the shear strength of discontinuities on the basis of relatively simple measurements was developed by Barton and his coworkers (Barton, 1976; Barton and Choubey, 1977; Barton and Bandis, 1990). According to the Barton model, the shear strength τ_f of a discontinuity subjected to an effective normal stress σ_n' is given by

$$\tau_f = \sigma'_n \tan\left[\text{JRC} \log\left(\frac{\text{JCS}}{\sigma'_n}\right) + \phi_r\right] \tag{7.37}$$

where JRC, JCS, and ϕ_r are, respectively, the joint (discontinuity) roughness coefficient, the joint (discontinuity) wall compressive strength, and the residual friction angle of the discontinuity, which can be estimated using the methods described in Section 6.3.

Eq. (7.37) suggests that there are three factors which control the shear strength of rock discontinuities: the geometrical component JRC, the asperity failure component expressed by the ratio JCS/σ_n', and the residual friction angle ϕ_r. Research results show that both JRC (geometrical component) and JCS (asperity failure component) decrease with increasing scale (Fig. 7.11). Based on extensive testing of discontinuities, discontinuity replicas, and a review of literature, Barton and Bandis (1982) proposed the scale corrections for JRC and JCS:

$$\text{JRC}_n = \text{JRC}_0\left[\frac{L_n}{L_0}\right]^{-0.02\text{JRC}_0} \tag{7.38}$$

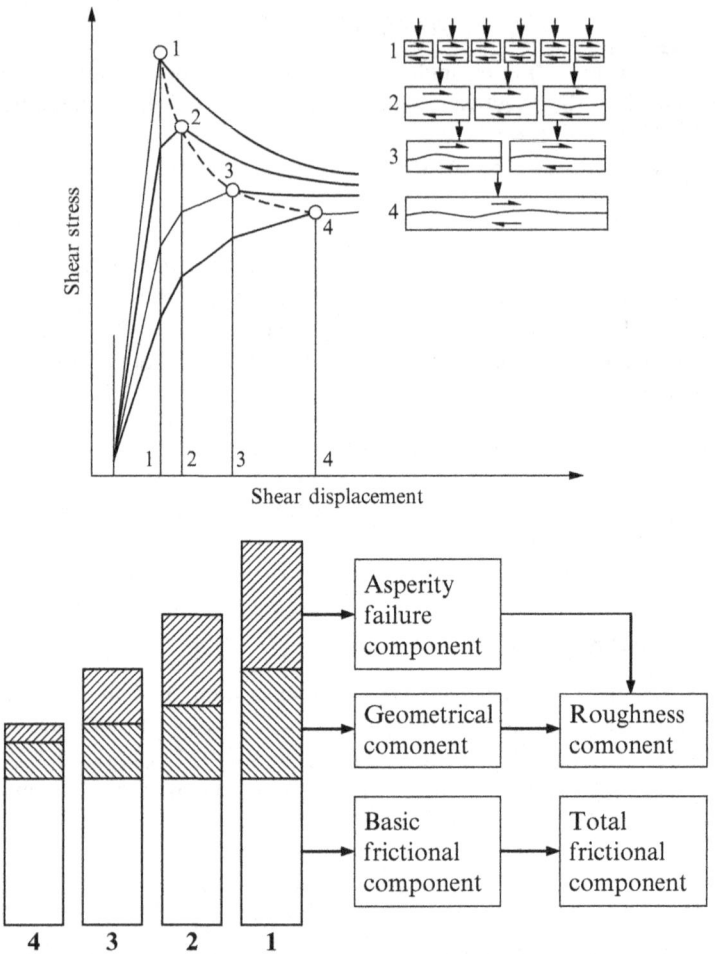

FIG. 7.11 Influence of scale on the three components of the shear strength of a rough discontinuity. *(Based on Bandis, S.C., 1990. Mechanical properties of rock joints. In: Barton, N., Stephanson, O. (Eds.), Proc. Int. Symp. on Rock Joints, Loen, Norway. Balkema, Rotterdam, pp. 125–140; Barton, N., Bandis, S. C., 1990. Review of predictive capabilities of JRC-JCS model in engineering practice. In: Barton, N., Stephansson, O. (Eds.), Proc. Int. Symp. on Rock Joints, Loen, Norway. Balkema, Rotterdam, pp. 603–610.)*

$$\mathrm{JCS_n = JCS_0} \left[\frac{L_n}{L_0} \right]^{-0.03\mathrm{JCS_0}} \tag{7.39}$$

where $\mathrm{JRC_0}$, $\mathrm{JCS_0}$, and L_0 (length) refer to 100 mm laboratory scale samples and $\mathrm{JRC_n}$, $\mathrm{JCS_n}$, and L_n refer to in situ block sizes.

It is worth noting two important limitations on the use of the Barton model for estimating the shear strength of discontinuities. Barton and Choubey (1977) suggest that the curves should be truncated such that the maximum allowable shear strength for design purposes is given by arctan $\left(\tau / \sigma_n' \right) = 70°$. For

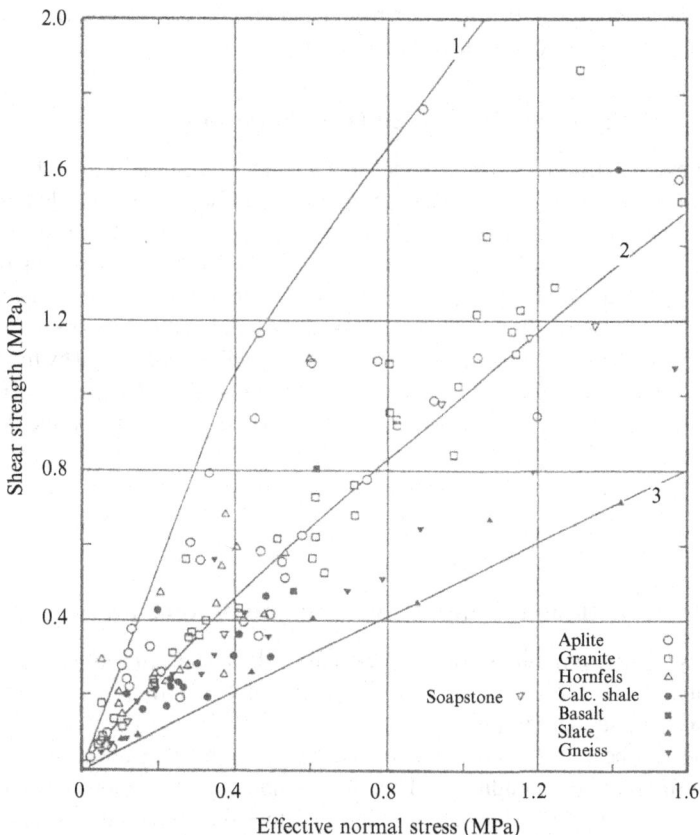

FIG. 7.12 Range of peak shear strength for 136 joints representing eight different rock types. Curves 1, 2, and 3 are evaluated using Eq. (7.37). *(Based on Barton, N., Choubey, V., 1977. The shear strength of rock joints in theory and practice. Rock Mech. 10, 1–54.)*

example, curve 1 in Fig. 7.12 has a linear "cut-off" representing the maximum suggested design value of 70 degree for the total frictional angle. Barton (1976) cautioned that when the effective normal stress exceeds the unconfined compressive strength of the rock material, the measured shear strength is always appreciably higher than that predicted by Eq. (7.37). Noting that this discrepancy was probably due to the effect of confining stresses that increase the strength of asperities, Barton proposed that a high stress version of Eq. (7.37) could be obtained by replacing JCS by $(\sigma'_1 - \sigma'_3)$, ie,

$$\tau = \sigma'_n \tan \left[\text{JRC} \log \left(\frac{\sigma'_1 - \sigma'_3}{\sigma'_n} \right) + \phi_r \right] \qquad (7.40)$$

where σ_1' is the effective axial stress required to yield the rock material under an effective confining stress σ_3'. The failure stress σ_1' can either be determined experimentally or can be estimated from an appropriate yield criterion such as the Hoek-Brown criterion described earlier.

7.3.4   Shear Strength of Filled Discontinuities

The previous sections deal with the shear strength of discontinuities in which rock wall contact occurs over the entire length of the surface under consideration. If the discontinuity contains a filling material such as clay gouge, the shear strength of the discontinuity will be influenced by the thickness and properties of the filling material. If the thickness of the filling material is more than about 25–50% of the amplitude of the asperities, there will be little or no rock-to-rock contact and the shear strength of the discontinuity will be controlled by the shear strength properties of the filling material (Goodman, 1970). The peak and residual shear strength of filled discontinuities can be expressed by the Mohr-Coulomb model (Eq. 7.33). Table 7.15 lists the shear strength parameters of filled discontinuities and filling materials summarized by Hoek and Bray (1981).

7.4   STRENGTH OF ROCK MASS

7.4.1   Unconfined Compressive Strength of Rock Mass

Because of discontinuities, jointed rock mass will have much lower unconfined compressive strength than intact rock. It is difficult to determine the unconfined compressive strength of jointed rock masses in the laboratory because the samples need to be undisturbed and sufficiently large so that they are representative of the discontinuity conditions. Therefore, empirical correlations considering the effect of discontinuities are usually used to estimate the unconfined compressive strength of rock masses.

7.4.1.1   Methods relating unconfined compressive strength with RQD

Kulhawy and Goodman (1987) suggested that, as a first approximation, the unconfined compressive strength σ_{cm} of rock masses be taken as $0.33\sigma_c$ when RQD is less than about 70% and then linearly increasing to $0.8\sigma_c$ when RQD increases from 70% to 100% (Fig. 7.13), where σ_c is the unconfined compressive strength of the intact rock. The Standard Specifications for Highway Bridges adopted by AASHTO (1996) suggest that σ_{cm} be estimated using the following expressions:

$$\sigma_{cm} = \alpha_\sigma \sigma_c \qquad (7.41a)$$

$$\alpha_\sigma = 0.0231 RQD - 1.32 \geq 0.15 \qquad (7.41b)$$

The variation of σ_{cm}/σ_c with RQD based on Eqs. (7.41a), (7.41b) is also shown in Fig. 7.13. It can be seen that the general trend of these two relations between σ_{cm}/σ_c and RQD is about the same: σ_{cm}/σ_c is constant when RQD is

TABLE 7.15 Shear Strength of Filled Discontinuities and Filling Materials

Rock	Description	Peak c_j (MPa)	Peak ϕ_j (degree)	Residual c_j (MPa)	Residual ϕ_j (degree)
Basalt	Clayey basaltic breccia, wide variation from clay to basalt content	0.24	42		
Bentonite	Bentonite seam in chalk	0.015	7.5		
	Thin layers	0.09–0.12	12–17		
	Triaxial tests	0.06–0.1	9–13		
Bentonitic shale	Triaxial tests	0–0.27	8.5–29		
	Direct shear tests			0.03	8.5
Clays	Overconsolidated, slips, joints, and minor shears	0–0.18	12–18.5	0–0.003	10.5–16
Clay shale	Triaxial tests	0.06	32		
	Stratification surfaces			0	19–25
Coal measure rocks	Clay mylonite seams, 10–25 mm	0.012	16	0	11–11.5
Dolomite	Altered shale bed, ±150 mm thick	0.04	14.5	0.02	17
Diorite, granodiorite and porphyry	Clay gouge (2% clay, PI = 17%)	0	26.5		
Granite	Clay filled faults	0–0.1	24–25		
	Sandy loam fault filling	0.05	40		
	Tectonic shear zone, schistose and broken granites, disintegrated rock and gouge	0.24	42		

Continued

TABLE 7.15 Shear Strength of Filled Discontinuities and Filling Materials—cont'd

Rock	Description	Peak c_j (MPa)	Peak ϕ_j (degree)	Residual c_j (MPa)	Residual ϕ_j (degree)
Greywacke	1–2 mm clay in bedding planes			0	21
Limestone	6 mm clay layer			0	13
	10–20 mm clay fillings	0.1	13–14		
	<1 mm clay filling	0.05–0.2	17–21		
Limestone, marl, and lignites	Interbedded lignite layers	0.08	38		
	Lignite/marl contact	0.1	10		
Limestone	Marlaceous joints, 20 mm thick	0	25	0	15–24
Lignite	Layer between lignite and clay	0.014–0.03	15–17.5		
Montmorillonite	80 mm seams of bentonite (montmorillonite) clay in chalk	0.36	14	0.08	11
Bentonite clay		0.016–0.02	7.5–11.5		
Schists, quartzites and siliceous schists	100–150 mm thick clay filling	0.03–0.08	32		
	Stratification with thin clay	0.61–0.74	41		
	Stratification with thick clay	0.38	31		
Slates	Finely laminated and altered	0.05	33		
Quartz/kaolin/pyrolusite	Remolded triaxial tests	0.042–0.09	36–38		

Based on Hoek, E., Bray, J.W., 1981. Rock Slope Engineering, third ed. Institution of Mining and Metallurgy, London.

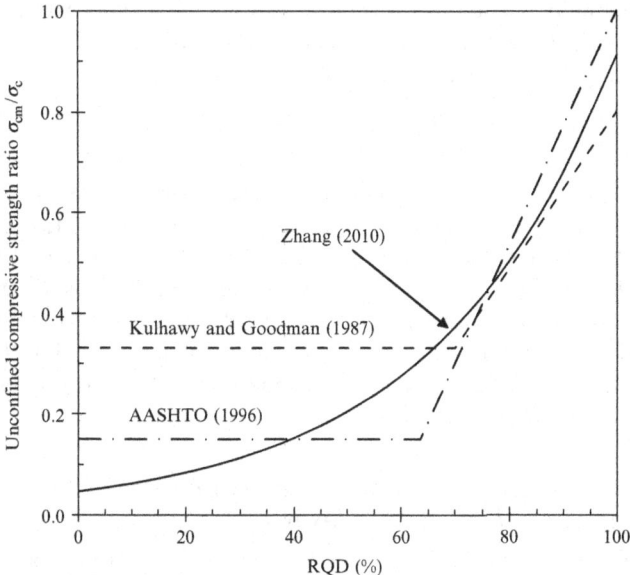

FIG. 7.13 Comparison of σ_{cm}/σ_c versus RQD relations by Kulhawy and Goodman (1987), AASHTO (1996), and Zhang (2010), respectively.

smaller than a certain value and then linearly increases when RQD increases. Obviously it is inappropriate to assume that σ_{cm}/σ_c is constant when RQD varies from 0 to a certain value (70% for the relation of Kulhawy and Goodman and 64% for the relation of AASHTO). For example, for a very poor-quality rock mass (RQD < 25%) and a fair-quality rock mass (RQD = 50% to 75%), different σ_{cm}/σ_c values should be expected.

It is also noted that the σ_{cm}/σ_c versus RQD relation in Eqs. (7.41a), (7.41b) is the same as the E_m/E_r versus RQD relation in Eq. (6.57), which may not be appropriate. Researchers have studied the relation between σ_{cm}/σ_c and E_m/E_r and found that they can be related approximately by the following equation (Ramamurthy, 1993; Singh et al., 1998; Singh and Rao, 2005):

$$\frac{\sigma_{cm}}{\sigma_c} = \alpha_\sigma = \left(\frac{E_m}{E_r}\right)^q = (\alpha_E)^q \qquad (7.42)$$

in which the power q varies from 0.5 to 1.0 and is most likely in the range of 0.61–0.74 with an average of 0.7. It can be seen that the AASHTO method (Eqs. 7.41a, 7.41b) uses the upper bound value of $q = 1.0$.

It needs to be noted that the relation between σ_{cm}/σ_c and E_m/E_r (Eq. 7.42) was derived based only on triaxial test data on jointed rock mass specimens with different joint frequencies, orientations, and conditions (Ramamurthy, 1993; Singh et al., 1998; Singh and Rao, 2005) and has not been tested against field cases. The power q in Eq. (7.42) may vary significantly for different rock types and discontinuity conditions. Nevertheless, using the average value of $q=0.7$ and the average E_m/E_r versus RQD relation in Eq. (6.60c), Zhang (2010) derived the following σ_{cm}/σ_c versus RQD relation:

$$\frac{\sigma_{cm}}{\sigma_c} = \alpha_\sigma = 10^{0.013RQD-1.34} \tag{7.43}$$

As shown in Fig. 7.13, relation (7.43) covers the entire range of $0 \leq RQD \leq 100\%$ continuously. For $RQD > 70\%$, relation (7.43) is in good agreement with the suggestions of Kulhawy and Goodman (1987) and AASHTO (1996). For $RQD < 70\%$, however, relation (7.43) is different from the suggestions of Kulhawy and Goodman (1987) and AASHTO (1996), with relation (7.43) considering the continuous variation of σ_{cm}/σ_c with RQD while the suggestions of Kulhawy and Goodman (1987) and AASHTO (1996) both assume constant σ_{cm}/σ_c values.

7.4.1.2 Methods relating unconfined compressive strength with RMR or geological strength index (GSI)

A large number of empirical correlations have been developed for estimating the rock mass unconfined compressive strength σ_{cm} based on RMR or GSI (Table 7.16). Fig. 7.14 shows a comparison of some of the empirical correlations with the in situ test data of Aydan and Dalgiç (1998).

It is noted that some of the correlations in Table 7.16 are derived from their corresponding strength criteria. As an example, the following shows the derivation of the correlation by Hoek (1994) and Hoek et al. (1995). With the Hoek-Brown strength criterion for rock masses (Eq. 7.51 in Section 7.4.3), the unconfined compressive strength can be expressed as:

$$\sigma_{cm} = \sqrt{s}\sigma_c \tag{7.44}$$

where s is a constant that depends on the characteristics of the rock mass, which can be estimated from RMR or GSI (see Section 7.4.3 for details). If Eq. (7.55b) is used to estimate s, Eq. (7.44) can be rewritten as:

$$\frac{\sigma_{cm}}{\sigma_c} = e^{\frac{GSI-100}{18}} \tag{7.45}$$

which is just what shown in Table 7.16.

7.4.1.3 Methods relating unconfined compressive strength with Q

Grimstad and Bhasin (1996) proposed the following relation for estimating the rock mass unconfined compressive strength σ_{cm} based on Q:

$$\sigma_{cm} = 7\gamma f_c Q^{1/3} \text{ (MPa)} \tag{7.46}$$

where $f_c = \sigma_c/100$ for $Q > 10$ and $\sigma_c > 100$ MPa, otherwise $f_c = 1$; and γ is the unit weight of the rock mass in g/cm^3.

TABLE 7.16 Empirical Correlations for Estimating Rock Mass Unconfined Compressive Strength σ_{cm} Based on RMR or GSI

Reference	Correlation
Yudhbir et al. (1983)	$\dfrac{\sigma_{cm}}{\sigma_c} = e^{\frac{7.65(RMR-100)}{100}}$
Laubscher (1984), Singh and Goel (1999)	$\dfrac{\sigma_{cm}}{\sigma_c} = \dfrac{RMR - Rating\ for\ \sigma_c}{106}$
Ramamurthy et al. (1985), Ramamurthy (1996)	$\dfrac{\sigma_{cm}}{\sigma_c} = e^{\frac{RMR-100}{18.75}}$
Trueman (1988), Asef et al. (2000)	$\sigma_{cm} = 0.5 e^{0.06 RMR}$ (MPa)
Kalamaras and Bieniawski (1993)	$\dfrac{\sigma_{cm}}{\sigma_c} = e^{\frac{RMR-100}{24}}$
Hoek (1994), Hoek et al. (1995)	$\dfrac{\sigma_{cm}}{\sigma_c} = e^{\frac{GSI-100}{18}}$
Sheorey (1997)	$\dfrac{\sigma_{cm}}{\sigma_c} = e^{\frac{RMR-100}{20}}$
Aydan and Dalgiç (1998)	$\dfrac{\sigma_{cm}}{\sigma_c} = \dfrac{RMR}{RMR + 6(100 - RMR)}$
Hoek et al. (2002)	$\dfrac{\sigma_{cm}}{\sigma_c} = e^{\frac{GSI-100}{9-3D}\left[\frac{1}{2} + \frac{1}{6}\left(e^{-\frac{GSI}{15}} - e^{-\frac{20}{3}}\right)\right]}$
Hoek (2004)	$\dfrac{\sigma_{cm}}{\sigma_c} = 0.036 e^{\frac{GSI}{30}}$

Notes: σ_c is the unconfined compressive strength of intact rock; RMR is the rock mass rating (note that the empirical relations before 1989 use RMR76, which is smaller than RMR89 by 5); GSI is the geological strength index; and D is the disturbance factor.

Barton (2002) also proposed a relation similar to Eq. (7.46) for estimating the rock mass unconfined compressive strength σ_{cm} based on Q:

$$\sigma_{cm} = 5\gamma(Q\sigma_c/100)^{1/3} \text{ (MPa)} \qquad (7.47)$$

where γ is the unit weight of the rock mass in g/cm^3.

7.4.1.4   Comments

Although the empirical methods are widely used in practice to estimate rock mass unconfined compressive strength, there are limitations for them:

(1) The anisotropy of the rock mass caused by discontinuities is not considered. If the classification index properties in a single direction are used, the estimated unconfined compressive strength may not be representative of the unconfined compressive strength in other directions. This will be further discussed later in this chapter.

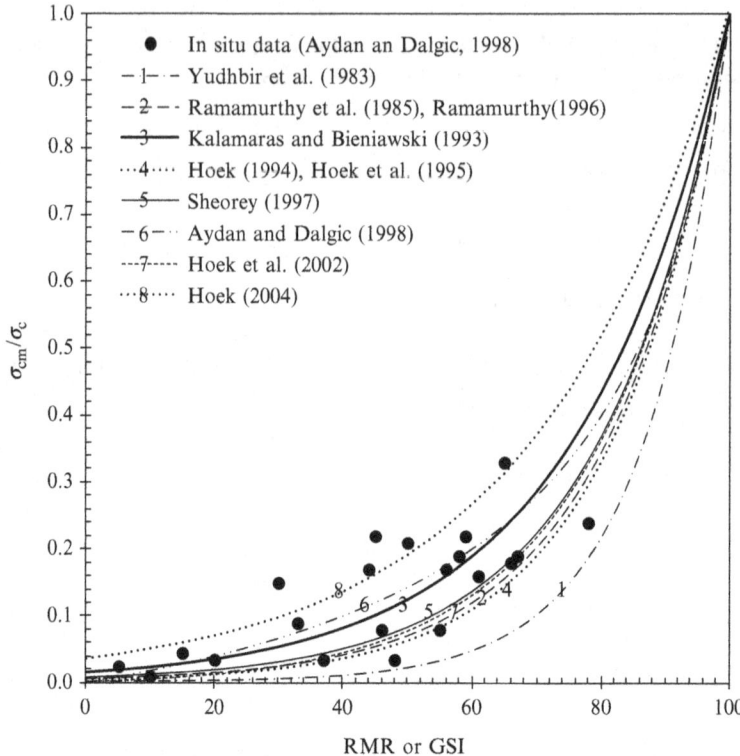

FIG. 7.14 Variation of ratio of rock mass unconfined compressive strength σ_{cm} to intact rock unconfined compressive strength σ_c with RMR or GSI ratings.

(2) Different empirical relations often give very different unconfined compressive strength values even for rock masses at the same site, as can be clearly seen from Fig. 7.15, which shows the predicted σ_{cm} values of the 13 rock masses listed in Table 6.12. So it is important that the estimation of rock mass unconfined compressive strength should not rely on a single empirical relation. Instead, various empirical relations should be used to get an idea on the possible range of the rock mass unconfined compressive strength.

7.4.2 Tensile Strength of Rock Mass

The tensile strength of a rock mass can also be estimated using empirical relations such as the one below by Singh and Goel (1999):

$$\sigma_{tm} = -0.029\gamma f_c Q^{0.3} \ (\text{MPa}) \tag{7.48}$$

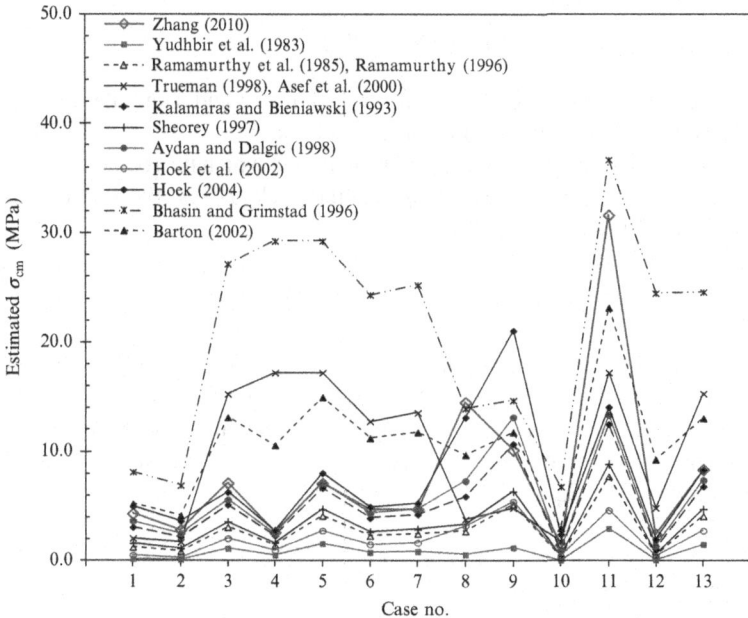

FIG. 7.15 Estimated rock mass unconfined compressive strength values from different empirical methods based on RQD, RMR, GSI, and/or Q.

where $f_c = \sigma_c/100$ for $Q > 10$ and $\sigma_c > 100$ MPa, otherwise $f_c = 1$; and γ is the unit weight of the rock mass in g/cm^3.

The strength criteria presented in Section 7.4.3 can also be used to obtain the tensile strength of a rock mass. For example, with the Hoek-Brown strength criterion for rock masses (Eq. 7.51), the tensile strength can be determined by

$$\sigma_{tm} = 0.5\sigma_c \left[m_b - \left(m_b^2 + 4s \right)^{0.5} \right] \tag{7.49}$$

where m_b is the material constant for the rock mass and s is a constant that depends on the characteristics of the rock mass (see Section 7.4.3 about the determination of m_b and s).

7.4.3   Empirical Strength Criteria of Rock Mass

There are different empirical strength criteria for rock masses. The four empirical strength criteria of rock masses, corresponding to those of intact rock presented in Section 7.2.3, are described in this section.

7.4.3.1   Hoek-Brown criterion

For jointed rock masses, the general form of the Hoek-Brown criterion, which incorporates both the original and the modified form, is given by

$$\sigma_1' = \sigma_3' + \sigma_c \left(m_b \frac{\sigma_3'}{\sigma_c} + s \right)^a \tag{7.50}$$

where m_b is the material constant for the rock mass; and s and a are constants that depend on the characteristics of the rock mass.

For most rocks of good to reasonable quality in which the rock mass strength is controlled by tightly interlocking angular rock pieces, the failure can be defined by setting $a = 0.5$ in Eq. (7.50), ie,

$$\sigma_1' = \sigma_3' + \sigma_c \left(m_b \frac{\sigma_3'}{\sigma_c} + s \right)^{0.5} \tag{7.51}$$

For poor-quality rock masses in which the tight interlocking has been partially destroyed by shearing or weathering, the rock mass has no tensile strength or "cohesion" and specimens will fall apart without confinement. For such rock masses the following modified criterion is more appropriate and it is obtained by putting $s = 0$ in Eq. (7.50):

$$\sigma_1' = \sigma_3' + \sigma_c \left(m_b \frac{\sigma_3'}{\sigma_c} \right)^a \tag{7.52}$$

For practical applications of Eqs. (7.50)–(7.52), it is important to determine the material constants m_b, s, and a. Hoek and Brown (1988) proposed a set of relations between the parameters m_b, s, and a and the 1976 version of Bieniawski's rock mass rating (RMR), assuming completely dry conditions and a very favorable (according to the RMR rating system) discontinuity orientation:

(i) disturbed rock masses

$$m_b = \exp\left(\frac{RMR - 100}{14} \right) m_i \tag{7.53a}$$

$$s = \exp\left(\frac{RMR - 100}{6} \right) \tag{7.53b}$$

$$a = 0.5 \tag{7.53c}$$

(ii) undisturbed or interlocking rock masses

$$m_b = \exp\left(\frac{RMR - 100}{28} \right) m_i \tag{7.54a}$$

$$s = \exp\left(\frac{RMR - 100}{9} \right) \tag{7.54b}$$

$$a = 0.5 \tag{7.54c}$$

Eqs. (7.53a)–(7.53c) and (7.54a)–(7.54c) are acceptable for rock masses with RMR values greater than about 25, but they do not work for very poor rock masses since the minimum value which RMR can assume is 18 for the 1976 RMR system and 23 for the 1989 RMR system (see Chapter 5 for details). To overcome this limitation, Hoek (1994) and Hoek et al. (1995) proposed the relations between parameters m_b, s, and a and the GSI are as follows:

(i) For GSI > 25, ie, rock masses of good to reasonable quality

$$m_b = \exp\left(\frac{GSI - 100}{28}\right) m_i \qquad (7.55a)$$

$$s = \exp\left(\frac{GSI - 100}{9}\right) \qquad (7.55b)$$

$$a = 0.5 \qquad (7.55c)$$

(ii) For GSI < 25, ie, rock masses of very poor quality

$$m_b = \exp\left(\frac{GSI - 100}{28}\right) m_i \qquad (7.56a)$$

$$s = 0 \qquad (7.56b)$$

$$a = 0.65 - \frac{GSI}{200} \qquad (7.56c)$$

Hoek et al. (2002) proposed new relationships between parameters m_b, s, and a and GSI by introducing a new parameter D, which is a factor that depends on the degree of disturbance due to blast damage and stress relaxation. The values of D range from 0 for undisturbed in situ rock masses to 1 for very disturbed rock masses. The guidelines for selecting D can be found by Hoek (2006).

$$m_b = \exp\left(\frac{GSI - 100}{28 - 14D}\right) m_i \qquad (7.57a)$$

$$s = \exp\left(\frac{GSI - 100}{9 - 3D}\right) \qquad (7.57b)$$

$$a = 0.5 + \frac{1}{6}[\exp(-GSI/15) - \exp(-20/3)] \qquad (7.57c)$$

Water has a great effect on the strength of rock masses. Many rocks show a significant strength decrease with increasing moisture content. Therefore, it is important to conduct laboratory tests at moisture contents which are as close as possible to those in the field. A more important effect of water is the strength reduction which occurs as a result of water pressures in the pore spaces in the rock. This is why the effective not the total stresses are used in the Hoek-Brown strength criterion.

The Hoek-Brown strength criterion was originally developed for intact rock and then extended to rock masses. The process used by Hoek and Brown in deriving their strength criterion for intact rock (Eq. 7.23) was one of pure trial and error (Hoek et al., 1995). Apart from the conceptual starting point provided by the Griffith theory, there is no fundamental relationship between the empirical constants included in the criterion and any physical characteristics of the rock. The justification for choosing this particular criterion (Eq. 7.23) over the numerous alternatives lies in the adequacy of its predictions of the observed rock fracture behavior, and the convenience of its application to a range of typical engineering problems (Hoek, 1983). The material constants m_i is derived based upon analyses of published triaxial test results of intact rock (Hoek, 1983; Doruk, 1991; Hoek et al., 1992). The strength criterion for rock masses is just an empirical extension of the criterion for intact rock. Since it is practically impossible to determine the material constants m_b and s using triaxial tests on rock masses, empirical relations are suggested to estimate them from RMR or GSI. The RMR and GSI rating systems are also empirical. For these reasons, the Hoek-Brown empirical rock mass strength criterion must be used with extreme care. In discussing the limitations in the use of their strength criterion, Hoek and Brown (1988) emphasized that it is not applicable to anisotropic rocks or to elements of rock masses that behave anisotropically by virtue of containing only a few discontinuities. Alternative empirical approaches and further developments of the Hoek-Brown criterion which seek to account for some of its limitations are given by Amadei (1988), Pan and Hudson (1988), Ramamurthy and Arora (1991), Amadei and Savage (1993), Ramamurthy (1993), and Saroglou and Tsiambaos (2008).

7.4.3.2 Bieniawski-Yudhbir criterion

Based on tests of jointed gypsum-celite specimens, Yudhbir et al. (1983) revised the strength criterion for intact rock (Eq. 7.24) to the form of

$$\frac{\sigma_1'}{\sigma_c} = a + b \left(\frac{\sigma_3'}{\sigma_c} \right)^{0.65} \tag{7.58}$$

to fit rock masses. Yudhbir et al. (1983) recommended that the parameter a in Eq. (7.58) be determined from

$$a = 0.0176 Q^{0.65} \quad \text{or} \quad a = \exp\left[7.65 \left(\frac{\text{RMR} - 100}{100} \right) \right] \tag{7.59}$$

where Q is the classification index of Barton et al. (1974) and RMR is Bieniawski's 1976 rock mass rating (Bieniawski, 1976). The parameter b in Eq. (7.58) is determined from Table 7.11.

Kalamaras and Bieniawski (1993) suggested that both a and b should be varied with RMR for better results and proposed the criterion of Table 7.17 specifically for coal seams.

TABLE 7.17 Rock Mass Criterion for Coal Seams by Kalamaras and Bieniawski (1993)

Equation	Parameters
$\dfrac{\sigma_1'}{\sigma_c} = a + 4\left(\dfrac{\sigma_3'}{\sigma_c}\right)^{0.6}$	$a = \exp\left(\dfrac{RMR - 100}{14}\right)$
$\dfrac{\sigma_1'}{\sigma_c} = a + b\left(\dfrac{\sigma_3'}{\sigma_c}\right)^{0.6}$	$a = \exp\left(\dfrac{RMR - 100}{12}\right), \quad b = \exp\left(\dfrac{RMR + 20}{52}\right)$

7.4.3.3 Johnston criterion

For rock masses, Johnston (1985) proposed the following strength criterion:

$$\sigma_{1n}' = \left(\frac{M}{B}\sigma_{3n}' + s\right)^{B} \qquad (7.60)$$

where σ_{1n}' and σ_{3n}' are the normalized effective principal stresses at failure, obtained by dividing the effective principal stresses, σ_1' and σ_3', by the relevant unconfined compressive strength, σ_c; B and M are intact material constants as described in Section 7.2.3; and s is a constant to account for the strength of discontinuous soil and rock masses in a manner similar to that used for the Hoek-Brown criterion.

7.4.3.4 Ramamurthy criterion

For rock masses, the strength criterion has the same form as that for intact rock (Ramamurthy et al., 1985; Ramamurthy, 1986, 1993), ie,

$$\frac{\sigma_1' - \sigma_3'}{\sigma_3'} = B_m\left(\frac{\sigma_{cm}}{\sigma_3'}\right)^{\alpha_m} \qquad (7.61)$$

where σ_{cm} is the rock mass unconfined compressive strength; B_m is a material constant for rock masses; and α_m is the slope of the plot between $(\sigma_1' - \sigma_3')/\sigma_3'$ and σ_{cm}/σ_3', which can be assumed to be 0.8 for rock masses as well. σ_{cm} and B_m can be obtained by

$$\sigma_{cm} = \sigma_c \exp\left(\frac{RMR - 100}{18.75}\right) \qquad (7.62a)$$

$$B_m = B_r \exp\left(\frac{RMR - 100}{75.5}\right) \qquad (7.62b)$$

in which σ_c is the unconfined compressive strength of intact rock; and B_r is a material constant for intact rock, as in Eq. (7.30).

7.4.3.5 *Comments*

Besides the four empirical strength criteria of rock masses described above, there are many others. All these criteria are purely empirical and thus it is impossible to say which one is correct or which one is not. However, the Hoek-Brown strength criterion is the most widely referred and used. Since its advent in 1980, considerable application experience has been gained by its authors as well as by others. As a result, this criterion has been modified several times to meet the needs of users who have applied it to conditions which were not visualized when it was originally developed.

It is noted that all the empirical strength criteria described above for rock masses have the following limitations:

1. The criteria are not applicable to anisotropic rock masses. So they can be used only when the rock masses are approximately isotropic, ie, when the discontinuity orientation does not have a dominant effect on failure. To consider the anisotropy, either the strength criterion parameters are modified as done by Saroglou and Tsiambaos (2008) or the equivalent continuum method in Section 7.4.5 is used.
2. The influence of the intermediate principal stress, which in some cases is important, is not considered. This will be further discussed in Section 7.5.

7.4.4   Mohr-Coulomb Parameters of Rock Mass

Since many of the numerical models and limit equilibrium analyses used in rock mechanics and rock engineering are expressed in terms of the Mohr-Coulomb failure criterion, it is important to estimate the cohesion and friction angle parameters of rock masses:

$$\tau_f = c_m + \sigma'_n \tan \phi_m \qquad (7.63)$$

where τ_f is the shear strength of the rock mass; c_m and ϕ_m are, respectively, the cohesion and internal friction angle of the rock mass; and σ'_n is the effective normal stress on the sliding plane.

Estimation of c_m and ϕ_m can be done using one of the empirical strength criteria presented in Section 7.4.3 and based on the solution published by Balmer (1952) in which the normal and shear stresses are expressed in terms of the corresponding principal stresses as follows:

$$\sigma'_n = \sigma'_3 + \frac{\sigma'_1 - \sigma'_3}{\dfrac{\partial \sigma'_1}{\partial \sigma'_3} + 1} \qquad (7.64)$$

$$\tau_f = (\sigma'_n - \sigma'_3) \sqrt{\frac{\partial \sigma'_1}{\partial \sigma'_3}} \qquad (7.65)$$

For example, for the Hoek-Brown strength criterion, one can have
For GSI > 25, when $a = 0.5$:

$$\frac{\partial \sigma_1'}{\partial \sigma_3'} = 1 + \frac{m_b \sigma_c}{2(\sigma_1' - \sigma_3')} \tag{7.66}$$

For GSI < 25, when $s = 0$:

$$\frac{\partial \sigma_1'}{\partial \sigma_3'} = 1 + a m_b^a \left(m_b \frac{\sigma_3'}{\sigma_c} \right)^{a-1} \tag{7.67}$$

Once a set of (σ_n', τ_f) values have been calculated from Eqs. (7.64), (7.65), the equivalent Mohr envelope defined by the following expression can be used to fit the (σ_n', τ_f) data:

$$\tau_f = A \sigma_c \left(\frac{\sigma_n' - \sigma_{tm}}{\sigma_c} \right)^B \tag{7.68}$$

where A and B are material constants which can be determined by fitting analysis; σ_n' is the effective normal stress; and σ_{tm} is the tensile strength of the rock mass which can be determined from Eq. (7.49).

After A and B are determined, the friction angle at a specified effective normal stress can be obtained by

$$\phi_m = \arctan \left[AB \left(\frac{\sigma_n' - \sigma_{tm}}{\sigma_c} \right)^{B-1} \right] \tag{7.69}$$

and the corresponding cohesion can then be determined by

$$c_m = \tau_f - \sigma_n' \tan \phi_m \tag{7.70}$$

The cohesion c_m given by Eq. (7.70) is an upper bound value and needs to be reduced to about 75% of the calculated value for practical applications (Hoek, 2006).

The values of c_m and ϕ_m obtained from the above method are very sensitive to the range of the minor principal stress σ_3' used to generate the (σ_n', τ_f) data sets. According to Hoek (2006), the most consistent results can be obtained when eight equally spaced values of σ_3' are used in the range of $0 < \sigma_3' < \sigma_c$. Fig. 7.16 shows the cohesive strength and friction angles of rock masses for different GSI and m_i values from Hoek (2006).

7.4.5 Equivalent Continuum Method for Estimating Rock Mass Strength

The equivalent continuum method treats the jointed rock mass as an equivalent anisotropic continuum with strength properties that are directional and reflect the properties of both the intact rock and the discontinuities. The discontinuities

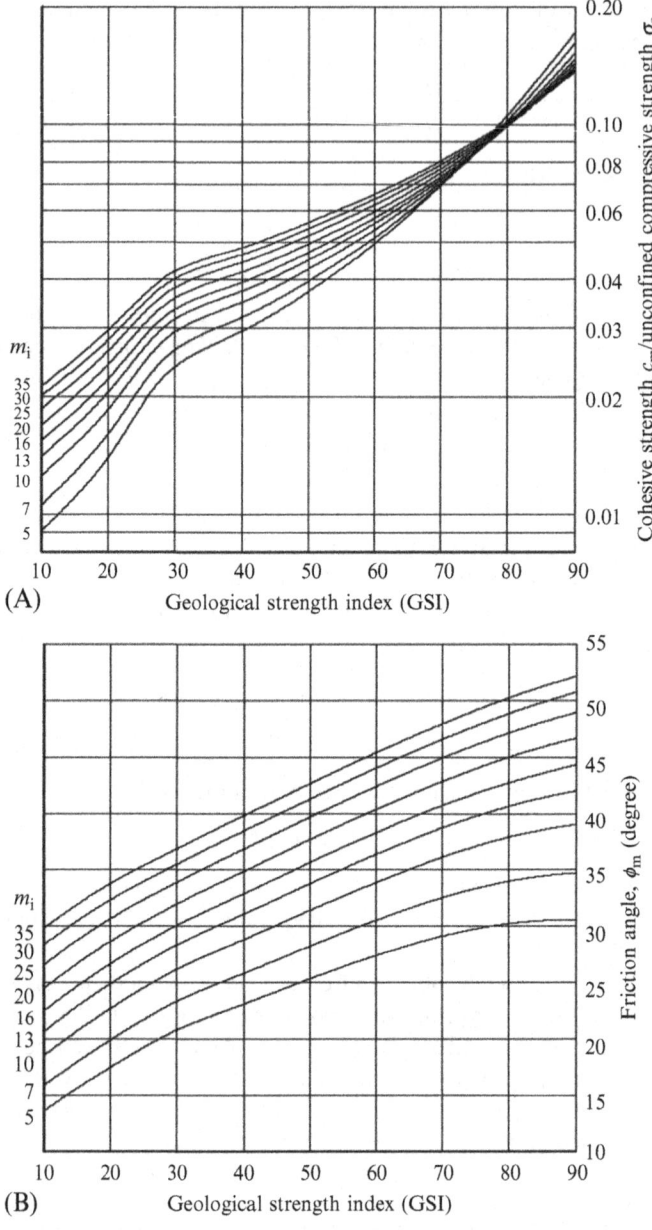

FIG. 7.16 (A) Cohesive strength c_m; and (B) friction angle ϕ_m at different GSI and m_i values. *(From Hoek, E., 2006. Rock engineering. http://www.rocscience.com.)*

are characterized without reference to their specific locations. Jaeger (1960) and Jaeger and Cook (1979) presented an equilibrium continuum strength model for jointed rock masses under axisymmetric loading condition. In their model, the strength of both the intact rock and the discontinuities are described by the Mohr-Coulomb criterion. Since the effect of the intermediate principal stress is not considered in the model of Jaeger (1960), Jaeger and Cook (1979), Amadei (1988), and Amadei and Savage (1989, 1993) derived solutions for the strength of a jointed rock mass under a variety of multiaxial stress states. As in the model of Jaeger (1960) and Jaeger and Cook (1979), the modeled rock mass is cut by a single discontinuity set. In the formulations of Amadei (1988) and Amadei and Savage (1989, 1993), however, the intact rock strength is described by the Hoek-Brown strength criterion and the discontinuity strength is modeled using a Mohr-Coulomb criterion with a zero tensile strength cut-off.

7.4.5.1 Model of Jaeger (1960) and Jaeger and Cook (1979)

Fig. 7.17A shows a cylindrical rock mass specimen subjected to an axial major principal stress σ_1' and a lateral minor principal stress σ_3'. The rock mass is cut by well-defined parallel discontinuities inclined at an angle β to the major principal stress σ_1'. The strength of both the intact rock and the discontinuities is described by the Mohr-Coulomb criterion, ie,

$$\tau_i = c_i + \sigma_n' \tan\phi_i \qquad (7.71)$$

$$\tau_j = c_j + \sigma_n' \tan\phi_j \qquad (7.72)$$

where τ_i and τ_j are, respectively, the shear strength of the intact rock and the discontinuities; c_i and ϕ_i are, respectively, the cohesion and internal friction angle of the intact rock; c_j and ϕ_j are, respectively, the cohesion and internal friction angle of the discontinuities; and σ_n' is the effective normal stress on the shear plane.

For the applied stresses on the rock mass cylinder, the effective normal stress σ_n' and the shear stress τ on a plane which makes an angle β' to the σ_1' axis are, respectively, given by

$$\sigma_n' = \frac{1}{2}(\sigma_1' + \sigma_3') - \frac{1}{2}(\sigma_1' - \sigma_3')\cos 2\beta' \qquad (7.73)$$

$$\tau = \frac{1}{2}(\sigma_1' - \sigma_3')\sin 2\beta' \qquad (7.74)$$

If shear failure occurs on the discontinuity plane, the effective normal stress σ_n' and the shear stress τ on the discontinuity plane can be obtained by replacing β' in Eqs. (7.73), (7.74) by β. Adopting the obtained stresses on the discontinuity plane to substitute for σ_n' and τ_j in Eq. (7.72) and then rearranging, the effective major principal stress required to cause shear failure along the discontinuity can be obtained as follows:

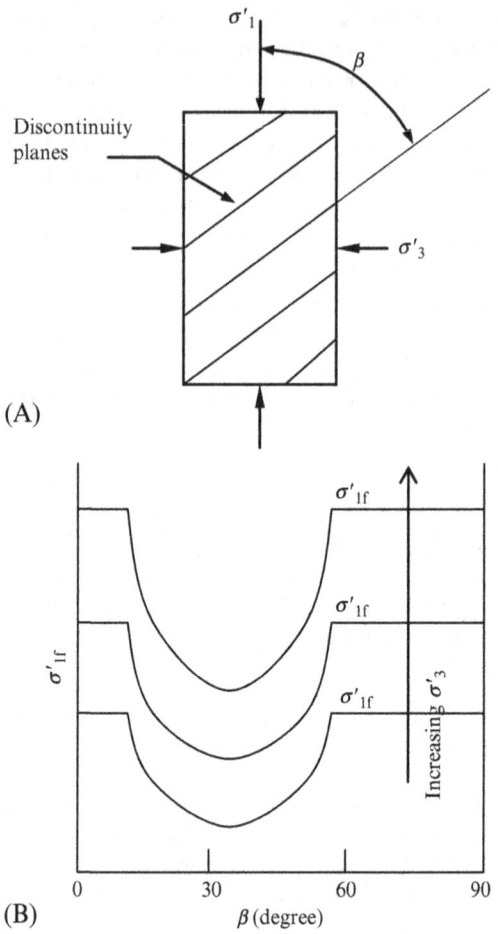

FIG. 7.17 Variation of compressive strength with angle β of the discontinuity plane. *(Based on Jaeger, J.C., Cook, N.G.W., 1979. Fundamentals of Rock Mechanics, third ed. Chapman and Hall, London.)*

$$\sigma'_{1f} = \sigma'_3 + \frac{2\left(c_j + \sigma'_3 \tan\phi_j\right)}{\sin 2\beta \left(1 - \tan\phi_j \tan\beta\right)} \tag{7.75}$$

Similarly, if the shear failure occurs in the intact rock, the minimum effective major principal stress can be obtained by

$$\sigma'_{1f} = 2c_i \tan\left(\frac{\pi}{4} + \frac{\phi_i}{2}\right) + \sigma'_3 \tan^2\left(\frac{\pi}{4} + \frac{\phi_i}{2}\right) \tag{7.76}$$

The model of Jaeger (1960) and Jaeger and Cook (1979) assumes that failure during compressive loading of a rock mass cylinder under a lateral stress σ'_3 (Fig. 7.17A) will occur when σ'_1 exceeds the smaller of the σ'_{1f} values given by Eqs. (7.75), (7.76). Fig. 7.17B shows the variation of σ'_{1f} with β, from which one can clearly see the anisotropy of the rock mass strength caused by the discontinuities.

7.4.5.2 Model of Amadei (1988) and Amadei and Savage (1989, 1993)

The principle used by Amadei (1988) and Amadei and Savage (1989, 1993) to derive the expressions of jointed rock mass strength is the same as that used by Jaeger (1960) and Jaeger and Cook (1979). However, since the effect of the intermediate principal stress is included and the nonlinear Hoek-Brown strength criterion is used, the derivation process and the final results are much more complicated. Due to the limit of space, only some of the typical results of Amadei and Savage (1989, 1993) are shown here.

Consider a jointed rock mass cube under stresses of σ'_x, σ'_y, and σ'_z. The orientation of the discontinuity plane is defined by two angles β and Ψ with respect to the xyz coordinate system (Fig. 7.18). Let nst be another coordinate system attached to the discontinuity plane such that the n-axis is along the discontinuity upward normal and the s- and t-axes are in the discontinuity plane. The t-axis is in the xz plane. The upward unit vector $\mathbf{n}$ has direction cosines as follows:

$$\bar{x} = \sin\psi\cos\beta; \quad \bar{y} = \cos\psi; \quad \bar{z} = \sin\psi\sin\beta \tag{7.77}$$

Defining $m = \sigma'_y/\sigma'_x$ and $n = \sigma'_z/\sigma'_x$ and introducing two functions

$$F_f = \tau^2 - \sigma'^2_n\tan\phi^2_j \quad \text{and} \quad F_n = \sigma'_n \tag{7.78}$$

where σ'_n and τ are, respectively, the normal and shear stresses acting on the discontinuity; and ϕ_j is the friction angle of the discontinuity, the limiting equilibrium (incipient slip) condition of the discontinuity can be derived as:

$$\begin{aligned} F_f = \sigma'^2_x\{ &[\bar{x}^2 - \bar{x}^4(1+\tan^2\phi_j)] + m^2[\bar{y}^2 - \bar{y}^4(1+\tan^2\phi_j)] \\ &+ m^2[\bar{z}^2 - \bar{z}^4(1+\tan^2\phi_j)] - 2m\bar{x}^2\bar{y}^2(1+\tan^2\phi_j) \\ &- 2n\bar{x}^2\bar{z}^2(1+\tan^2\phi_j) - 2mn\bar{y}^2\bar{z}^2(1+\tan^2\phi_j)\} = 0 \end{aligned} \tag{7.79}$$

The nonnegative normal stress condition of the discontinuity is

$$F_n = \sigma'_x(\bar{x}^2 + m\bar{y}^2 + n\bar{z}^2) \geq 0 \tag{7.80}$$

So for a discontinuity with orientation angles β and Ψ, the condition $F_f = 0$ corresponds to impending slip. No slip takes place when F_f is negative. Fig. 7.19 shows a typical set of failure surfaces $F_f(m, n) = 0$ for Ψ equal to 40 or 80 degrees and β ranging between 0 and 90 degrees. In this figure the ranges $F_f(m, n) > 0$ are shaded and $F_n = 0$ is represented as a dashed straight line. The positive normal stress condition ($F_n > 0$) is shown as the region on either side of the line $F_n = 0$ depending on the sign of σ_x.

Depending on the ordering of σ'_x, σ'_y, and σ'_z, the Hoek-Brown strength criterion for intact rock (Eq. 7.23) assumes six possible forms as shown in

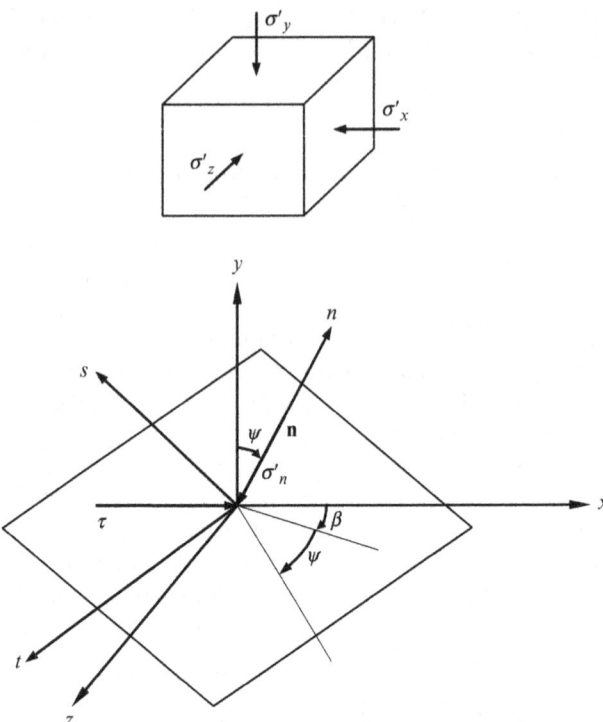

FIG. 7.18 Discontinuity plane in a triaxial stress field. *(Based on Amadei, B., Savage, W.Z., 1993. Effect of joints on rock mass strength and deformability. In: Hudson, J.A. (Ed.), Comprehensive Rock Engineering—Principle, Practice and Projects, vol. 1. Pergamon, Oxford, pp. 331–365.)*

Table 7.18. Using $m_i = 7$ and $\sigma_c = 42$ MPa, the intact rock failure surfaces for different values of σ_x'/σ_c can be obtained as shown in Fig. 7.20.

The failure surfaces of the jointed rock mass can then be obtained by superposition of the discontinuity failure surfaces and the intact rock failure surfaces. Fig. 7.21 is obtained by superposition of the failure surfaces in Figs. 7.19 and 7.20. The following remarks can be made about the diagrams shown in Fig. 7.21:

1. In general, for a given value of σ_x'/σ_c, the size of the stable domain enclosed by the intact rock failure surface is reduced because of the discontinuities. The symmetry of the intact rock failure surface with respect to the $m = n$ axis in the m, n space is lost. The strength of the jointed rock mass is clearly anisotropic.
2. The strength reduction associated with the discontinuities is more pronounced for discontinuities with orientation angles β and Ψ for which the discontinuity failure surface in the m, n space is ellipse than when it is a hyperbola or a parabola.

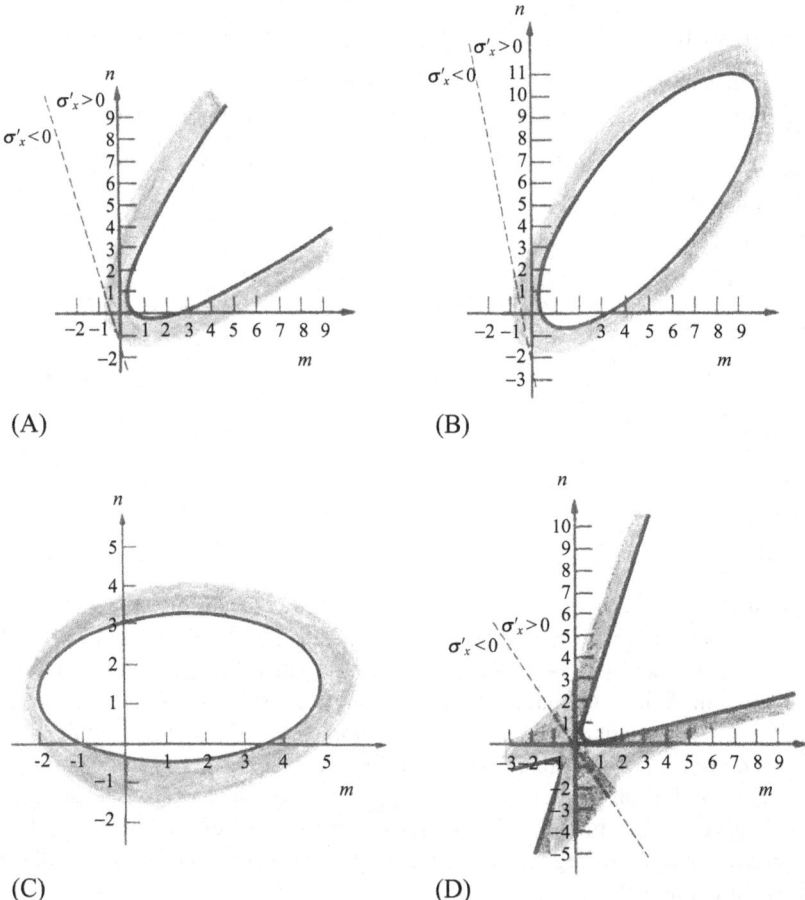

FIG. 7.19 Shape of failure surface $F_f(m, n)=0$ in the $m=\sigma'_y/\sigma'_x$, $n=\sigma'_z/\sigma'_x$ space for (A) $\beta=38.935$ degree, $\Psi=40$ degree; (B) $\beta=30$ degree, $\Psi=40$ degree; (C) $\beta=20$ degree, $\Psi=80$ degree; and (D) $\beta=70$ degree, $\Psi=40$ degree. The region $F_n>0$ is above the dashed line $F_n=0$ when σ'_x is compressive and below that line when σ'_x is tensile. Friction angle $\phi_j=30$ degree. *(Based on Amadei, B., Savage, W.Z., 1993. Effect of joints on rock mass strength and deformability. In: Hudson, J.A. (Ed.), Comprehensive Rock Engineering—Principle, Practice and Projects, vol. 1. Pergamon, Oxford, pp. 331–365.)*

3. Despite the zero discontinuity tensile strength and the strength reduction associated with the discontinuities, jointed rock masses can be stable under a wide variety of states of stress σ'_x, $\sigma'_y=m\sigma'_x$, $\sigma'_z=n\sigma'_x$. These states of stress depend on the values of discontinuity orientation angles β and Ψ and the stress ratio σ'_x/σ_c.

TABLE 7.18 Forms of Eq. (7.23) for Different Orderings of σ'_x, σ'_y, and σ'_z

Principal Stress Ordering	Major Stress σ'_1	Minor Stress σ'_3	Forms of Eq. (7.23)
$\sigma'_x > \sigma'_y > \sigma'_z$	σ'_x	σ'_z	$\sigma'_x = \sigma'_z + \sigma_c\sqrt{m_i\sigma'_z/\sigma_c + 1}$
$\sigma'_x > \sigma'_z > \sigma'_y$	σ'_x	σ'_y	$\sigma'_x = \sigma'_y + \sigma_c\sqrt{m_i\sigma'_y/\sigma_c + 1}$
$\sigma'_y > \sigma'_x > \sigma'_z$	σ'_y	σ'_z	$\sigma'_y = \sigma'_z + \sigma_c\sqrt{m_i\sigma'_z/\sigma_c + 1}$
$\sigma'_y > \sigma'_z > \sigma'_x$	σ'_y	σ'_x	$\sigma'_y = \sigma'_x + \sigma_c\sqrt{m_i\sigma'_x/\sigma_c + 1}$
$\sigma'_z > \sigma'_x > \sigma'_y$	σ'_z	σ'_y	$\sigma'_z = \sigma'_y + \sigma_c\sqrt{m_i\sigma'_y/\sigma_c + 1}$
$\sigma'_z > \sigma'_y > \sigma'_x$	σ'_z	σ'_x	$\sigma'_z = \sigma'_x + \sigma_c\sqrt{m_i\sigma'_x/\sigma_c + 1}$

7.4.5.3 Comments

The model of Jaeger (1960) and Jaeger and Cook (1979) discussed earlier considers a rock mass with one discontinuity set. If the model of Jaeger (1960) and Jaeger and Cook (1979) is applied to a rock mass with several discontinuity sets, the strength of the rock mass can be obtained by considering the effect of each discontinuity set. For example, consider a simple case of two discontinuity sets A and B in Fig. 7.22A, with the angle between them being α. The corresponding variation of the compressive strength $\sigma'_{1\beta}$, if the two discontinuity sets are present singly, is shown in Fig. 7.22B. As the angle β_a of discontinuity set A is changed from 0 to 90 degrees, the angle β_b of discontinuity set B with the major stress direction will be

$$\beta_b = |\alpha - \beta_a| \quad \text{for} \quad a \leq 90° \tag{7.81}$$

When β_a is varied from 0 to 90 degrees, the resultant strength variation for $\alpha = 60$ and 90 degrees will be as in Fig. 7.22C, by choosing the smaller of $\sigma'_{1\beta a}$ and $\sigma'_{1\beta b}$ from the curves in Fig. 7.22B.

Hoek and Brown (1980) have shown that with three or more discontinuity sets, all sets having identical strength characteristics, the rock mass will exhibit an almost flat strength variation (Fig. 7.23). This means that for highly jointed rock masses, it is possible to adopt one of the empirical isotropic rock mass failure criteria presented in Section 7.4.3.

It should be noted that, in the models of the equivalent continuum method, discontinuities are assumed to be persistent and all discontinuities in one set have the same orientation. In reality, however, discontinuities are usually nonpersistent and the discontinuities in one set have orientation distributions.

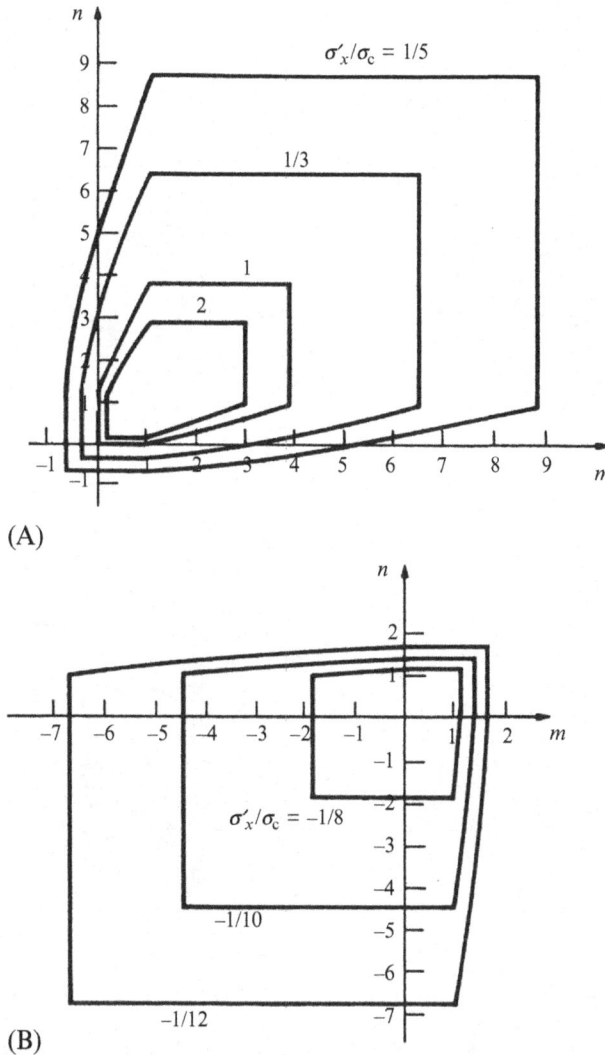

FIG. 7.20 Geometrical representation of the Hoek-Brown failure surface for intact rock in the $m = \sigma'_y / \sigma'_x$, $n = \sigma'_z / \sigma'_x$ space for different values of σ'_x / σ_c with $m_i = 7$ and $\sigma_c = 42$ MPa: (A) σ'_x is compressive and (B) σ'_x is tensile. *(Based on Amadei, B., Savage, W.Z., 1993. Effect of joints on rock mass strength and deformability. In: Hudson, J.A. (Ed.), Comprehensive Rock Engineering— Principle, Practice and Projects, vol. 1. Pergamon, Oxford, pp. 331–365.)*

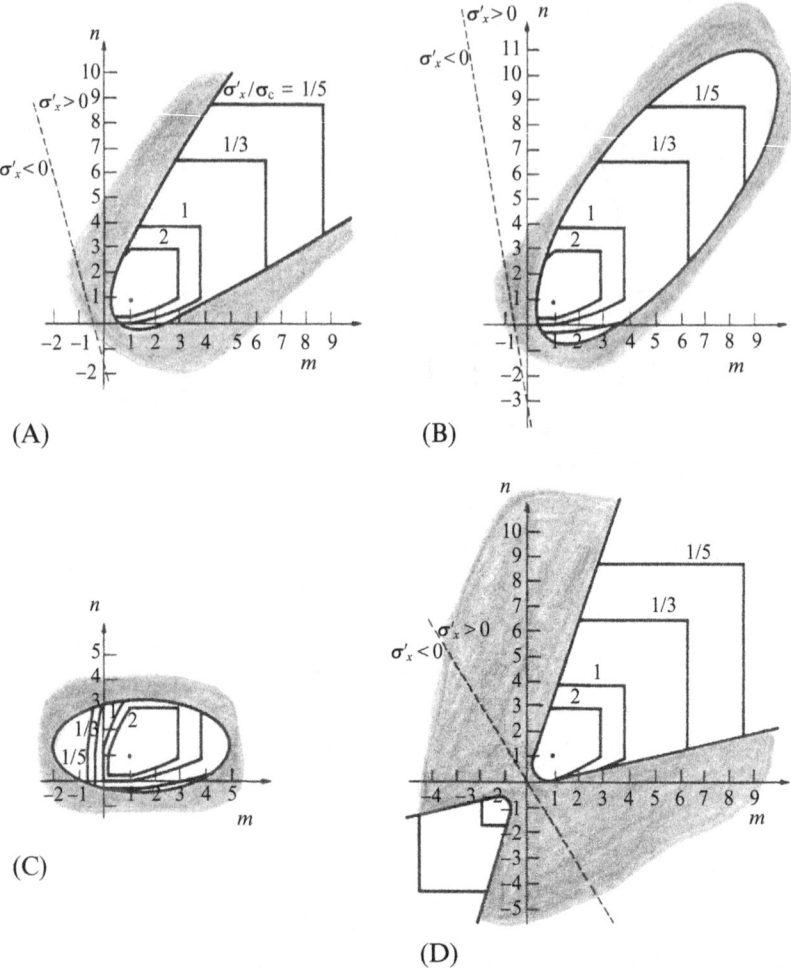

FIG. 7.21 Superposition of the joint failure surface with $\phi_j = 30$ degree and the intact rock failure surface with $m_i = 7$ and $\sigma_c = 42$ MPa in the $m = \sigma'_y/\sigma'_x$, $n = \sigma'_z/\sigma'_x$ space for (A) $\beta = 38.935$ degree, $\Psi = 40$ degree; (B) $\beta = 30$ degree, $\Psi = 40$ degree; (C) $\beta = 20$ degree, $\Psi = 80$ degree; and (D) $\beta = 70$ degree, $\Psi = 40$ degree. *(Based on Amadei, B., Savage, W.Z., 1993. Effect of joints on rock mass strength and deformability. In: Hudson, J.A. (Ed.), Comprehensive Rock Engineering— Principle, Practice and Projects, vol. 1. Pergamon, Oxford, pp. 331–365.)*

7.5   THREE-DIMENSIONAL STRENGTH CRITERIA OF ROCK

The strength criteria presented in Sections 7.2.3 and 7.4.3 do not take account of the influence of the intermediate principal stress although much evidence has indicated that the intermediate principal stress does influence the rock strength in many instances. Therefore, researchers have developed different 3D strength criteria for rock, such as the modified Wiebols and Cook criterion

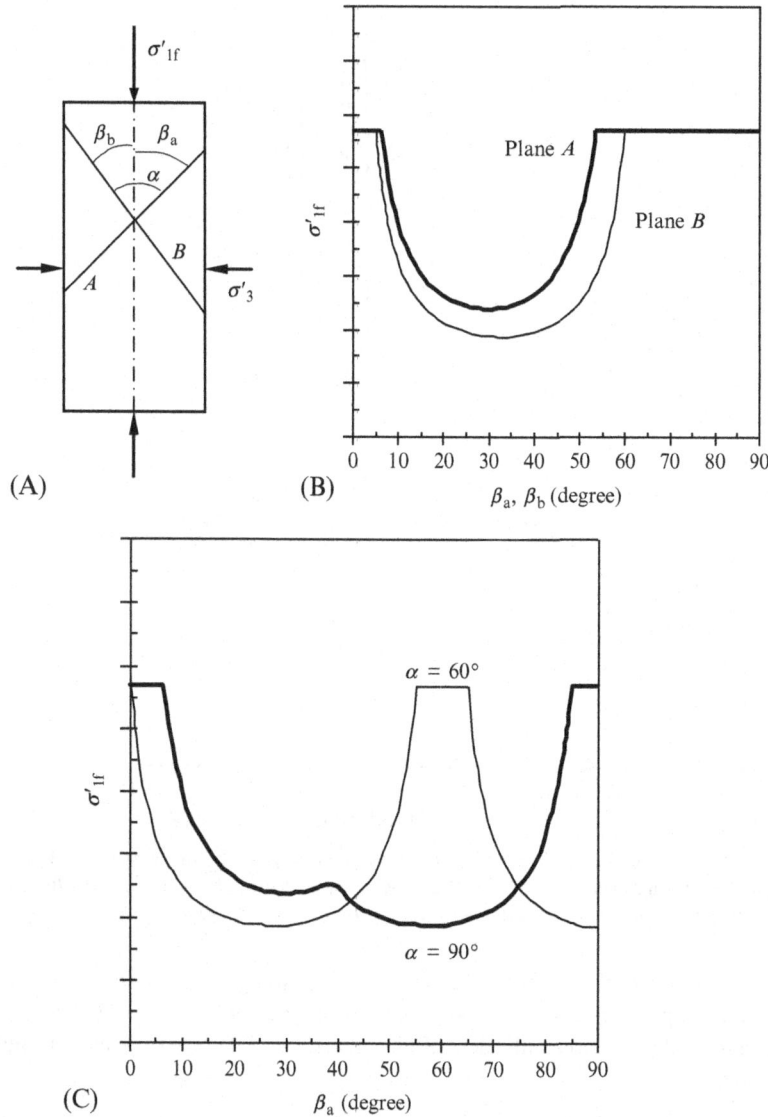

FIG. 7.22   (A) Rock mass with two discontinuity sets A and B; (B) strength variation with β if the discontinuity sets are present singly; and (C) strength variation when both discontinuity sets are present.

by Zhou (1994), the modified Lade criterion by Ewy (1999), the Mogi-Coulomb criterion by Al-Ajmi and Zimmerman (2005), the generalized 3D criterion by Jaiswal and Shrivastva (2012), and the 3D versions of the Hoek-Brown strength criterion by Pan and Hudson (1988), Priest (2005), Zhang and Zhu (2007), and Zhang (2008). The 3D versions of the Hoek-Brown strength criterion have

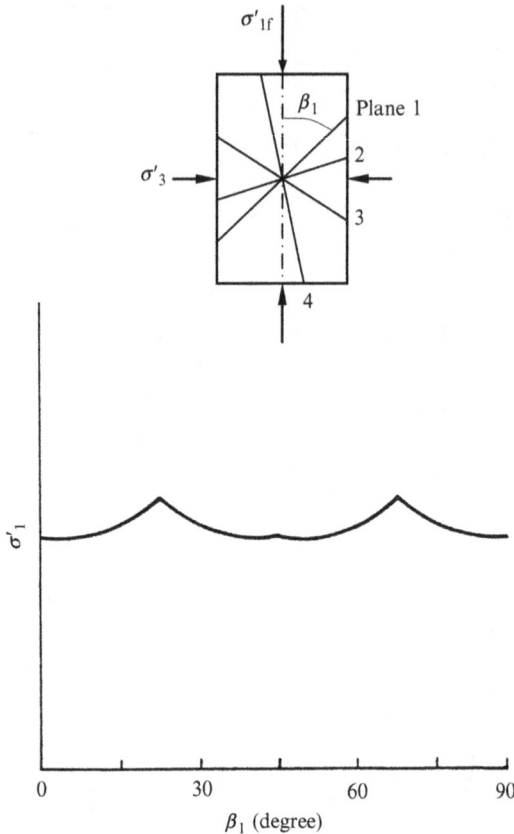

FIG. 7.23 Strength variation with angle β_1 of discontinuity plane 1 in the presence of 4 discontinuity sets, the angle between two adjoining planes being 45 degree. *(Based on Hoek, E., Brown, E. T., 1980. Empirical strength criterion for rock masses. J. Geotech. Eng. ASCE 106, 1013–1035.)*

advantages over other non Hoek-Brown type 3D strength criteria in that they use the same input parameters as the widely used Hoek-Brown strength criterion. Therefore, this section only describes some of the 3D Hoek-Brown strength criteria.

7.5.1 Pan-Hudson Criterion

Pan and Hudson (1988) proposed a 3D version of the original Hoek-Brown strength criterion for rock mass (Eq. 7.51), which is expressed as

$$\frac{9}{2\sigma_c}\tau_{oct}^2 + \frac{3}{2\sqrt{2}}m_b\tau_{oct} - m_b\frac{I_1}{3} = s\sigma_c \tag{7.82}$$

where τ_{oct} and I_1 are, respectively, the octahedral shear stress and the first stress invariant defined by

$$\tau_{\text{oct}} = \frac{1}{3}\sqrt{(\sigma_1' - \sigma_2')^2 + (\sigma_2' - \sigma_3')^2 + (\sigma_3' - \sigma_1')^2} \qquad (7.83)$$

$$I_1 = \sigma_1' + \sigma_2' + \sigma_3' \qquad (7.84)$$

in which σ_1', σ_2', and σ_3' are, respectively, the major, intermediate, and minor effective principal stresses.

7.5.2 Generalized Priest Criterion

Priest (2005) developed a 3D version of the generalized Hoek-Brown strength criterion (Eq. 7.50) by combining it with the Drucker-Prager criterion, which is expressed by

$$J_2^{1/2} = A J_1 + B \qquad (7.85)$$

where A and B are empirical parameters; $J_1 = I_1/3$ is the mean effective stress; and J_2 is the second deviatoric stress invariant defined by

$$J_2 = \frac{1}{6}\left[(\sigma_1' - \sigma_2')^2 + (\sigma_2' - \sigma_3')^2 + (\sigma_3' - \sigma_1')^2\right] \qquad (7.86)$$

Since the 3D stress state satisfying the generalized Hoek-Brown strength criterion (Eq. 7.50) is (σ_1', σ_3', σ_3'), the strategy is to determine parameters A and B for the Drucker-Prager failure surface that intersects the Hoek-Brown failure point (σ_1', σ_3', σ_3'). With parameters A and B determined, the strength σ_{zf}' for an element of material subjected to a general state of 3D stresses $\sigma_x' \neq \sigma_y'$ / $= \sigma_{zf}'$ can then be predicted. This process is equivalent to the well-known process identifying the Drucker-Prager parameters that give the circumscribed fit for the Coulomb strength criterion. Priest (2005) originally suggested a numerical procedure for determining parameters A and B and strength σ_{zf}'. Melkoumian et al. (2009) later developed an explicit solution for determining strength σ_{zf}' at intermediate and minor principal stresses σ_x' and σ_y', which is summarized as follows:

$$\sigma_{zf}' = 3\sigma_p' + \sigma_c\left[\frac{m_b\sigma_p'}{\sigma_c} + s\right]^a - \left(\sigma_x' + \sigma_y'\right) \qquad (7.87)$$

in which

$$\sigma_p' = \frac{\sigma_x' + \sigma_y'}{2} + \frac{-E + \sqrt{E^2 - F\left(\sigma_x' - \sigma_y'\right)^2}}{2F} \qquad (7.88)$$

where $E = 2C^a\sigma_c$, $F = 3 + 2aC^{a-1}m_b$, and $C = s + \dfrac{m_b\left(\sigma_x' + \sigma_y'\right)}{2\sigma_c}$.

7.5.3   Generalized Zhang-Zhu Criterion

Zhang and Zhu (2007) proposed a 3D version of the original Hoek-Brown strength criterion for rock mass (Eq. 7.51) by combining the general Mogi (1971) criterion and the Hoek-Brown strength criterion, which is expressed as:

$$\frac{9}{2\sigma_c}\tau_{oct}^2 + \frac{3}{2\sqrt{2}}m_b\tau_{oct} - m_b\sigma'_{m,2} = s\sigma_c \qquad (7.89)$$

where τ_{oct} is the octahedral shear stress as defined by Eq. (7.83), and $\sigma'_{m,2}$ is the mean effective stress defined by

$$\sigma'_{m,2} = \frac{\sigma'_1 + \sigma'_3}{2} \qquad (7.90)$$

By modifying Eq. (7.89), Zhang (2008) proposed a 3D version of the generalized Hoek-Brown strength criterion (Eq. 7.50):

$$\frac{1}{\sigma_c^{(1/a-1)}}\left(\frac{3}{\sqrt{2}}\tau_{oct}\right)^{1/a} + \frac{m_b}{2}\left(\frac{3}{\sqrt{2}}\tau_{oct}\right) - m_b\sigma'_{m,2} = s\sigma_c \qquad (7.91)$$

which reduces to Eq. (7.89) when $a = 0.5$.

7.5.4   Evaluation of 3D Versions of Hoek-Brown Strength Criterion

Fig. 7.24 shows the π-plane plot of the Hoek-Brown strength criterion and the 3D versions of the Hoek-Brown strength criterion by Pan and Hudson (1988), Priest (2005), Zhang and Zhu (2007), and Zhang (2008). The (generalized) 3D Zhang-Zhu criterion predicts the same strength as the 2D Hoek-Brown strength criterion at both triaxial compression $(\sigma'_2 = \sigma'_3)$ and extension $(\sigma'_1 = \sigma'_2)$ stress states; but the 3D Pan-Hudson criterion does not predict the same strength as the Hoek-Brown criterion at either triaxial compression or extension stress state. Since the 3D Priest criterion is derived by identifying the Drucker-Prager parameters that give the circumscribed fit for the Hoek-Brown criterion, it predicts the same strength as the Hoek-Brown strength criterion at triaxial compression stress state but not at triaxial extension stress state. So the (generalized) 3D Zhang-Zhu criterion tends to provide better strength predictions, especially at the triaxial compression and extension stress states. However, it is noted that the (generalized) 3D Zhang-Zhu criterion envelope is not smooth at either the triaxial compression or extension stress state and concave at the triaxial extension stress state, which may lead to problems with some stress paths and cause inconvenience in numerical applications.

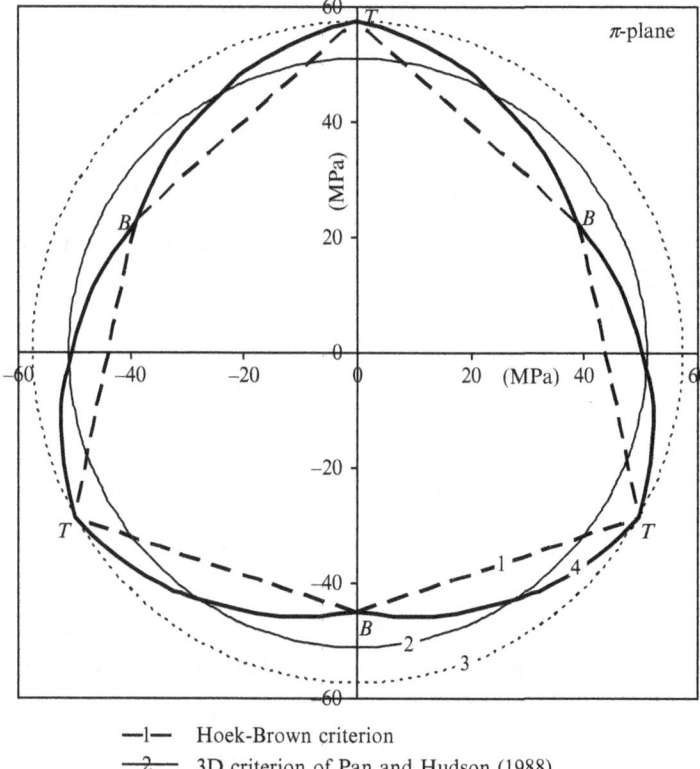

—1— Hoek-Brown criterion
—2— 3D criterion of Pan and Hudson (1988)
······3······ 3D criterion of Priest (2005)
—4— 3D criterion of Zhang and Zhu (2007), Zhang (2008)

FIG. 7.24 π-Plane plot of Hoek-Brown criterion and 3D criteria of Pan and Hudson (1988), Priest (2005), Zhang and Zhu (2007), and Zhang (2008) at $I_1 = 100$ MPa for a rock mass with $\sigma_c = 50$ MPa, $m_b = 5$, $s = 0.01$, and $a = 0.5$.

7.5.5 Modified Generalized 3D Zhang-Zhu Criterion

To address the nonsmoothness and nonconvexity problem for the generalized 3D Zhang-Zhu criterion, Zhang et al. (2013) used the following three Lode dependences with smoothness and convexity characteristics to replace the Lode dependence of the generalized 3D Zhang-Zhu criterion:

(a) The dependence proposed by William and Warnke (1975), which uses an elliptical approximation of the deviatoric section (E-D):

$$L(\theta_\sigma)_{\text{E-D}}$$
$$= \frac{2\left(1-\delta^2\right)\cos\left(\pi/6+\theta_\sigma\right) + (2\delta-1)\sqrt{4\left(1-\delta^2\right)\cos^2\left(\pi/6+\theta_\sigma\right) + \delta(5\delta-4)}}{4\left(1-\delta^2\right)\cos^2\left(\pi/6+\theta_\sigma\right) + (2\delta-1)^2}$$

$$(7.92)$$

(b) The hyperbolic dependence (H-D) proposed by Yu (1990):

$$L(\theta_\sigma)_{\mathrm{H-D}}$$

$$= \frac{2\delta(1-\delta^2)\cos(\pi/6-\theta_\sigma) + \delta(\delta-2)\sqrt{4(\delta^2-1)\cos^2(\pi/6-\theta_\sigma)+(5-4\delta)}}{4(1-\delta^2)\cos^2(\pi/6-\theta_\sigma)-(\delta-2)^2}$$

$$(7.93)$$

(c) The dependence proposed by Matsuoka and Nakai (1974) based on the spatial mobilized plane (S-D):

$$L(\theta_\sigma)_{\mathrm{S-D}} = \frac{\sqrt{3}\delta}{2\sqrt{\delta^2-\delta+1}} \frac{1}{\cos\gamma}$$

$$\begin{cases} \gamma = \dfrac{1}{6}\arccos\left(-1 + \dfrac{27\delta^2(1-\delta)^2}{2(\delta^2-\delta+1)^3}\sin^2(3\theta_\sigma)\right) & \text{for } \theta_\sigma \le 0 \\[4mm] \gamma = \dfrac{\pi}{3} - \dfrac{1}{6}\arccos\left(-1 + \dfrac{27\delta^2(1-\delta)^2}{2(\delta^2-\delta+1)^3}\sin^2(3\theta_\sigma)\right) & \text{for } \theta_\sigma > 0 \end{cases} \quad (7.94)$$

In Eqs. (7.92), (7.93) and (7.94), δ is the aspect ratio and θ_σ is the Lode angle, which can be determined by

$$\delta = \sqrt{J_{\min}/J_{\max}} = \sqrt{J_2(-\pi/6)/J_2(\pi/6)} \qquad (7.95)$$

$$\theta_\sigma = \tan^{-1}\left[\frac{\sigma_1+\sigma_3-2\sigma_2}{\sqrt{3}(\sigma_1-\sigma_3)}\right] \qquad (7.96)$$

Since σ_1', σ_2', and σ_3' can be expressed as:

$$\begin{bmatrix} \sigma_1' \\ \sigma_2' \\ \sigma_3' \end{bmatrix} = -\frac{2}{\sqrt{3}}\sqrt{J_2}\begin{bmatrix} \sin\left(\theta_\sigma - \dfrac{2}{3}\pi\right) \\ \sin(\theta_\sigma) \\ \sin\left(\theta_\sigma + \dfrac{2}{3}\pi\right) \end{bmatrix} + \begin{bmatrix} I_1/3 \\ I_1/3 \\ I_1/3 \end{bmatrix} \qquad (7.97)$$

where I_1 and J_2 are, respectively, the first stress invariant and the second deviatoric stress invariant as defined earlier, the generalized 3D Zhang-Zhu criterion (Eq. 7.91) can be expressed in terms of I_1, J_2, θ_σ as:

$$\frac{1}{\sigma_c^{(1/a-1)}}\left(\sqrt{3J_2}\right)^{1/a} + \left(\frac{\sqrt{3}}{2} - \frac{\sin\theta_\sigma}{\sqrt{3}}\right)m_b\sqrt{J_2} - m_b\frac{I_1}{3} - s\sigma_c = 0 \qquad (7.98)$$

Using one of these Lode dependences from Eqs. (7.92), (7.93) and (7.94) to replace the Lode dependence of the generalized 3D Zhang-Zhu criterion, the modified criteria can be expressed as:

$$\sqrt{J_2} = L(\theta_\sigma)_{X-D} \sqrt{J_{max}} \tag{7.99}$$

where X is E, H, or S, respectively for the elliptical dependence, the hyperbolic dependence or the spatial mobilized plane dependence.

When $a = 0.5$, J_{max}, J_{min} and the aspect ratio δ (Eq. 7.95) can be determined using the following explicit expressions

$$\sqrt{J_{max}} = \frac{1}{6\sqrt{3}} \left(\sqrt{m_b^2 \sigma_c^2 + 12 m_b \sigma_c I_1 + 36 s \sigma_c^2} - m_b \sigma_c \right) \tag{7.100}$$

$$\sqrt{J_{min}} = \frac{1}{3\sqrt{3}} \left(\sqrt{m_b^2 \sigma_c^2 + 3 m_b \sigma_c I_1 + 9 s \sigma_c^2} - m_b \sigma_c \right) \tag{7.101}$$

$$\delta = \sqrt{\frac{J_{min}}{J_{max}}} = \frac{2\sqrt{m_b^2 \sigma_c^2 + 3 m_b \sigma_c I_1 + 9 s \sigma_c^2} - 2 m_b \sigma_c}{\sqrt{m_b^2 \sigma_c^2 + 12 m_b \sigma_c I_1 + 36 s \sigma_c^2} - m_b \sigma_c} \tag{7.102}$$

and Eq. (7.99) can then be directly used to determine the strength with given σ_2' and σ_3'. When $a \neq 0.5$, however, an explicit expression for J_{max}, J_{min}, and δ cannot be derived and thus a numerical solution method below needs to be used:

- Assume an initial $(\sigma_1')_a$ with known σ_2' and σ_3', say $(\sigma_1')_a = \sigma_2'$.
- Calculate I_1 with Eq. (7.84).
- Calculate J_{max} and J_{min} from Eq. (7.98) with known m_b, s, a and using $\theta_\sigma = \pi/6$ for J_{max} and $\theta_\sigma = -\pi/6$ for J_{min}.
- Calculate aspect ratio δ with Eq. (7.95).
- Determine σ_1' with Eq. (7.99) using the Lode dependence from Eqs. (7.92), (7.93), or (7.94) and denote it as $(\sigma_1')_n$.
- Check the error defined as follows:

$$|(\sigma_1')_n - (\sigma_1')_a| < \varepsilon \tag{7.103}$$

where ε is the prescribed convergence limit (eg, 0.001 MPa).
- If Eq. (7.103) is satisfied, the obtained $(\sigma_1')_n$ is the right σ_1' at given σ_2' and σ_3' from the modified criteria. If Eq. (7.103) is not satisfied, the obtained $(\sigma_1')_n$ is used as $(\sigma_1')_a$ and the previous steps are repeated until Eq. (7.103) is satisfied.

Fig. 7.25 shows the π-plane plots of the generalized 3D Zhang-Zhu (Z-Z) criterion and the three modified criteria with, respectively, the elliptical dependence (E-D), hyperbolic dependence (H-D) and spatial mobilized plane dependence (S-D), for a rock mass with $m_i = 15$, GSI $= 75$, $D = 0$ and the corresponding m_b, s, a determined from Eqs. (7.57a) to (7.57c). The smoothness

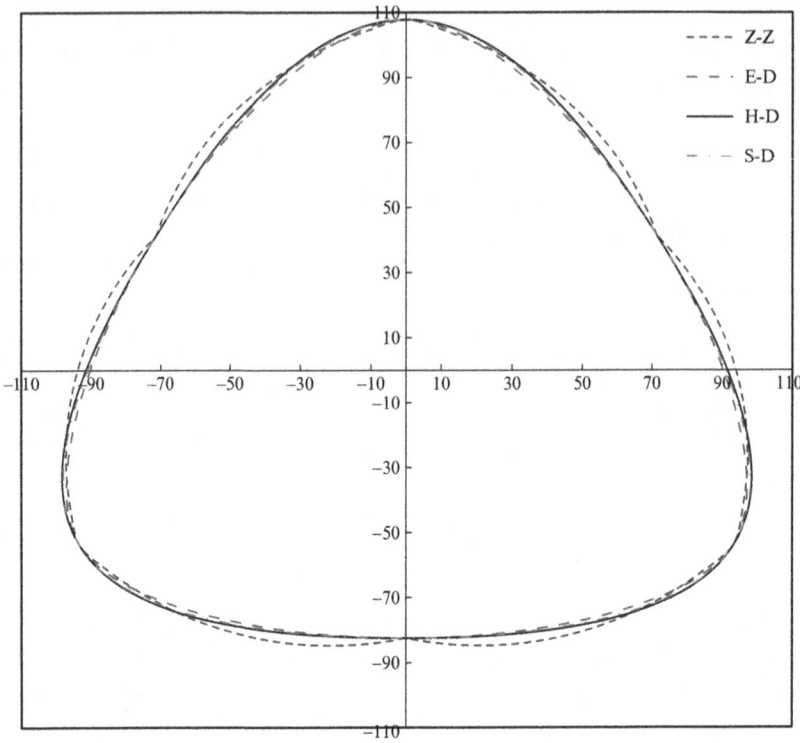

FIG. 7.25 π-Plane plot of generalized Zhang-Zhu (Z-Z) criterion and three modified criteria with elliptical dependence (E-D), hyperbolic dependence (H-D), and spatial mobilized plane dependence (S-D) at $I_1 = 300$ MPa for a rock mass with $m_i = 15$, GSI $= 75$, and $D = 0$. *(From Zhang, Q., Zhu, H., Zhang, L., 2013. Modification of a generalized three-dimensional Hoek-Brown strength criterion. Int. J. Rock Mech. Min. Sci. 59, 80–96.)*

and convexity of the three modified criteria can be obviously seen from the figure. The three modified criteria also predict the same strength as the Z-Z criterion at both triaxial compression and extension states. But the three modified criteria predict slightly higher strength at the zone near the triaxial compression state and slightly lower strength at the zone near the triaxial extension state than the Z-Z criterion.

7.6   RESIDUAL STRENGTH OF ROCK

The postpeak behavior of rock is important in the design of many engineering structures in or on rock because it can have a significant influence on the stability of the structure. In general, rock masses exhibit a strain-softening postpeak behavior and thus the residual strength parameters are lower than

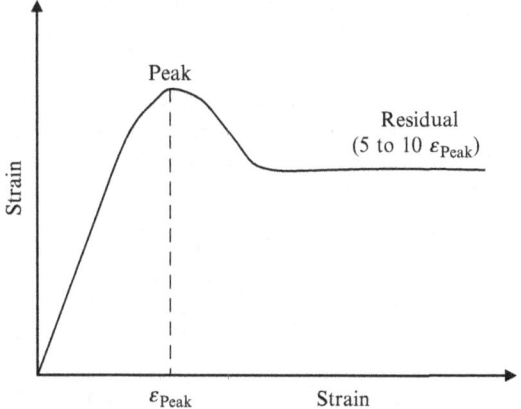

FIG. 7.26 Strain-softening of rock.

the peak parameters (Fig. 7.26). To design structures in or on rock properly, it is important to evaluate the residual strength of rock masses.

Several methods have been proposed for estimating the residual strength of rock masses by reducing the GSI from the in-situ value to a lower value GSI_r corresponding to the residual state of the rock mass. For example, Russo et al. (1998) proposed that the residual GSI_r is 36% of the peak GSI value. This simple relation may underestimate the residual GSI_r values for poor-quality rock masses but overestimate the residual GSI_r values for very good-quality rock masses and thus should be used with caution.

Cai et al. (2007) proposed a method for estimating the residual GSI_r by using the same Eq. (5.16) but with the block volume V_b^r and joint condition factor J_C^r of the rock mass at the residual state, ie,

$$GSI_r = \frac{26.5 + 8.79 \ln J_C^r + 0.9 \ln V_b^r}{1 + 0.015 \ln J_C^r - 0.0253 \ln V_b^r} \qquad (7.104)$$

The residual block volume V_b^r can be obtained as follows:

(1) $V_b^r = 10 \, \text{cm}^3$ if the peak block volume V_b is $>10 \, \text{cm}^3$.
(2) $V_b^r = V_b$ if the peak block volume V_b is $>10 \, \text{cm}^3$.

The residual joint condition factor J_C^r is calculated using the same Eq. (5.15) for the peak joint condition factor J_C but with the large-scale waviness J_W^r, small-scale smoothness J_S^r and joint alteration factor J_A^r at the residual state, ie,

$$J_C^r = \frac{J_W^r J_S^r}{J_A^r} \qquad (7.105)$$

The residual J_W^r, J_S^r, and J_A^r are determined as follows:

$$\text{If } J_W/2 < 1, \quad J_W^r = 1; \quad \text{Otherwise,} \quad J_W^r = J_W/2 \qquad (7.106a)$$

$$\text{If } J_S/2 < 0.75, \quad J_S^r = 0.75; \quad \text{Otherwise,} \quad J_S^r = J_S/2 \qquad (7.106b)$$

$$J_A^r = J_A, \quad \text{ie, there is no reduction for } J_A \qquad (7.106c)$$

With the obtained GSI_r, the Hoek-Brown strength parameters of the rock mass at the residual state can then be determined using the same Eqs. (7.57a)–(7.57c) for the peak strength but with GSI_r, ie,

$$m_r = \exp\left(\frac{\text{GSI}_r - 100}{28 - 14D}\right) m_i \qquad (7.107a)$$

$$s_r = \exp\left(\frac{\text{GSI}_r - 100}{9 - 3D}\right) \qquad (7.107b)$$

$$a_r = 0.5 + \frac{1}{6}[\exp(-\text{GSI}_r/15) - \exp(-20/3)] \qquad (7.107c)$$

Because the rock mass is damaged at the residual state, $D = 0$ should be used in Eqs. (7.107a)–(7.107c) for determining the residual Hoek-Brown strength parameters m_r, s_r, and a_r. With the obtained m_r, s_r, and a_r, the residual strength of the rock mass can then be determined using the Hoek-Brown strength criterion as described earlier.

7.7   SCALE EFFECT ON ROCK STRENGTH

Research results (see, eg, Heuze, 1980; Hoek and Brown, 1980; Medhurst and Brown, 1996) indicate that rock masses show strong scale-dependent mechanical properties. In the following, the scale effect on the strength of rock masses is briefly discussed.

Experimental results show that rock strength decreases significantly with increasing sample size. Based upon analyses of published data, Hoek and Brown (1980) suggested that the unconfined compressive strength σ_{cd} of a rock specimen with a diameter of d mm is related to the unconfined compressive strength σ_{c50} of a 50 mm diameter specimen by

$$\sigma_{cd} = \sigma_{c50} \left(\frac{50}{d}\right)^{0.18} \qquad (7.108)$$

This relationship, together with the data upon which it was based, is illustrated in Fig. 7.27. Hoek and Brown (1997) suggested that the reduction in strength is due to the greater opportunity for failure through and around grains, the "building blocks" of intact rock, as more and more of these grains are included in the test sample. Eventually, when a sufficiently large number of grains are included in the sample, the strength reaches a constant value.

Medhurst and Brown (1996) reported the results of laboratory triaxial tests on 61, 101, 146, and 300 mm diameter samples of a highly cleated mid-brightness coal from the Moura mine in Australia. The results of these tests are as summarized in Table 7.19 and Fig. 7.28. It can be seen that the strength decreases

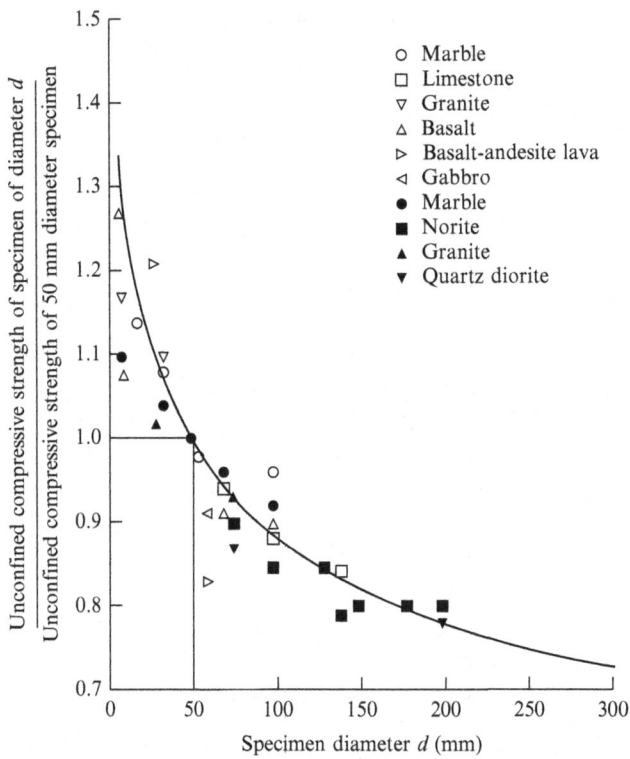

FIG. 7.27 Influence of specimen size on the strength of intact rock. *(From Hoek, E., Brown, E.T., 1980. Empirical strength criterion for rock masses. J. Geotech. Eng. ASCE 106, 1013–1035.)*

TABLE 7.19 Peak Strength of Moura Coal in Terms of the Parameters in Eq. (7.50), Based Upon a Value of $\sigma_c = 32.7$ MPa

Diameter (mm)	m_b	s	a
61	19.4	1	0.5
101	13.3	0.555	0.5
146	10.0	0.236	0.5
300	5.7	0.184	0.6
Mass	2.6	0.052	0.65

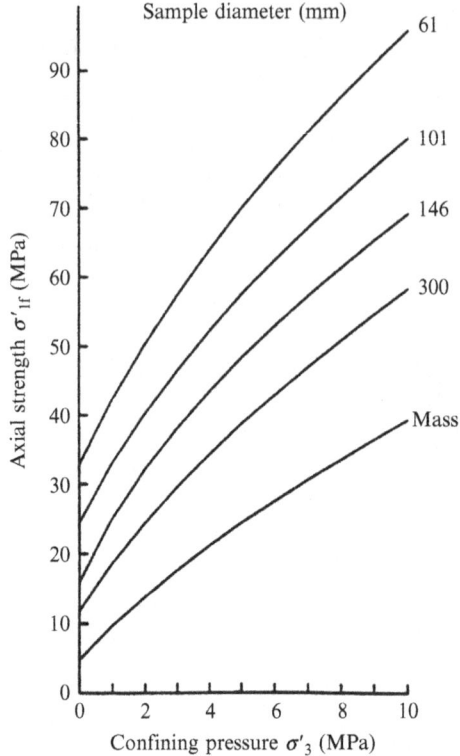

FIG. 7.28   Peak strength for Australian Moura coal. *(From Medhurst, T.P., Brown, E.T., 1996. Large scale laboratory testing of coal. In: Proc. 7th Australian-New Zealand Conf. on Geomech, Canberra, Australia, pp. 203–208.)*

significantly with increasing specimen size. This is attributed to the effects of cleat spacing. For this coal, the persistent cleats are spaced at 0.3–1.0 m while nonpersistent cleats within vitrain bands and individual lithotypes define blocks of 1 cm or less. This cleating results in a "critical" sample size of about 1 m above which the strength remains constant. Heuze (1980) conducted an extensive literature search and found results of 77 plate tests as shown in Fig. 7.29. The test volume shown in this figure is calculated in the following way:

(1) For a circular plate, the test volume is taken as that of a sphere having a diameter of four times the diameter of the plate.
(2) For a rectangular or square plate of given area, the diameter of a circle of equal area is first calculated, and the test volume is then determined using the equivalent diameter.

The number shown next to the open triangles in the figure indicates the number of tests performed; the mean value of these test results is plotted as the triangle.

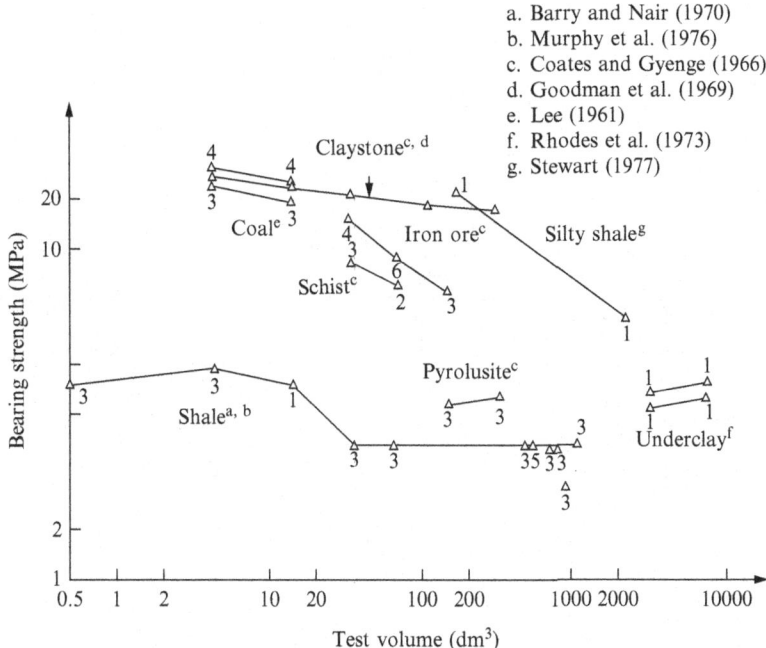

a. Barry and Nair (1970)
b. Murphy et al. (1976)
c. Coates and Gyenge (1966)
d. Goodman et al. (1969)
e. Lee (1961)
f. Rhodes et al. (1973)
g. Stewart (1977)

FIG. 7.29   Effect of test volume on measured bearing strength of rock masses. The number next to the triangle indicates the number of tests performed. *(Based on Heuze, F.E., 1980. Scale effects in the determination of rock mass strength and deformability. Rock Mech. 12, 167–192.)*

The test results (except those of Coates and Gyenge (1966) and Rhodes et al. (1973)) show that the strength decreases with increasing test volume.

Fig. 7.30 shows a simplified representation of the influence of the relation between the discontinuity spacing and the size of the problem domain on the selection of a rock mass behavior model (Hoek-Brown strength criterion). As the problem domain enlarges, the corresponding rock behavior changes from that of the isotropic intact rock, through that of a highly anisotropic rock mass in which failure is controlled by one or two discontinuities, to that an isotropic heavily jointed rock mass.

7.8   ANISOTROPY OF ROCK STRENGTH

Some intact rocks, such as those composed of parallel arrangements of flat minerals like mica, chlorite, and clay, show strong strength anisotropy. Fig. 7.31 shows the anisotropy of compressive strength recorded for a series of tests performed on a slate. The maximum strength is generally found when the major principal stress is nearly perpendicular or parallel to the stratification plane. The minimum strength is obtained when the angle β between the major principal stress and the stratification plane is at 30–60 degrees. The degree of strength

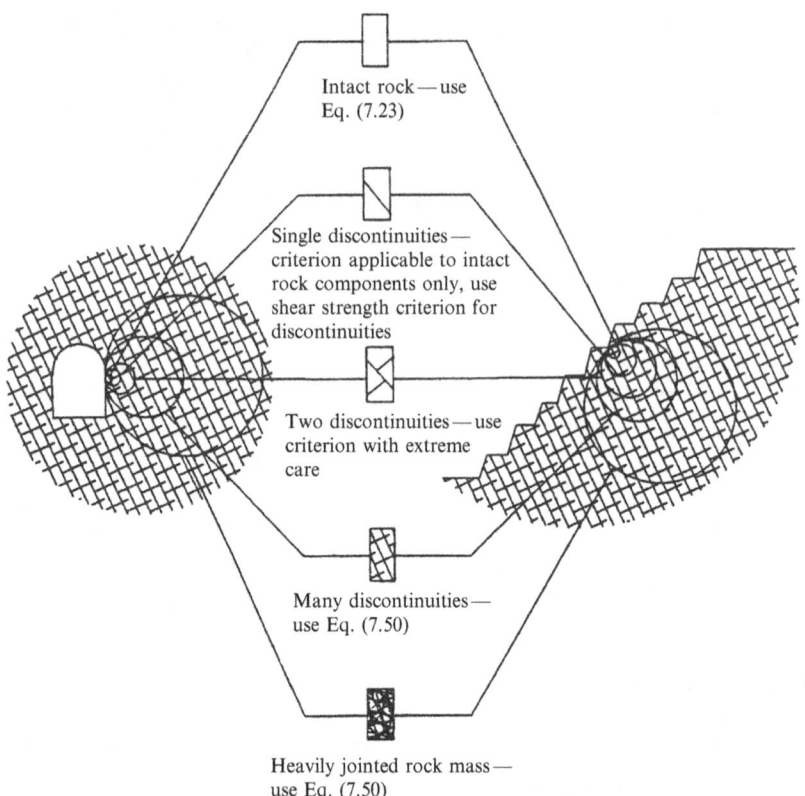

Intact rock—use
Eq. (7.23)

Single discontinuities—
criterion applicable to intact
rock components only, use
shear strength criterion for
discontinuities

Two discontinuities—use
criterion with extreme
care

Many discontinuities—
use Eq. (7.50)

Heavily jointed rock mass—
use Eq. (7.50)

FIG. 7.30   Simplified representation of the influence of scale on the type of rock mass behavior. *(Based on Hoek, E., Kaiser, P.K., Bawden, W.F., 1995. Support of Underground Excavations in Hard Rock. Balkema, Rotterdam.)*

anisotropy is commonly quantified by the strength anisotropy ratio R_c defined as follows:

$$R_c = \frac{\sigma_{cmax}}{\sigma_{cmin}} \qquad (7.109)$$

where σ_{cmax} and σ_{cmin} are, respectively, the maximum and minimum compressive strengths at a given confining pressure. Table 7.20 lists the values of R_c for different rocks at the conditions of unconfined compression. According to Ramamurthy (1993), the strength anisotropy of intact rocks can be classified as in Table 7.21.

It is noted that the strength anisotropy ratio R_c decreases when the confining pressure is higher (Fig. 7.32). So the effect of strength anisotropy will be reduced when the confining pressure is increased.

The degree of strength anisotropy can also be quantified by the point load strength anisotropy index $I_{a(50)}$ defined as follows:

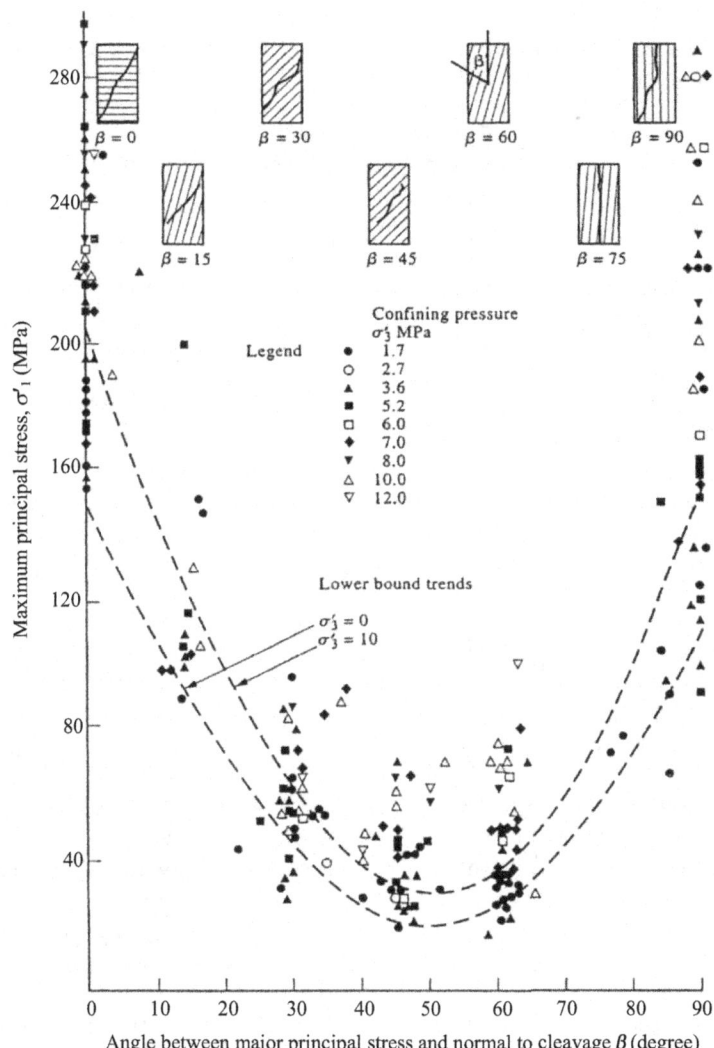

FIG. 7.31 Compressive strength anisotropy in dark gray slate. *(Based on Brown, E.T., Richards, L. R., Barr, M.V., 1977. Shear strength characteristics of the Delabole Slates. In: Proc. Conf. Rock Eng. Newcastle University, pp. 33–51.)*

$$I_{a(50)} = \frac{I_{s(50)v}}{I_{s(50)h}} \tag{7.110}$$

where $I_{s(50)v}$ and $I_{s(50)h}$ are the point load strength index values perpendicular and parallel to the stratification planes, respectively. Table 7.22 shows the anisotropy classification based on $I_{a(50)}$ suggested by Tsidzi (1990).

TABLE 7.20 Strength Anisotropy Ratio R_c for Different Rocks at Unconfined Compression

Rock	β for σ_{cmax} (degree)	Anisotropy Ratio R_c	Reference
Yeoncheon schost	90	18.60	Cho et al. (2012)
Angers schist	90	13.48	Duveau et al. (1998)
Martinsburg slate	90	13.46	Donath (1964)
Fractured sandstone	90	6.37	Horino and Ellickson (1970)
Chamera phyllites-micaceous	90	6.00	Singh (1988)
Barnsley hard coal	90	5.18	Pomeroy et al. (1971)
Penrhyn slate	90	4.85	Attewell and Sandford (1974)
Diatomite	90	3.74	Allirot and Boehler (1979)
Siltshale	90	3.70	Ajalloeian and Lashkaripour (2000)
South African slate	0	3.68	Hoek (1964)
Claystone	90	3.04	Colak and Unlu (2004)
Mudshale	90	3.01	Ajalloeian and Lashkaripour (2000)
Texas slate	90	3.00	McLamore and Gray (1967)
Asan gneiss	90	2.60	Cho et al. (2012)
Boryeong shale	90	2.60	Cho et al. (2012)
Permian shale	90	2.33	Chenevert and Gatlin (1965)
Siltstone-B	90	2.30	Colak and Unlu (2004)
Travertine	90	2.29	Yilmaz and Yucel (2014)
Crystalline schist	90	2.24	Mogi et al. (1978)
Chamera phyllites-quartizitic	90	2.19	Singh (1988)
Chamera phyllites-carbonaceous	90	2.19	Singh (1988)
Siltstone-A	90	1.94	Colak and Unlu (2004)

Continued

TABLE 7.20 Strength Anisotropy Ratio R_c for Different Rocks at Unconfined Compression—cont'd

Rock	β for σ_{cmax} (degree)	Anisotropy Ratio R_c	Reference
Sandstone-A (fine grained)	90	1.75	Colak and Unlu (2004)
Sandstone-B (fine grained)	90	1.62	Colak and Unlu (2004)
Green River shale I	0	1.62	McLamore and Gray (1967)
Green River shale II	0	1.41	McLamore and Gray (1967)
Green River shale	0, 90	1.37	Chenevert and Gatlin (1965)
Sandstone-D (medium grained)	90	1.34	Colak and Unlu (2004)
Sandstone-E (medium grained)	90	1.23	Colak and Unlu (2004)
Sandstone-C (fine grained)	90	1.15	Colak and Unlu (2004)
Kota sandstone	0	1.12	Rao (1984)
Arkansas sandstone	0	1.10	Chenevert and Gatlin (1965)

TABLE 7.21 Classification of Strength Anisotropy of Intact Rocks

Anisotropy Ratio R_c	Class	Rock Types
$1.0 < R_c \leq 1.1$	Isotropic	⎫ Sandstone
$1.1 < R_c \leq 2.0$	Low anisotropy	⎭
$2.0 < R_c \leq 4.0$	Medium anisotropy	⎱ Shale
$4.0 < R_c \leq 6.0$	High anisotropy	⎱ Slate, phyllite
$6.0 < R_c$	Very high anisotropy	⎭

Based on Ramamurthy, T., 1993. Strength and modulus responses of anisotropic rocks. In: Hudson, J. A. (Ed.), Comprehensive Rock Engineering—Principle, Practice and Projects, vol. 1. Pergamon, Oxford, pp. 313–329.

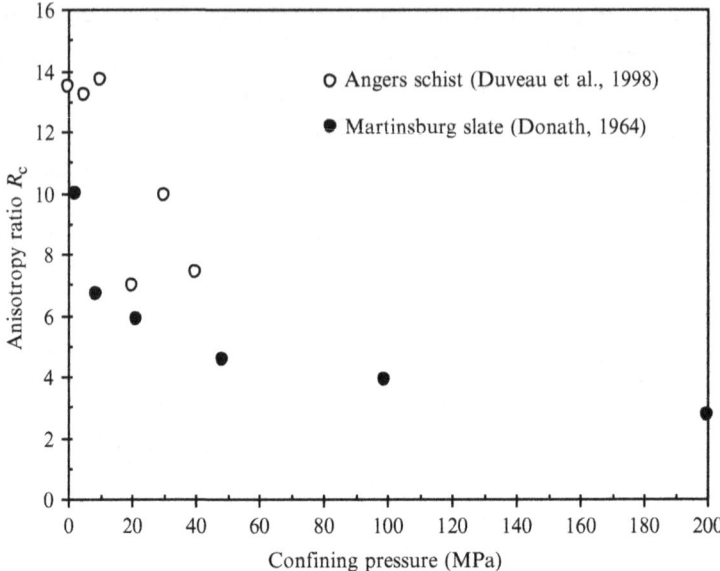

FIG. 7.32   Variation of strength anisotropy ratio with confining pressure.

TABLE 7.22 Classification of Foliated Rocks Based on Point Load Strength Anisotropy Index $I_{a(50)}$

Nature of Rock	$I_{a(50)}$	Class
Very weakly foliated or nonfoliated	$I_{a(50)} \leq 1.1$	Quasiisotropic
Weakly foliated	$1.1 < I_{a(50)} \leq 1.5$	Fairly anisotropic
Moderately foliated	$1.5 < I_{a(50)} \leq 2.5$	Moderately anisotropic
Strongly foliated	$2.5 < I_{a(50)} \leq 3.5$	Highly anisotropic
Very strongly foliated	$3.5 < I_{a(50)}$	Very highly anisotropic

Based on Tsidzi, K.E.N., 1990. The influence of foliation on point load strength anisotropy of foliated rocks. Eng. Geol. 29, 49–58.

Rock masses containing discontinuities also display strength anisotropy. The equivalent continuum models presented in Section 7.4.5 clearly show the variation of compressive strength with the direction of the principal stresses.

REFERENCES

AASHTO, 1996. Standard Specifications for Highway Bridges, 16th ed. American Association of State Highway and Transportation Officials, Washington, DC.

Adachi, T., Yoshida, N., 2002. In situ investigation on mechanical characteristics of soft rocks. In: Sharma, V.M., Saxena, K.R. (Eds.), In-Situ Characterization of Rocks. Balkema, Lisse, pp. 131–186.

Ajalloeian, R., Lashkaripour, G.R., 2000. Strength anisotropies in mudrocks. Bull. Eng. Geol. Environ. 59, 195–199.

Al-Ajmi, A.M., Zimmerman, R.W., 2005. Relation between the Mogi and the Coulomb failure criteria. Int. J. Rock Mech. Min. Sci. 42 (3), 431–439.

Al-Jassar, S.H., Hawkins, A.B., 1979. Geotechnical properties of the carboniferous limestone of the Bristol area—the influence of petrography and chemistry. In: 4th ISRM Cong., Montreau. International Society for Rock mechanics, vol. 1, pp. 3–14.

Allirot, D., Boehler, J.-P., 1979. Evolution of mechanical properties of a stratified rock under confining pressure. In: Proc. 4th Int. Cong. Rock Mech., Montreux. vol. 1. Balkema, Rotterdam, pp. 15–22 (in French).

Amadei, B., 1988. Strength of a regularly jointed mass under biaxial and axisymmetric loading conditions. Int. J. Rock Mech. Min. Sci. Geomech. Abstr. 25, 3–13.

Amadei, B., Savage, W.Z., 1989. Anisotropic nature of jointed rock mass strength. J. Geotech. Eng. ASCE 115, 525–542.

Amadei, B., Savage, W.Z., 1993. Effect of joints on rock mass strength and deformability. In: Hudson, J.A. (Ed.), Comprehensive Rock Engineering—Principle, Practice and Projects, vol. 1. Pergamon, Oxford, pp. 331–365.

Asef, M.R., Reddish, D.J., Lloyd, P.W., 2000. Rock-support interaction analysis based on numerical modeling. Geotech. Geol. Eng. 18, 23–37.

Atkinson, R.H., 1993. Hardness tests for rock characterization. In: Hudson, J.A. (Ed.), Compressive Rock Engineering—Rock Testing and Site Characterization, vol. 3. Pergamon, Oxford, pp. 105–117.

Attewell, P.B., Sandford, M.R., 1974. Intrinsic shear strength of a brittle anisotropic rock I, experimental and mechanical interpretation. Int. J. Rock Mech. Min. Sci. Geomech. Abstr. 11, 423–430.

Aufmuth, R.E., 1973. A systematic determination of engineering criteria for rocks. Bull. Assoc. Eng. Geol. 11, 235–245.

Aydan, Ö., Dalgiç, S., 1998. Prediction of deformation behavior of 3-lanes Bolu tunnels through squeezing rocks of North Anatolian fault zone (NAFZ). In: Proc. Regional Symp. Sedimentary Rock Eng., Taipei, pp. 228–233.

Aydan, Ö., Sato, A., Yagi, M., 2014. The inference of geo-mechanical properties of soft rocks and their degradation from needle penetration tests. Rock Mech. Rock. Eng. 47, 1867–1890.

Balmer, G., 1952. A general analytical solution for Mohr's envelope. Am. Soc. Test. Mater. 52, 1260–1271.

Barry, A.J., Nair, O.B., 1970. In situ tests of bearing capacity of roof and floor in selected bituminous coal mines. A Progress Report—Longwall Mining, US Bureau of Mines, RI 7406.

Barton, N., 1976. The shear strength of rock and rock joints. Int. J. Rock Mech. Min. Sci. Geomech. Abstr. 13, 255–279.

Barton, N., 2002. Some new Q value correlations to assist in site characterization and tunnel design. Int. J. Rock Mech. Min. Sci. Geomech. Abstr. 39, 185–216.

Barton, N., Bandis, S.C., 1982. Effects of block size on the shear behavior of jointed rock. In: Proc. 23rd U.S. Symp. on Rock Mech., Berkeley, CA, pp. 739–760.

Barton, N., Bandis, S.C., 1990. Review of predictive capabilities of JRC-JCS model in engineering practice. In: Barton, N., Stephansson, O. (Eds.), Proc. Int. Symp. on Rock Joints. Loen, Norway. Balkema, Rotterdam, pp. 603–610.

Barton, N., Choubey, V., 1977. The shear strength of rock joints in theory and practice. Rock Mech. 10, 1–54.

Barton, N., Lien, R., Lunde, J., 1974. Engineering classification of rock masses for the design of tunnel support. Rock Mech. 6, 189–236.

Basu, A., Aydin, A., 2006. Predicting uniaxial compressive strength by point load test: significance of cone penetration. Rock Mech. Rock. Eng. 39, 483–490.

Basu, A., Kamran, M., 2010. Point load test on schistose rocks and its applicability in predicting uniaxial compressive strength. Int. J. Rock Mech. Min. Sci. 47, 823–828.

Beverly, B.E., Schoenwolf, D.A., Brierly, G.S., 1979. Correlations of Rock Index Values With Engineering Properties and the Classification of Intact Rock. FHWA, Washington, DC.

Bieniawski, Z.T., 1974. Geomechanics classification of rock masses and its application in tunneling. In: Proc. 3rd Int. Cong. on Rock Mech., Denver, CO., International Society for Rock Mechanics vol. IIA, pp. 27–32.

Bieniawski, Z.T., 1975. The point load test in geotechnical practice. Eng. Geol. 9, 1–11.

Bieniawski, Z.T., 1976. Rock mass classification in rock engineering. In: Bieniawski, Z.T. (Ed.), Proc. Symp. on Exploration for Rock Eng. vol. 1. Balkema, Cape Town, pp. 97–106.

Bowden, A.J., Spink, T.W., Mortmore, R.N., 2002. Engineering description of chalk: its strength, hardness and density. Q. J. Eng. Geol. Hydrogeol. 35, 355–361.

Broch, E., Franklin, J.A., 1972. Point load strength test. Int. J. Rock Mech. Min. Sci. 9, 669–697.

Brook, N., 1993. The measurement and estimation of basic rock strength. In: Hudson, J.A. (Ed.), Rock Testing and Site Characterization—Comprehensive Rock Engineering, vol. 3, Pergamon, Oxford, pp. 41–66.

Bruno, G., Vessia, G., Bobbo, L., 2013. Statistical method for assessing the uniaxial compressive strength of carbonate rock by Schmidt hammer tests performed on core samples. Rock Mech. Rock. Eng. 46, 199–206.

Cai, M., Kaiser, P.K., Tasaka, Y., Minami, M., 2007. Determination of residual strength parameters of jointed rock masses using the GSI system. Int. J. Rock Mech. Min. Sci. 44, 247–265.

Cargill, J.S., Shakoor, A., 1990. Evaluation of empirical methods for measuring the uniaxial compressive strength. Int. J. Rock Mech. Min. Sci. 27, 495–503.

Cha, Y.H., Kang, J.S., Jo, C.-H., 2006. Application of linear-array microtremor surveys for rock mass classification in urban tunnel design. Explor. Geophys. 37, 108–113.

Chang, C., Haimson, B., 2005. Non-dilatant deformation and failure mechanism in two Long Valley Caldera rocks under true triaxial compression. Int. J. Rock Mech. Min. Sci. 42, 402–414.

Chau, K.T., Wong, R.H.C., 1996. Uniaxial compressive strength and point load strength of rocks. Int. J. Rock Mech. Min. Sci. Geomech. Abstr. 33, 183–188.

Chen, Y.-L., Ni, J., Shao, W., Azzam, R., 2012. Experimental study on the influence of temperature on the mechanical properties of granite under uni-axial compression and fatigue loading. Int. J. Rock Mech. Min. Sci. 56, 62–66.

Chenevert, M.E., Gatlin, C., 1965. Mechanical anisotropies of laminated sedimentary rocks. Soc. Pet. Eng. J. 5, 67–77.

Cho, J.-W., Kim, H., Jeon, S., Min, K.-B., 2012. Deformation and strength anisotropy of Asan gneiss, Boryeong shale, and Yeoncheon schist. Int. J. Rock Mech. Min. Sci. 50, 158–169.

Coates, D.F., Gyenge, M., 1966. Plate-load testing on rock for deformation and strength properties. ASTM Spec. Tech. Publ. 402, 19–35.

Colak, K., Unlu, T., 2004. Effect of transverse anisotropy on the Hoek-Brown strength parameter "m_i" for intact rocks. Int. J. Rock Mech. Min. Sci. 41, 1045–1052.

Colback, P.S.B., Wild, B.L., 1965. The influence of moisture content on the compressive strength of rock. In: Proc. 3rd Canadian Rock Mech. Symp, Toronto, pp. 65–83.

D'Andrea, D.V., Fischer, R.L., Fogelson, D.E., 1965. Prediction of compressive strength of rock from other properties. US Bureau of Mines Rep. Investigation, No. 6702.

Das, B.M., 1985. Evaluation of the point load strength for soft rock classification. In: Proceedings of the Fourth International Conference Ground Control in Mining, Morgantown, WV, pp. 220–226.

Deere, D.U., Miller, R.P., 1966. Engineering classification and index properties for intact rock. Tech. Rep. No. AFWL-TR-65-116, Air Force Weapons Lab, Kirtland Air Force Base, Albuquerque, NM.

Del Potro, R., Hürlimann, M., 2009. A comparison of different indirect techniques to evaluate volcanic intact rock strength. Rock Mech. Rock. Eng. 42, 931–938.

Dincer, I., Acar, A., Cobanoglu, I., Uras, Y., 2004. Correlation between Schmidt hardness, uniaxial compressive strength and Young's modulus for andesites, basalts and tuffs. Bull. Eng. Geol. Environ. 63, 141–148.

Donath, F.A., 1964. Strength variation and deformation behavior of anisotropic rocks. In: Judd, W.R. (Ed.), State of Stress in the Earth's Crust. Elsevier, New York, NY, pp. 281–318.

Doruk, P., 1991. Analysis of the laboratory strength data using the original and modified Hoek-Brown failure criterion. MS thesis, Department of Civil Engineering, University of Toronto, Toronto.

Duveau, G., Shao, J.F., Henry, J.P., 1998. Assessment of some failure criteria for strongly anisotropic geomaterials. Mech. Cohesive-Frictional Mat. 3, 1–26.

Dwivedi, R.D., Goel, R.K., Prasad, V.R.R., Sinha, A., 2008. Thermo-mechanical properties of Indian and other granites. Int. J. Rock Mech. Min. Sci. 45, 303–315.

Dyke, C.J., Dobereiner, L., 1991. Evaluating the strength and deformability of sandstones. Q. J. Eng. Geol. 24, 123–134.

Erguler, Z.A., Ulusay, R., 2007. Estimation of uniaxial compressive strength of clay-bearing weak rocks using needle penetration resistance. In: Proceedings of 11th Congress on Int. Soc. Rock Mech., Lisbon., ISRM, vol. 1. pp. 265–268.

Erguler, Z.A., Ulusay, R., 2009. Water-induced variations in mechanical properties of clay-bearing rocks. Int. J. Rock Mech. Min. Sci. 46, 355–370.

Ewy, R., 1999. Wellbore-stability predictions by use of a modified Lade criterion. SPE Drill. Complet. 14, 85–91.

Fener, M., Kahraman, S., Bilgil, A., Gunaydin, O., 2005. A comparative evaluation of indi-rect methods to estimate the compressive strength of rocks. Rock Mech. Rock. Eng. 38, 329–343.

Forster, I.R., 1983. The influence of core sample geometry on the axial point-load test. Int. J. Rock Mech. Min. Sci. Geomech. Abstr. 20, 291–295.

Freyburg, E., 1972. Der untere und mittlere Buntsandstein SW-Thüringens in seinen gesteinstechnischen Eigenschaften. Ber. Dte. Ges. Geol. Wiss. A, Berlin 17, 911–919.

Ghose, A.K., Chakraborti, S., 1986. Empirical strength indices of Indian coals an investigation. In: Proc. 27th US Symp. on Rock Mech, Balkema, Rotterdam, pp. 56–61.

Ghosh, D.K., Srivastava, M., 1991. Point-load strength: an index for classification of rock material. Bull. Int. Assoc. Eng. Geol. 44, 27–33.

Göktan, R.M., 1988. Theoretical and practical analysis of rock rippability. Ph.D. thesis, Istanbul Technical University.

Golubev, A., Rabinovich, G.J., 1976. Resultaty primenenia apparatury akusticeskogo karotasa dlja predelenia procnostych svoistv gornych porod na mestorosdeniach tverdych iskopaemych. Prikladnaja Geofizika Moskva 73, 109–116.

Goodman, R.E., 1970. The deformability of joints. Determination of the In-Situ Modulus of Deformation of Rock, ASTM Special Technical Publication No. 477, ASTM International, pp. 174–196.

Goodman, R.E., Van, T.K., Heuze, F.E., 1969. The measurement of rock deformability in boreholes. In: Proc. 10th U.S. Symp. on Rock Mech., Austin, TX, pp. 523–555.

Gorjainov, N.N., Ljachovickij, F.M., 1979. Seismiceskie metody v insenernoi geologii, Izdat. Nedra, Moskva.

Grasso, P., Xu, S., Mahtab, A., 1992. Problems and promises of index testing of rocks. In: Proc. of 33rd US Symp. of Rock Mechanics. Santa Fe, New Mexico, Balkema, Rotterdam, pp. 879–888.

Grimstad, E., Bhasin, R., 1996. Stress strength relationships and stability in hard rock. Proc. Conf. on Recent Advances in Tunnelling Technology. New Delhi, India 1, 3–8.

Gunsallus, K.L., Kulhawy, F.H., 1984. A comparative evaluation of rock strength measures. Int. J. Rock Mech. Min. Sci. Geomech. Abstr. 21, 233–248.

Haramy, K.Y., DeMarco, M.J., 1985. Use of the Schmidt hammer for rock and coal testing. In: 26th US Symp. on Rock Mechanics, Rapid City, SD, pp. 549–555.

Hassani, F.P., Scoble, M.J., Whittaker, B.N., 1980. Application of the point load index test to strength determination of rock and proposals for a new size-correction chart. In: The State of the Art in Rock Mechanics. Proc. 21st US Symp. on Rock Mech., Rolla, Missouri, pp. 543–553.

Hawkins, A.B., McConnel, B.J., 1992. Sensitivity of sandstone strength and deformability to changes in moisture content. Q. J. Eng. Geol. 62, 115–130.

Hawkins, A.B., Olver, J.A.G., 1986. Point load tests: correlation factor and contractual use. An example from the Corallian at Weymouth. In: Hawkins, A.B. (Ed.), Site Investigation Practice: Assessing BS 5930. Geological Society, London, pp. 269–271.

Heard, H.C., 1980. Thermal expansion and inferred permeability of climax quartz monazite to 300 °C and 27.6 MPa. Int. J. Rock Mech. Min. Sci. Geomech. Abstr. 17, 289–296.

Heuze, F.E., 1980. Scale effects in the determination of rock mass strength and deformability. Rock Mech. 12, 167–192.

Hoek, E., 1964. Fracture of anisotropic rock. J. S. Afr. Inst. Min. Metall. 64, 510–518.

Hoek, E., 1983. Strength of jointed rock masses—23rd Rankine lecture. Geotechnique 33 (3), 187–223.

Hoek, E., 1994. Strength of rock and rock masses. ISRM News J. 2, 4–16.

Hoek, E., 2004. Personal communication.

Hoek, E., 2006. Rock engineering. http://www.rocscience.com.

Hoek, E., Bray, J.W., 1981. Rock Slope Engineering, third ed. Institution of Mining and Metallurgy, London.

Hoek, E., Brown, E.T., 1980. Empirical strength criterion for rock masses. J. Geotech. Eng. ASCE 106, 1013–1035.

Hoek, E., Brown, E.T., 1988. The Hoek-Brown criterion—a 1988 update. In: Proc. 15th Can. Rock Mech. Symp. University of Toronto, Toronto, pp. 31–38.

Hoek, E., Brown, E.T., 1997. Practical estimates of rock mass strength. Int. J. Rock Mech. Min. Sci. 34, 1165–1186.

Hoek, E., Wood, D., Shah, S., 1992. A modified Hoek-Brown criterion for jointed rock masses. In: Hudson, J.A. (Ed.), Proc. Rock Characterization Symp. ISRM: Eurock'92, Chester, pp. 209–214.

Hoek, E., Kaiser, P.K., Bawden, W.F., 1995. Support of Underground Excavations in Hard Rock. Balkema, Rotterdam.

Hoek, E., Carranza-Torres, C., Corkum, B., 2002. Hoek-Brown failure criterion—2002 edition. In: Proc. 5th North American Rock Mechanics Symposium & 17th Tunneling Association of Canada Conference: NARMS-TAC 2002. Mining Innovation and Tech., Toronto, pp. 267–273

Homand-Etienne, H., Houpert, R., 1989. Thermally induced micro-cracking in granites: characterization and analysis. Int. J. Rock Mech. Min. Sci. Geomech. Abstr. 26, 125–134.

Horino, F.G., Ellickson, M.L., 1970. A method of estimating the strength of rock containing planes of weakness. U.S. Bureau of Mines, Report Investigation 7449.

Inoue, M., Ohomi, M., 1981. Relation between uniaxial compressive strength and elastic wave velocity of soft rock. In: Int. Symp. on Weak Rock, Tokyo, pp. 9–13.

Irfan, T.Y., Dearman, W.R., 1978. Engineering classification and index properties of weathered granite. Bull. Int. Assoc. Eng. Geol. 17, 79–90.

ISRM, 1985. Suggested methods for determining the point load strength. International Society for Rock Mechanics, Commission on Standardization of Laboratory and Field Tests. Int. J. Rock Mech. Min. Sci. Geomech. Abstr. 22, 51–60.

Jaeger, J.C., 1960. Shear failure of anisotropic rocks. Geol. Mag. 97, 65–72.

Jaeger, J.C., 1971. Friction of rocks and the stability of rock slopes, Rankine Lecture. Geotechnique 21, 97–134.

Jaeger, J.C., Cook, N.G.W., 1979. Fundamentals of Rock Mechanics, third ed. Chapman and Hall, London.

Jaiswal, A., Shrivastva, B.K., 2012. A generalized three-dimensional failure criterion for rock masses. J. Rock Mech. Geotech. Eng. 4, 333–343.

Johnston, I.W., 1985. Strength of intact geomechanical materials. J. Geotech. Eng. ASCE 111, 730–749.

Kahraman, S., 2001. Evaluation of simple methods for assessing the uniaxial compressive strength of rock. Int. J. Rock Mech. Min. Sci. 38, 981–994.

Kahraman, S., Gunaydin, O., Fener, M., 2005. The effect of porosity on the relation between uniaxial compressive strength and point load index. Int. J. Rock Mech. Min. Sci. 42, 584–589.

Kalamaras, G.S., Bieniawski, Z.T., 1993. A rock mass strength concept for coal seams. In: Proc. 12th Conf. Ground Control in Mining, Morgantown, pp. 274–283.

Karakul, H., Ulusay, R., 2013. Empirical correlations for predicting strength properties of rocks from P-wave velocity under different degrees of saturation. Rock Mech. Rock. Eng. 46, 981–999.

Katz, O., Reches, Z., Roegiers, J.-C., 2000. Evaluation of mechanical rock properties using a Schmidt hammer. Int. J. Rock Mech. Min. Sci. 37, 723–728.

Kidybinski, A., 1980. Bursting liability indices of coal. Int. J. Rock Mech. Min. Sci. Geomech. Abstr. 17, 167–171.

Koncagül, E.C., Santi, P.M., 1999. Predicting the unconfined compressive strength of the Breathitt shale using slake durability, shore hardness and rock structural properties. Int. J. Rock Mech. Min. Sci. 36, 139–153.

Konecny, P., Kozusnikova, A., Goel, R.K., Dwivedi, R.D., Swarup, A., Prasad, V.V.R., 2005. A comparative study of influence of temperature on granites from India and Czech Republic. In: Proceedings of EUROCK 2005, Brno, Czech Republic, pp. 265–267.

Kulhawy, F.H., Goodman, E.E., 1987. Foundations in rock. In: Bell, F.G. (Ed.), Ground Engineer's Reference Book. Butterworths, London, pp. 55/1–13.

Kwaśniewski, M., 2011. A new approach to the modeling of the pressure dependency of the strength of rocks. Rock Mech. Rock. Eng. 44, 103–111.

Kwaśniewski, M., Oitaben, P.R., 2009. Effect of water on the deformability of rocks under uniaxial compression. In: ISRM Regional Symposium EUROCK2009 Rock Engineering in Difficult Ground Conditions-Soft Rock and Karst, Dubrovnik, Cavtat, Croatia, pp. 271–276.

Lade, P.V., 1993. Roc strength criteria: the theories and the evidence. In: Hudson, J.A. (Ed.), Comprehensive Rock Engineering—Principle, Practice & Projects, vol. 1. Pergamon, Oxford, pp. 255–284.

Lashkaripour, G.R., 2002. Predicting mechanical properties of mudrock from index parameters. Bull. Eng. Geol. Environ. 61, 73–77.

Lashkaripour, G.R., Ghafoori, M., 2002. The engineering geology of the Tabarak Abad Dam. Eng. Geol. 66, 233–239.

Laubscher, D.H., 1984. Design aspects and effectiveness of support system in different mining conditions. Trans. Inst. Min. Metall. 93, A70–A81.

Lee, R.D., 1961. Testing mine floors. Colliery Eng. 38, 255–261.

Li, D., Wong, L.N.Y., 2013. Point load test on meta-sedimentary rocks and correlation to UCS and BTS. Rock Mech. Rock. Eng. 46, 889–896.

Matsuoka, H., Nakai, T., 1974. Stress-deformation and strength characteristics of soil under three different principal stress. In: Proceedings of Japan Society of Civil Engineers, Japan, pp. 59–70.

McLamore, R., Gray, K.E., 1967. The mechanical behavior of anisotropic sedimentary rocks. J. Eng. Ind. 89, 62–76.

McNally, G.H., 1987. Estimation of coal measures rock strength using sonic and neutron logs. Geoexploration 24, 381–395.

Medhurst, T.P., Brown, E.T., 1996. Large scale laboratory testing of coal. In: Proc. 7th Australian-New Zealand Conf. on Geomech., Canberra, Australia, pp. 203–208.

Melkoumian, N., Priest, S.D., Hunt, S.P., 2009. Further development of the three-dimensional Hoek-Brown yield criterion. Rock Mech. Rock. Eng. 42 (6), 835–847.

Militzer, H., Stoll, R., 1973. Einige Beiträge der Geophysik zur primärdatenerfassung im Bergbau. Neue Bergbautechnik, Leipzig 3, 21–25.

Mishra, D.A., Basu, A., 2012. Use of the block punch test to predict the compressive and tensile strengths of rocks. Int. J. Rock Mech. Min. Sci. 51, 119–127.

Mogi, K., 1971. Fracture and flow of rocks under high triaxial compression. J. Geophys. Res. 76 (5), 1255–1269.

Mogi, K., Kwasniewski, M., Mochizuki, H., 1978. Fracture of anisotropic rocks under general triaxial compression (abstract). Seismol. Soc. Japan Bull. 1 (D40), 25(in Japanese).

Morales, T., Uribe-Etxebarria, G., Uriarte, J.A., de Valderrama, I.F., 2004. Probabilistic slope analysis—state-of-play. Eng. Geol. 71, 343–362.

Murphy, D.K., Levay, J., Rancourt, M., 1976. The LG-2 underground power house. In: Proc. RETC Conf., Las Vegas, NV, pp. 515–533.

Nabaei, M., Shahbazi, K., Shadravan, A., 2010. Uncertainty analysis in unconfined rock compressive strength prediction. In: Proc. SPE Deep Gas Conference and Exhibition, Manama, Bahrain, Paper SPE 131719.

Naoto, U., Yoshitake, E., Hidehiro, O., Norihiko, M., 2004. Strength evaluation of deep mixing soil-cement by needle penetration test. J. Jpn. Soc. Soil Mech. Found. Eng. 52 (7), 23–25 (in Japanese).

Norbury, D.R., 1986. The point load test. In: Hawkins, A.B. (Ed.), Site Investigation Practice: Assessing BS 5930. Geological Society, London, pp. 325–329.

O'Rourke, J.E., 1988. Rock index properties for geo engineering design in underground development. SME preprint 88–48, Society for Mining, Metallurgy & Exploration, 5 p.

Okada, S., Izumiya, Y., Iizuka, Y., Horiuchi, S., 1985. The estimation of soft rock strength around a tunnel by needle penetration test. J. Jpn. Soc. Soil Mech. Found. Eng. 33 (2), 35–38 (in Japanese).

Palassi, M., Emami, V., 2014. A new nail penetration test for estimation of rock strength. Int. J. Rock Mech. Min. Sci. 66, 124–127.

Palchik, V., 1999. Influence of porosity and elastic modulus on uniaxial compressive strength in soft brittle porous sandstones. Rock Mech. Rock. Eng. 32, 303–309.

Palchik, V., Hatzor, Y.H., 2004. Influence of porosity on tensile and compressive strength of porous chalks. Rock Mech. Rock. Eng. 37, 331–341.

Pan, X.D., Hudson, J.A., 1988. A simplified three-dimensional Hoek-Brown yield criterion. In: Romana, M. (Ed.), Rock Mechanics and Power Plants. Balkema, Rotterdam, pp. 95–103.

Pappalardo, G., 2015. Correlation between P-wave velocity and physical–mechanical properties of intensely jointed dolostones, Peloritani Mounts, NE Sicily. Rock Mech. Rock Eng. 48, 1711–1721.

Patton, F.D., 1966. Multiple modes of shear failure in rock. In: Proc. 1st Int. Cong. on Rock Mech., Lisbon. vol. 1, pp. 509–515.

Peng, J., Rong, G., Cai, M., Wang, X., Zhou, C., 2014. An empirical failure criterion for intact rocks. Rock Mech. Rock. Eng. 47, 347–356.

Pomeroy, C.D., Hobbs, D.W., Mahmoud, A., 1971. The effect of weakness plane orientation on the fracture of Barnsley Hards by triaxial compression. Int. J. Rock Mech. Min. Sci. Geomech. Abstr. 8, 227–238.

Priest, S.D., 2005. Determination of shear strength and three-dimensional yield strength for the Hoek–Brown yield criterion. Rock Mech. Rock. Eng. 38, 299–327.

Quane, S.L., Russel, J.K., 2003. Rock strength as a metric of welding intensity in pyroclastic deposits. Eur. J. Mineral. 15, 855–864.

Rabbani, E., Sharif, F., KoolivandSalooki, M., Moradzadeh, A., 2012. Application of neural network technique for prediction of uniaxial compressive strength using reservoir formation properties. Int. J. Rock Mech. Min. Sci. 56, 100–111.

Rajabzadeh, M.A., Moosavinasab, G., Rakhshandehroo, G., 2012. Effects of rock classes and porosity on the relation between uniaxial compressive strength some rock properties for carbonate rocks. Rock Mech. Rock. Eng. 45, 113–122.

Ramamurthy, T., 1986. Stability of rock mass—eighth Indian Geotech. Soc. annual lecture. Indian Geotech. J. 16, 1–73.

Ramamurthy, T., 1993. Strength and modulus responses of anisotropic rocks. In: Hudson, J.A. (Ed.), Comprehensive Rock Engineering—Principle, Practice and Projects, vol. 1. Pergamon, Oxford, pp. 313–329.

Ramamurthy, T., 1996. Stability of rock mass—eighth Indian Geotechnical Society annual lecture. Indian Geotech. J. 16, 1–73.

Ramamurthy, T., Arora, V.K., 1991. A simple stress-strain model for jointed rocks. In: Wittke, W. (Ed.), Proc. 7th Int. Cong. on Rock Mech., Anchen. vol. 1. Balkema, Rotterdam, pp. 323–326.

Ramamurthy, T., Rao, G.V., Rao, K.S., 1985. A strength criterion for rocks. In: Proc. Indian Geotech. Conf., Roorkee. vol. 1. pp. 59–64.

Rao, K.S., 1984. Strength and deformation behavior of sandstones. PhD thesis, India Institute of Technology, Delhi.

Read, J.R.L., Thornton, P.N., Regan, W.M., 1980. A rational approach to the point load test. In: Proc. Aust.-N.Z. Geomech. Conf.vol. 2. pp. 35–39.

Rhodes, G.W., Stephenson, R.W., Rockaway, J.D., 1973. Plate bearing tests on coal underclay. In: Proc. 19th U.S. Symp. on Rock Mech., Stateline, NV., Univ. of Nevada Publ., vol. 2, pp. 16–27.

Robertson, A., 1970. The interpretation of geologic factors for use in slope theory. In: Proc. Symp. on the Theoretical Background to the Planning of Open Pit Mines, Johannesburg, South Africa, pp. 55–71.

Romana, M., 1999. Correlation between uniaxial compressive and point-load (Franklin test) strengths for different rock classes. In: 9th ISRM Congress. vol. 1. Balkema, Paris, pp. 673–676.

Rshewski, W.W., Novik, G.J., 1978. Osnovy fiziki gornich porod, Izdat. Nauka, Moskva.

Rusnak, J.A., Mark, C., 1999. Using the point load test to determine the uniaxial compressive strength of coal measure rock. In: Proceedings of 19th International Conference on Ground Control in Mining. Morgantown, WV, pp. 362–371.

Russo, G., Kalamaras, G.S., Grasso, P., 1998. A discussion on the concepts of geomechanical classes behavior categories and technical classes for an underground project. Gallerie e Grandi Opere Sotterranee, 54.

Sachpazis, C.I., 1990. Correlating Schmidt hardness with compressive strength and Young's modulus of carbonate rocks. Bull. Int. Assoc. Eng. Geol. 42, 75–84.

Saroglou, H., Tsiambaos, G., 2008. A modified Hoek-Brown failure criterion for anisotropic intact rock. Int. J. Rock Mech. Min. Sci. 45, 223–234.

Shalabi, F.I., Cording, E.J., Al-Hattamleh, O.H., 2007. Estimation of rock engineering properties using hardness tests. Eng. Geol. 90, 138–147.

Sheorey, P.R., 1997. Empirical Rock Failure Criteria. Balkema, Rotterdam.

Singh, D.P., 1981. Determination of some engineering properties of weak rocks. In: IAkai, K. (Ed.), Proc. Int. Sym. on Weak Rock. Balkema, Rotterdam, pp. 21–24.

Singh, J., 1988. Strength prediction of anisotropic rocks. PhD thesis, India Institute of Technology, Delhi.

Singh, B., Goel, R.K., 1999. Rock Mass Classifications—A Practical Approach in Civil Engineering. Elsevier, Amsterdam.

Singh, M., Rao, K.S., 2005. Empirical methods to estimate the strength of jointed rock masses. Eng. Geol. 77 (1–2), 127–137.

Singh, V.K., Singh, D.P., 1993. Correlation between point load index and compressive strength for quartzite rocks. Geotech. Geol. Eng. 11, 269–272.

Singh, B., Goel, R.K., Mehrotra, V.K., Garg, S.K., Allu, M.R., 1998. Effect of intermediate principal stress on strength of anisotropic rock mass. Tunn. Undergr. Space Technol. 13 (1), 71–79.

Singh, T.N., Kainthola, A., Venkatesh, A., 2012. Correlation between point load index and uniaxial compressive strength for different rock types. Rock Mech. Rock. Eng. 45, 259–264.

Smith, H.J., 1997. The point load test for weak rock in dredging applications. Int. J. Rock Mech. Min. Sci. 34 (3–4). Paper No. 295.

Sousa, L.M.O., del Rio, L.M.S., Calleja, L., de Argandona, V.G.R., Rey, A.R., 2005. Influence of microfractures and porosity on the physico-mechanical properties and weathering of ornamental granites. Eng. Geol. 77, 153–168.

Stewart, C.L., 1977. Subsurface rock mechanics instrumentation program for demonstration of shield-type long wall supports at York Canyon, Raton, New Mexico. In: Proc. 18th U.S. Symp. on Rock Mech., Keystone, CO, 1C-2, pp. 1–13.

Stimpson, B., Acott, C.P., 1983. Application of the NCB cone indenter to strength index testing of sedimentary rocks from Western Canada. Can. Geotech. J. 20, 532–535.

Szlavin, J., 1974. Relationships between some physical properties of rock determined by laboratory tests. Int. J. Rock Mech. Min. Sci. Geomech. Abstr. 11, 57–66.

Takahashi, K., Noto, K., Yokokawa, I., 1998. Strength characteristics of Kobe Formation in Akashi Strata (no. 1). In: Proc. of 10th Japan National Conf. on Geotech. Eng. The Japanese Geotechnical Society, Tokyo, pp. 1231–1232 (in Japanese).

Trueman, R., 1988. An evaluation of strata support techniques in dual life gateroads. Ph.D. thesis, University of Wales, Cardiff.

Tsiambaos, G., Sabatakakis, N., 2004. Considerations on strength of intact sedimentary rocks. Eng. Geol. 72, 261–273.

Tsidzi, K.E.N., 1990. The influence of foliation on point load strength anisotropy of foliated rocks. Eng. Geol. 29, 49–58.

Tsidzi, K.E.N., 1991. Point load-uniaxial compressive strength correlation. In: Wittke, W. (Ed.), Proc. 7th ISRM Cong. vol. 1. Balkema, Rotterdam, pp. 637–639.

Tullis, J., Yund, R.A., 1977. Experimental deformation of dry Westerly granite. J. Geophys. Res. 82, 5705–5718.

Tuğrul, A., 2004. The effect of weathering on pore geometry and compressive strength of selected rock types from Turkey. Eng. Geol. 75, 215–227.

Tuğrul, A., Zarif, I.H., 1999. Correlation of mineralogical and textural characteristics with engineering properties of selected granitic rocks from Turkey. Eng. Geol. 51, 303–317.

Ulusay, R., Erguler, Z.A., 2012. Needle penetration test: evaluation of its performance and possible uses in predicting strength of weak and soft rocks. Eng. Geol. 149–150, 47–56.

Ulusay, R., Tureli, K., Ider, M.H., 1994. Prediction of engineering properties of a selected litharenite sandstone from its petrographic characteristics using correlation and multivariate statistical techniques. Eng. Geol. 38, 135–157.

Vallejo, L.E., Welsh, R.A., Robinson, M.K., 1989. Correlation between unconfined compressive and point load strength for Appalachian rocks. In: Khair, A.W. (Ed.), Proc. 30th US Symp. Rock Mech. Balkema, Rotterdam, pp. 461–468.

Vasarhelyi, B., 2003. Some observations regarding the strength and deformability of sandstones in case of dry and saturated conditions. Bull. Eng. Geol. Environ. 62, 245–249.

Vasarhelyi, B., 2005. Statistical analysis of the influence of water content on the strength of the Miocene limestone. Rock Mech. Rock. Eng. 38, 69–76.

William, K.J., Warnke, E.P., 1975. Constitutive Model for the Triaxial Behavior of Concrete. International Association for Bridge and Structural Engineering, Bergamo.

Wuerker, R.G., 1953. The status of testing strength of rocks. Trans. Min. Eng. AIME, 1108–1113.

Xu, S., Grasso, P., Mahtab, A., 1990. Use of Schmidt hammer for estimating mechanical properties of weak rock. In: 6th Int. IAEG Cong. Balkema, Rotterdam, pp. 511–519.

Yamaguchi, Y., Ogawa, N., Kawasaki, M., Nakamura, A., 1997. Evaluation of seepage failure resistance potential of dam foundation with simplified tests. J.Jpn. Soc. Eng. Geol. 38 (3), 130–144.

Yasar, E., Erdoğan, Y., 2004a. Estimation of rock physicomechnical properties using hardness methods. Eng. Geol. 71, 281–288.

Yasar, E., Erdoğan, Y., 2004b. Correlating sound velocity with the density, compressive strength and Young's modulus of carbonate rocks. Int. J. Rock Mech. Min. Sci. 41, 871–875.

Yilmaz, I., 2009. A new testing method for indirect determination of the unconfined compressive strength of rocks. Int. J. Rock Mech. Min. Sci. 46, 1349–1357.

Yilmaz, I., 2010. Influence of water content on the strength and deformability of gypsum. Int. J. Rock Mech. Min. Sci. 47, 342–347.

Yilmaz, I., Sendir, H., 2002. Correlation of Schmidt hardness with unconfined compressive strength and Young's modulus in gypsum from Sivas (Turkey). Eng. Geol. 66, 211–219.

Yilmaz, I., Yucel, Ö., 2014. Use of the core strangle test for determining strength anisotropy of rocks. Int. J. Rock Mech. Min. Sci. 66, 57–63.

Yilmaz, I., Yuksek, G., 2009. Prediction of the strength and elasticity modulus of gypsum using multiple regression, ANN, and ANFIS models. Int. J. Rock Mech. Min. Sci. 46, 803–810.

Yu, M., 1990. New System of Strength Theory. Xian Jiaotong University, Xian.

Yudhbir, F., Lemanza, W., Prinzl, F., 1983. An empirical failure criterion for rock masses. In: Proc. 5th Int. Congress on Rock Mechanics. vol. 1. ISRM, Melbourne, pp. B1–B8.

Zorlu, K., Ulusay, R., Ocakoglu, F., Gokceoglu, C., Sonmez, H., 2004. Predicting intact rock properties of selected sandstones using petrographic thin-section data. Int. J. Rock Mech. Min. Sci. 41, 93–98.

Zhang, L., 2008. A generalized three-dimensional Hoek-Brown strength criterion. Rock Mech. Rock. Eng. 41, 893–915.

Zhang, L., 2010. Estimating the strength of jointed rock masses. Rock Mech. Rock. Eng. 43, 391–402.

Zhang, L., Zhu, H., 2007. Three-dimensional Hoek–Brown strength criterion for rocks. J. Geotech. Geoenviron. Eng. 133 (9), 1128–1135.

Zhang, Q., Zhu, H., Zhang, L., 2013. Modification of a generalized three-dimensional Hoek-Brown strength criterion. Int. J. Rock Mech. Min. Sci. 59, 80–96.

Zhang, L., Mao, X., Liu, R., Guo, X., Ma, D., 2014. The mechanical properties of mudstone at high temperatures: an experimental study. Rock Mech. Rock. Eng. 47, 1479–1484.

Zhou, S., 1994. A program to model the initial shape and extent of borehole breakout. Comput. Geosci. 20, 1143–1160.

Chapter 8

Permeability

8.1 INTRODUCTION

The permeability of a rock is a measure of its capacity for transmitting a fluid. Rock permeability is one of the most important parameters controlling project performance in energy, civil, and environmental engineering. Effective characterization of fluid flow and chemical transport in rock requires accurate determination of permeability. The coefficient of permeability (or hydraulic conductivity) is defined as the discharge velocity through a unit area under a unit hydraulic gradient and is dependent upon the properties of the medium, as well as the viscosity and density of the fluid. According to Darcy's law, the quantity of flow through a cross-sectional area of rock can be calculated by

$$q = KiA \tag{8.1}$$

where q is the quantity of flow; K is the permeability coefficient of the rock, having the units of a velocity; i is the hydraulic gradient (head loss divided by length over which the head loss occurs); and A is the cross-sectional area of flow.

The permeability coefficient of a rock varies for different fluids depending on their density and viscosity as follows:

$$K = k\frac{\rho g}{\mu} = k\frac{g}{\nu} \tag{8.2}$$

where k is the intrinsic (or specific) permeability of the rock, having the units of length squared; ρ, μ, and ν are, respectively, the density, viscosity, and dynamic viscosity of the fluid; and g is the gravitational acceleration (9.81 m/s^2). The intrinsic permeability k is independent of the properties of the fluid in the rock.

Very often in rock engineering water is the percolating fluid. In the following discussion, therefore, the fluid in a rock, if not specifically mentioned, will be water.

Because of the presence of discontinuities in the rock mass, the permeability of rock mass is controlled not only by the intact rock but also by the discontinuities separating the intact rock blocks.

Engineering Properties of Rocks. http://dx.doi.org/10.1016/B978-0-12-802833-9.00008-0

339

This chapter presents the representative values of the permeability of different rocks and describes various methods for estimating the permeability of rocks. The factors affecting the permeability of rocks are also discussed.

8.2   PERMEABILITY OF INTACT ROCK

The permeability of an intact rock is usually referred to as the primary permeability. The intact rock permeability is governed by the porosity, which varies with factors like the rock type, geological history, and in situ stress conditions. Figs. 8.1 and 8.2 are two examples of typical permeability k versus porosity n plots. The increase of permeability with growing porosity follows a quasilinear log k–n relationship (Van-Golf Racht, 1982):

$$\log k = a_1 n + a_2 \qquad (8.3)$$

where a_1 and a_2 are fitting constants. Since the porosity of intact rocks varies widely (see Table 3.6), the intact rock permeability varies in a great range: at least eight orders of magnitude. Fig. 8.3A shows the range of the intact rock permeability coefficient K of different rock types.

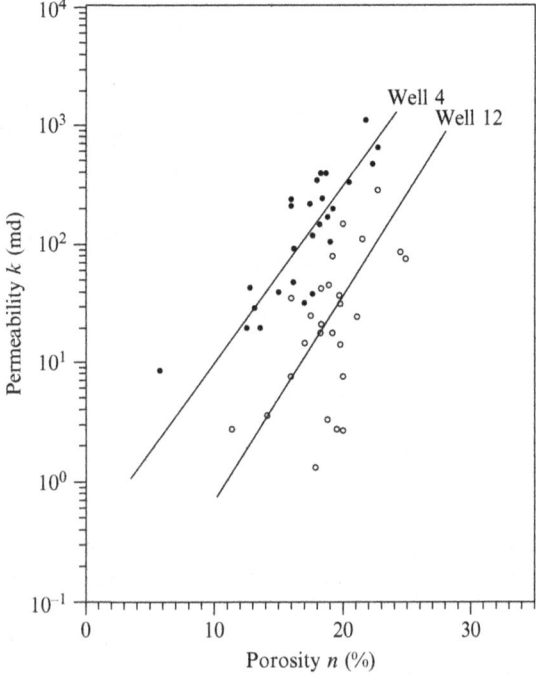

FIG. 8.1   Permeability versus porosity for Totliegent sandstone. *(From Schön, J.H., 1996. Physical Properties of Rocks—Fundamentals and Principles of Petrophysics. Pergamon, Oxford.)*

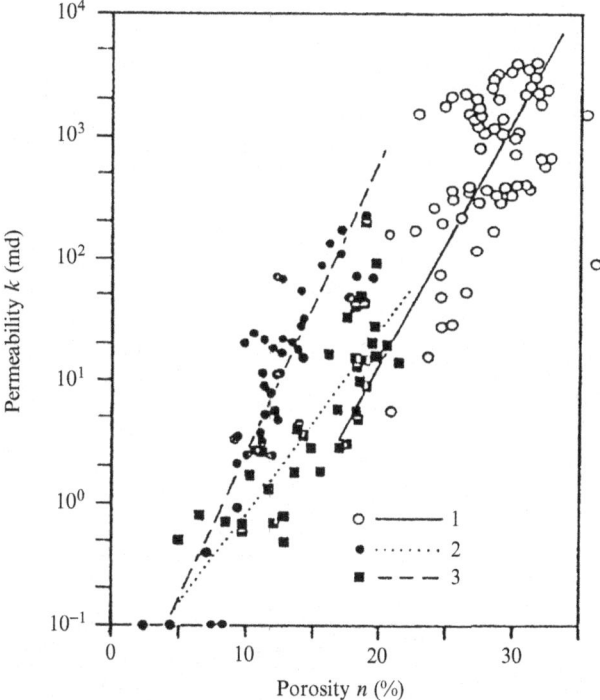

FIG. 8.2   Permeability versus porosity for three sandstones (1, Gulf Coast Field; 2, Colorado Field; 3, California Field). *(From Schön, J.H., 1996. Physical Properties of Rocks—Fundamentals and Principles of Petrophysics. Pergamon, Oxford.)*

The permeability of intact rocks also varies with the grain size, increasing with larger grain size. Fig. 8.4 shows a plot of permeability k versus grain size d. The trend of these data can be represented by

$$\log k = 2.221 \; \log d - 2.101 \tag{8.4}$$

where k is in md (millidarcy $\approx 10^{-15}$ m^2) and d is in μm (Schöpper, 1982).

Theoretical and experimental data show that at a given porosity, pore size distributions of rocks are controlled in the first order by grain size (Cade et al., 1994; Yang and Aplin, 2007, 2010). At a given porosity, the pore size and thus the permeability decrease as the grain size decreases. Using clay content (CF) as a lithological or grain size descriptor, Yang and Aplin (2010) derived the following expression relating permeability to both void ratio [$e = n/(1-n)$] and CF for mudstones:

$$\log k = \begin{aligned}&-69.59 - 26.79 \times CF + 44.07 \times CF^{0.5} \\ &+ \left(-53.61 - 80.03 \times CF + 132.78 \times CF^{0.5}\right) \times e \\ &+ \left(86.61 + 81.91 \times CF - 163.61 \times CF^{0.5}\right) \times e^{0.5}\end{aligned} \tag{8.5}$$

where k is in m^2 and CF in %.

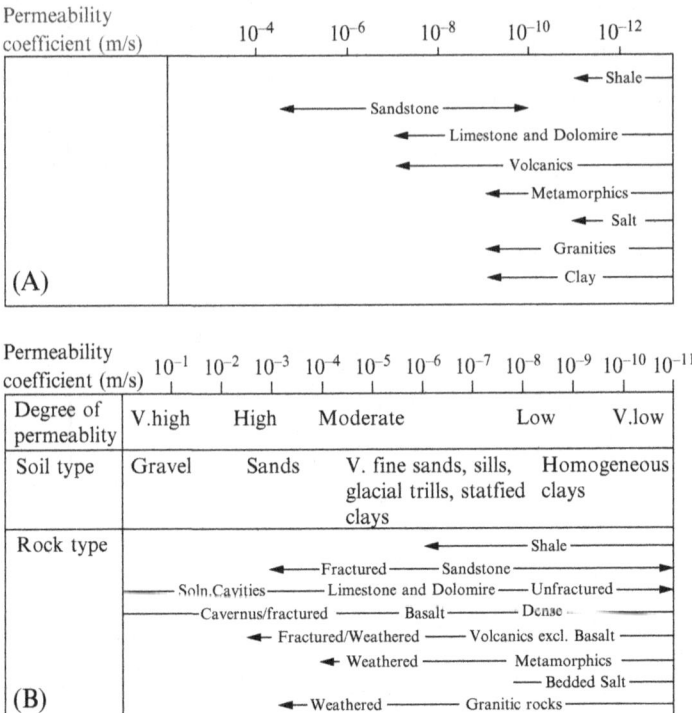

FIG. 8.3 Typical values of permeability coefficient for (A) intact rocks and (B) rock masses. *(Modified from Isherwood, D., 1979. Geoscience Data Base Handbook for Modeling Nuclear Waste Repository, vol. 1. NUREG/CR-0912 V1, UCRL-52719, V1.)*

8.3 PERMEABILITY OF DISCONTINUITIES

The flow of fluid through discontinuities in rock has been studied in great detail by different researchers, such as Huitt (1956), Snow (1968a,b), Louis (1969), Sharp (1970), Hoek and Bray (1981), Zhang et al. (2007), Fernandez and Moon (2010), and Zou et al. (2013). If discontinuities are infilled, the permeability of a discontinuity is simply that of the infilling material. For unfilled discontinuities, by modeling a discontinuity as an equivalent parallel plate conductor, the permeability coefficient along the discontinuity can be determined for the laminar flow by

$$K = \frac{ge^2}{12\nu C} \tag{8.6}$$

where e is the aperture of the discontinuity; ν is the kinematic viscosity of the fluid (which for water can be taken as 1.0×10^{-6} m^2/s); and C is a correction factor representing the discrepancy between the actual physical aperture of the discontinuity and its equivalent hydraulic aperture.

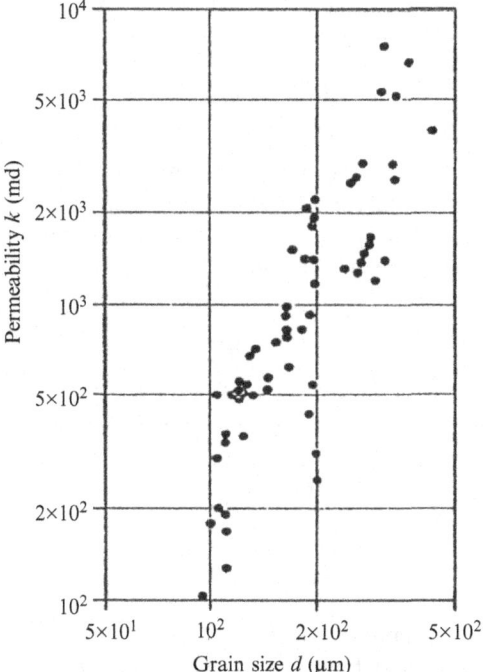

FIG. 8.4   Permeability versus grain size for Bentheim sandstone, Scherhorn oilfield, Germany. *(From Schön, J.H., 1996. Physical Properties of Rocks—Fundamentals and Principles of Petrophysics. Pergamon, Oxford.)*

If the equivalent hydraulic aperture is used, Eq. (8.6) can be rewritten as

$$K = \frac{g e_{h}^{2}}{12 \nu} \qquad (8.7)$$

where e_{h} is the equivalent hydraulic aperture of the discontinuity, which is related to its physical or mechanical aperture, e, as follows:

$$e_{h}^{2} = \frac{e^{2}}{C} \qquad (8.8)$$

Owing to the wall friction and the tortuosity, the mechanical aperture e is generally larger than the hydraulic aperture e_{h}. Hakami (1995) showed that the ratio of mechanical mean aperture to hydraulic aperture was 1.1–1.7 for discontinuities with a mean aperture of 100–500 μm. A study by Zimmerman and Bodvarsson (1996) concluded that the mechanic aperture is larger than hydraulic aperture by a factor that depends on the ratio of the mean value of the aperture to its standard deviation. Many researchers have evaluated the factor

C, including Lomize (1951), Louis (1969), and Quadros (1982). Their findings can be summarized by the following expression:

$$C = 1 + m \left(\frac{y}{2e}\right)^{1.5} \tag{8.9}$$

where $m = 17$ from Lomize (1951), $m = 8.8$ from Louis (1969), and $m = 20.5$ from Quadros (1982); y is the magnitude of the discontinuity surface roughness. For a smooth parallel discontinuity, y becomes zero and thus C becomes one and $e_h = e$.

Barton et al. (1985) related factor C to the joint (discontinuity) roughness coefficient (JRC) by

$$C = \frac{JRC^5}{e^2} \tag{8.10}$$

where C is dimensionless and e is in the unit of μm. The methods for determining JRC have been described in Chapter 6.

Combining Eqs. (8.8), (8.10) gives

$$e_h = \frac{e^2}{JRC^{2.5}} \tag{8.11}$$

The background data for Eq. (8.10) and thus Eq. (8.11) mainly comes from normal deformation fluid flow tests. Olsson and Barton (2001) found that Eq. (8.9) only applies to the case that the shear displacement u_s does not exceed 75% of the peak shear displacement u_{sp} (the shear displacement at peak shear stress). After the peak shear stress ($u_s \geq u_{sp}$), the hydraulic aperture e_h can be calculated from

$$e_h = e^{1/2} JRC_{mob} \quad (u_s \geq u_{sp}) \tag{8.12}$$

where JRC_{mob} is the mobilized value of JRC. In the phase of $u_s \geq u_{sp}$, the geometry of the discontinuity wall changes with increasing shear displacement and thus JRC_{mob} should be used. The value of JRC_{mob} is dependent on the strength of the discontinuity wall, on the applied normal stress and on the magnitude of the shear displacement. It is also dependent on the size of the discontinuity plane and on the residual friction angle of the discontinuity. For the calculation of u_{sp} and JRC_{mob}, the reader can refer to Olsson and Barton (2001).

Since the intermediate phase ($0.75u_{sp} < u_s < u_{sp}$) is difficult to define, Olsson and Barton (2001) recommended using a transition curve by connecting the two phases defined by Eqs. (8.11), (8.12).

For a set of parallel discontinuities, the permeability coefficient parallel to the discontinuities can be determined by

$$K = \frac{g(e_{avg})^3}{12\nu C_{avg} s_{avg}} \tag{8.13}$$

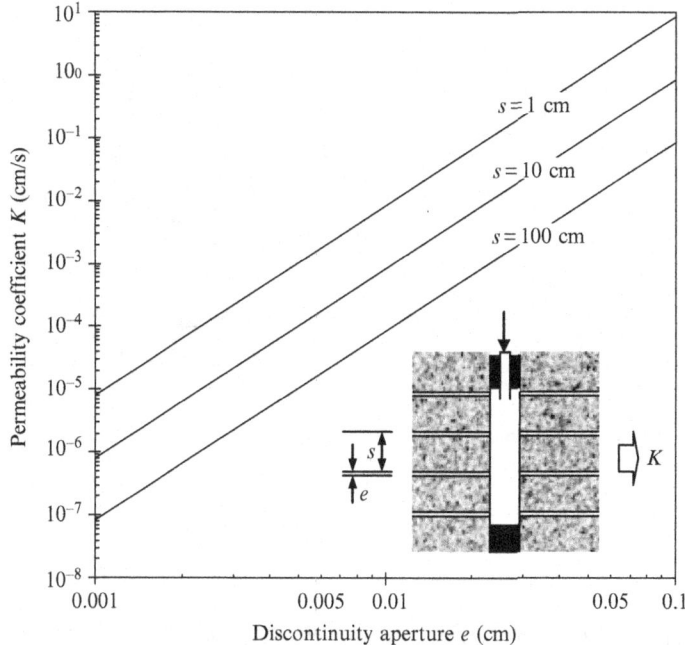

FIG. 8.5 Variation of permeability coefficient K of a set of smooth parallel discontinuities with discontinuity aperture e and spacing s. *(Modified from Hoek, E., Bray, J.W., 1981. Rock Slope Engineering. Institution of Mining and Metallurgy, London.)*

where e_{avg} is the average of individual values of e for discontinuities in the set under consideration; s_{avg} is the average of individual spacing s between discontinuities; and C_{avg} is estimated from Eq. (8.9) using $(y/e)_{avg}$ which is the average of the individual values of (y/e).

Fig. 8.5 shows the variation of permeability coefficient K of a set of smooth parallel discontinuities with the discontinuity aperture and the discontinuity spacing, based on Eq. (8.13). The permeability coefficient is very sensitive to small changes in aperture.

8.4   PERMEABILITY OF ROCK MASS

Fig. 8.3B illustrates the range of rock mass permeability coefficient for different rock types. It can be seen that the rock mass permeability coefficient varies in a very wide range: 11 orders of magnitude.

For a rock mass containing a single set of continuous discontinuities, as illustrated in Fig. 8.6A, the permeability coefficient of the rock mass in the direction of the discontinuities can be estimated as

$$K = \frac{ge_1^3}{12\nu C_1 s_1} + K_i(1 - e_1/s_1) \qquad (8.14)$$

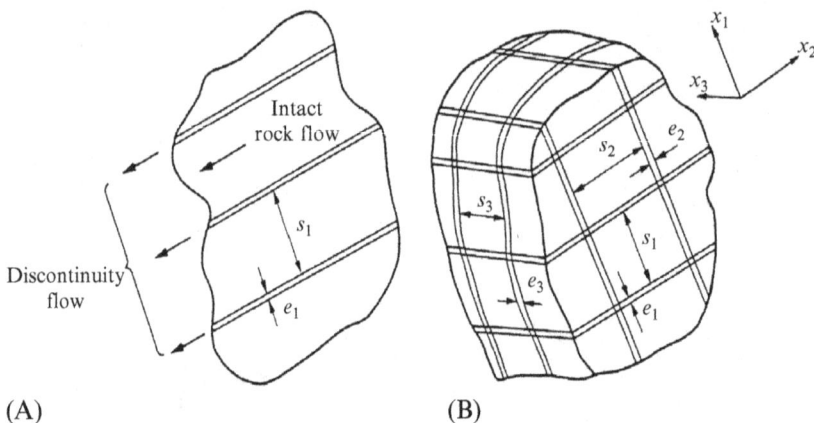

FIG. 8.6 Rock mass containing: (A) a single discontinuity set and (B) three orthogonal discontinuity sets.

where e_1, s_1, and C_1 are, respectively, the aperture, spacing, and correction factor of discontinuity set 1; and K_i is the permeability coefficient of the intact rock.

The permeability coefficient of the rock mass in the direction perpendicular to the discontinuities can be simply taken as that of the intact rock.

As further discontinuity sets are added to produce an orthogonal array, the principal magnitudes of permeability coefficient remain coincident with the lines of intersection of the discontinuity sets. The permeability coefficient, K_{11}, in the x_1 direction of Fig. 8.6B can be estimated as

$$K_{11} = \frac{g}{12\nu}\left(\frac{e_2^3}{C_2 s_2} + \frac{e_3^3}{C_3 s_3}\right) + K_i(1 - e_2/s_2)(1 - e_3/s_3) \qquad (8.15)$$

The permeability coefficients in the two other orthogonal directions may be determined from Eq. (8.15) through appropriate permutation. Where discontinuity apertures and spacings between discontinuities differ for each of the sets, the permeability of the rock mass will be anisotropic. Commonly, the discontinuity permeability dominates over the intact rock permeability. Consequently, the second term of Eqs. (8.14), (8.15) may often be neglected.

Fig. 8.7 shows the variation of rock mass permeability coefficient measured by conventional packer tests with the ratio of field rock mass wave velocity measured from seismic survey to that of rock core measured in laboratory, based on extensive studies at 89 concrete dam sites in the UK (Knill, 1969). All measurements were performed on saturated rocks. The field rock mass permeability coefficient increases by 10,000 times when the wave velocity ratio decreases from 1.0 to 0.5 due to fractures.

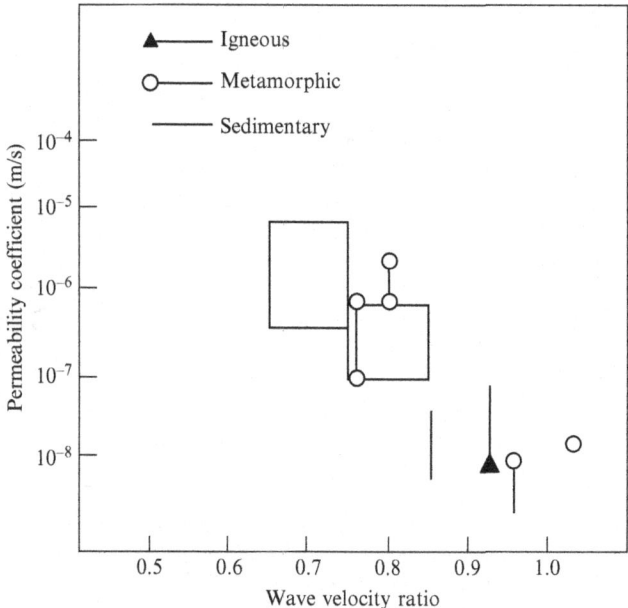

FIG. 8.7   Variation of field rock mass permeability coefficient with wave velocity ratio. *(Modified from Knill, J.L., 1969. The application of seismic methods in the prediction of grout taken in rock. In: Proc. Conf. In-Situ Investigations in Soils and Rocks, London, pp. 93–99.)*

Researchers have also related field measured rock mass permeability to rock mass quality indices such as rock quality designation (RQD), rock mass rating (RMR), and Q (El-Naqa, 2001; Cha et al., 2006; Barton, 2008; Jiang et al., 2009). Fig. 8.8 shows the measured permeability coefficient from packer tests versus RQD for the Cambrian sandstone rock mass in central Jordan. The following relation between permeability coefficient K and RQD can be derived using regression analysis (El-Naqa, 2001):

$$\log K = 2.300 - 0.0157 \text{RQD} \quad \left(r^2 = 0.64\right) \tag{8.16}$$

where K is in the unit of cm/day and r^2 is the determination coefficient. The permeability coefficient shows a progressive decrease with the increase of RQD. It needs to be noted that RQD does not include the information of discontinuity aperture. Since aperture is a main factor affecting the permeability of rock masses, caution should be taken when applying an empirical relation as Eq. (8.16) to a site different from the site where the empirical relation was derived. For example, the extensive study by Jiang et al. (2009) at a site composed of monzonitic granite, quartz monzonite, and quartz syenite in Eastern Shandong Province, China shows a similar relation to Eq. (8.16) between field

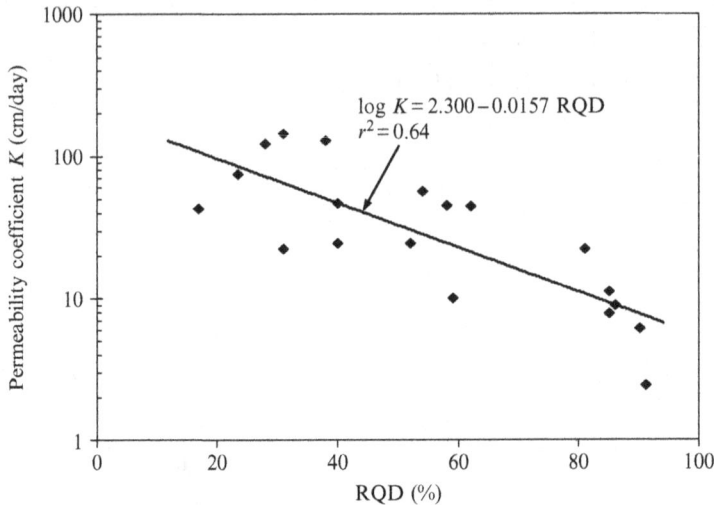

FIG. 8.8 Measured permeability coefficient versus RQD for Cambrian sandstone rock mass in central Jordan. *(From El-Naqa, A., 2001. The hydraulic conductivity of the fractures intersecting Cambrian sandstone rock masses, central Jordan. Environ. Geol. 40, 973–982.)*

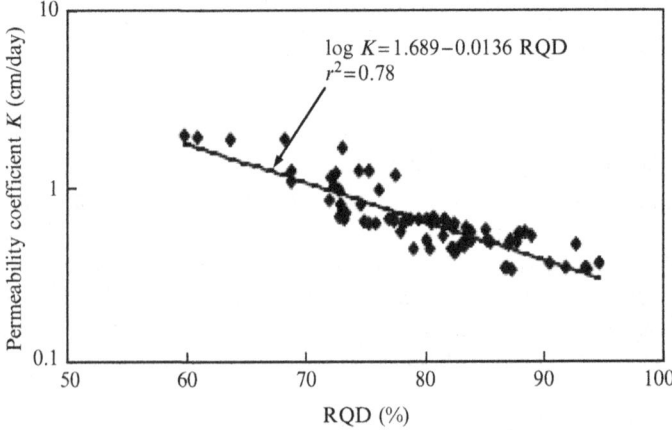

FIG. 8.9 Measured rock mass permeability coefficient versus RQD at a site composed of monzonitic granite, quartz monzonite, and quartz syenite in Eastern Shandong Province, China. *(From Jiang, X.-W., Wan, L., Wang, X.-S., Wu, X., Zhang, X., 2009. Estimation of rock mass deformation modulus using variations in transmissivity and RQD with depth. Int. J. Rock Mech. Min. Sci. 46, 1370–1377.)*

rock mass permeability coefficient K and RQD but the specific coefficients are different (Fig. 8.9):

$$\log K = 1.689 - 0.0236\text{RQD} \quad \left(r^2 = 0.78\right) \qquad (8.17)$$

where K is in the unit of cm/day and r^2 is the determination coefficient.

Sen (1996) derived analytical relations between rock mass permeability and RQD by considering the effect of discontinuity aperture. Fig. 8.10 is the permeability-RQD-aperture chart produced based on the analytical relations. It clearly shows that the discontinuity aperture has a major effect on rock permeability.

Fig. 8.11 shows the measured permeability coefficient from packer tests versus RMR for the Cambrian sandstone rock mass in central Jordan. The

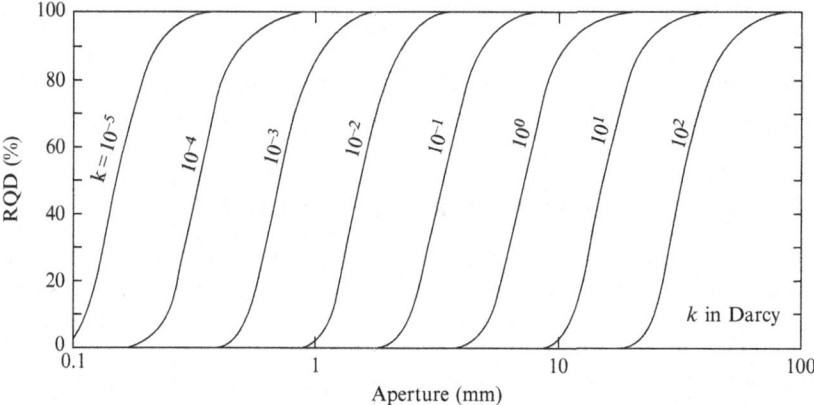

FIG. 8.10 Permeability-RQD-discontinuity aperture chart. *(From Sen, Z., 1996. Theoretical RQD-porosity-conductivity-aperture charts. Int. J. Rock Mech. Min. Sci. Geomech. Abstr. 33, 173–177.)*

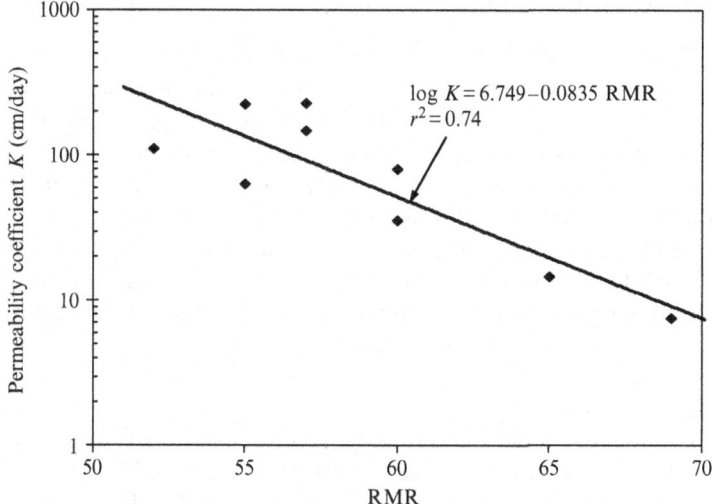

FIG. 8.11 Measured permeability coefficient versus RMR for Cambrian sandstone rock mass in central Jordan. *(Modified from El-Naqa, A., 2001. The hydraulic conductivity of the fractures intersecting Cambrian sandstone rock masses, central Jordan. Environ. Geol. 40, 973–982.)*

following relation between permeability coefficient K and RMR can be derived using regression analysis (El-Naqa, 2001):

$$\log K = 6.749 - 0.0835 \text{RMR} \quad (r^2 = 0.74) \tag{8.18}$$

where K is in the unit of cm/day and r^2 is the determination coefficient.

According to Barton (2008), the permeability coefficient K of an in situ rock mass can be related to its quality index Q as follows:

$$K = \left(\frac{0.002}{Q}\right)\left(\frac{100}{\text{JCS}}\right)\left(\frac{1}{H^{5/3}}\right) \quad (\text{m/s}) \tag{8.19}$$

where JCS is the joint wall compressive strength in MPa as described in Chapter 5 and H is the depth of the rock mass below ground surface.

8.5 EFFECT OF STRESS ON ROCK PERMEABILITY

Stress has a great effect on the permeability of both intact rocks and rock masses. A number of studies on the variation of rock permeability with stress can be found in the literature (Brace et al., 1968; Gangi, 1978; Kranz et al., 1979; Oda et al., 1989; Read et al., 1989; Jouanna, 1993; Azeemuddin et al., 1995; Indraratna et al., 1999; Ranjith, 2000; Zhang et al., 2000; Baghbanan and Jing, 2008; Ghabezloo et al., 2009; Jiang et al., 2010; Yang et al., 2010; Chen et al., 2011; Konecny and Kozusnikova, 2011; Meng et al., 2011; Metwally and Sondergeld, 2011; Kishida et al., 2011; Wong et al., 2013; Zou et al., 2013). Tiller (1953) found an empirical power relationship between the permeability of intact rock and the effective pressure:

$$K = A\sigma^{-m} \quad (\sigma > \sigma_{\text{threshold}}) \tag{8.20}$$

where A and m are constants and σ is the effective pressure (the difference between the exterior confining pressure and the pore-fluid pressure). Ghabezloo et al. (2009) proposed the same type of relation between permeability and effective pressure based on constant head permeability tests on limestone in a triaxial cell with different conditions of confining pressure and pore pressure.

Louis et al. (1977) and other many researchers (Seidle et al., 1992; David et al., 1994; Liu and Rutqvist, 2010; Meng et al., 2011; Kawano et al., 2011) used the following negative exponential expression to describe the relationship between rock permeability k and effective stress σ:

$$k = k_0 e^{-a\sigma} \tag{8.21}$$

where k_0 is the permeability at zero effective stress condition and a is an empirical coefficient. For example, based on best fitting analysis, Meng et al. (2011) related the field measured permeability in the Southern Qinshui coalbed methane reservoir in China to the in situ stresses as follows:

$$k = 58.135 e^{-0.435(\sigma_v - p_p)} \tag{8.22a}$$

$$k = 2.6357e^{-0.193\left(\sigma_{\text{hmax}} - p_{\text{p}}\right)} \tag{8.22b}$$

$$k = 4.6752e^{-0.446\left(\sigma_{\text{hmin}} - p_{\text{p}}\right)} \tag{8.22b}$$

where k is the permeability in md; σ_{v}, σ_{vmax}, and σ_{vmin} are the vertical, maximum horizontal, and minimum horizontal stresses, respectively, in MPa; and p_{p} is the pore water pressure in MPa.

Based on the Hertz theory of deformation of spheres, Gangi (1978) derived the following expression illustrating the effect of confining pressure on the intact rock permeability:

$$k = k_0 \left[1 - C_0 \left(\frac{\sigma + \sigma_i}{p_0} \right)^{2/3} \right]^4 \tag{8.23}$$

where k_0 is the initial permeability of the loose-grain packing; C_0 is a constant depending on the packing and is of the order of 2; σ is the confining pressure; σ_i is the equivalent pressure due to the cementation and permanent deformation of the grains; and p_0 is the effective elastic modulus of the grains and is of the order of the grain material bulk modulus.

Fig. 8.12 shows the variation of the intrinsic permeability of the intact Westerly granite rock with the confining pressure. The intact rock permeability decreases significantly when the confining pressure increases.

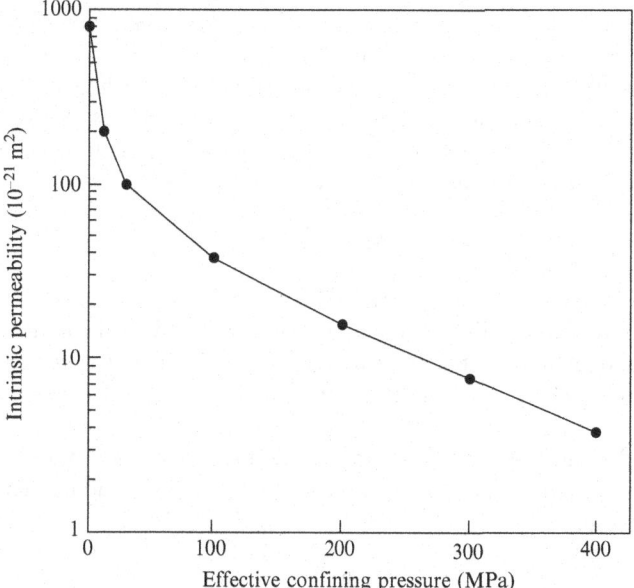

FIG. 8.12 Effect of confining pressure on the permeability of Westerly granite rock. *(From Indraratna, B., Ranjith, P., 2001. Hrdromechanical Aspects and Unsaturated Flow in Jointed Rock. Balkema, Lisse.)*

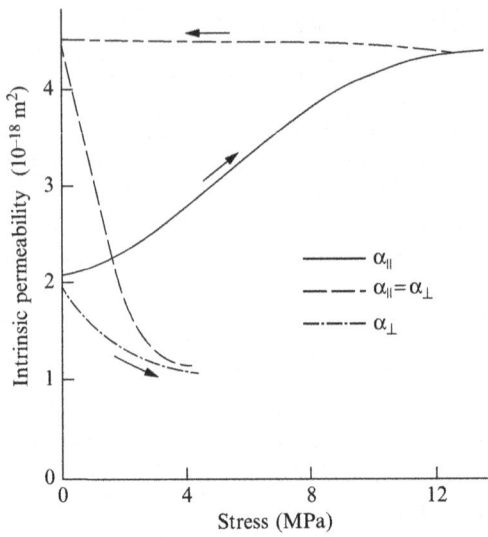

FIG. 8.13 Effect of load acting parallel and perpendicular to a discontinuity on the permeability. *(From Brace, W.F., 1978. A note on permeability changes in geological material due to stress. Pure Appl. Geophys. 116, 627–633.)*

The effect of stresses on rock mass permeability depends on their direction with respect to the discontinuity orientation. According to Brace (1978), a stress parallel to the discontinuities increases the permeability, while a stress perpendicular to the discontinuities decreases the permeability (Fig. 8.13).

Snow (1968a) presented the following empirical relation between the discontinuity permeability and the normal stress:

$$K = K_0 + k_n \frac{g e_h^2}{\nu s} (\sigma - \sigma_0) \tag{8.24}$$

where K is the discontinuity permeability at normal stress σ; K_0 is the initial discontinuity permeability at initial normal stress σ_0; k_n is the normal stiffness of the discontinuity; ν is the dynamic viscosity of the fluid; g is the gravitational acceleration (9.81 m/s²); s is the discontinuity spacing; and e_h is the hydraulic aperture.

Based on the test results on carbonate rocks, Jones (1975) proposed the following empirical relation between the discontinuity permeability and the confining pressure σ:

$$K = c_0 \left[\log \left(\frac{\sigma_h}{\sigma} \right) \right]^3 \tag{8.25}$$

where σ_h is the confining pressure at which the permeability is zero; and c_0 is a constant that depends on the discontinuity surface and the initial aperture.

By a "bed of nails" model for the asperities of a discontinuity, Gangi (1978) derived the following relation between discontinuity permeability and effective confining stress σ:

$$K = K_0 \left[1 - \left(\frac{\sigma}{p_1} \right)^m \right]^3 \tag{8.26}$$

where K_0 is the zero pressure permeability; m is a constant $(0 < m < 1)$ which characterizes the distribution function of the asperity lengths; and p_1 is the effective modulus of the asperities and is of the order of one-tenth to one-hundredth of the asperity material bulk modulus.

Nelson (1975) proposed the following general expression for the permeability of discontinuities:

$$K = A + B\sigma^{-m} \tag{8.27}$$

where σ is the effective confining stress and A, B, and m are constants determined by regression analysis of test results. These constants vary with the rock type, and even for the same rock type, change with the discontinuity surface.

Based on the simple model of Walsh and Grosenbaugh (1979) for describing the deformation of discontinuities, Walsh (1981) derived the following relation between permeability and confining pressure σ:

$$K = K_0 \left[1 - \left(2\frac{h}{a_0} \ln \frac{\sigma}{\sigma_0} \right)^{0.5} \right]^3 \left[\frac{1 - b(\sigma - u_w)}{1 + b(\sigma - u_w)} \right] \tag{8.28}$$

where K_0 is the permeability at reference confining pressure σ_0; h is the root mean square value of the height distribution of the discontinuity surface; a_0 is the half aperture at the reference confining pressure; u_w is the pore fluid pressure; and

$$b = \left[\frac{3\pi f}{E(1 - \nu^2)h} \right]^{0.5} \tag{8.29}$$

where f is the autocorrelation distance and E and ν are the elastic modulus and Poisson's ratio of the rock, respectively.

8.6   VARIATION OF ROCK PERMEABILITY WITH DEPTH

Because in situ rock stress increases with depth (see Chapter 2 for the detailed description of in situ rock stress), the permeability of field rock mass decreases with depth. Figs. 8.14 and 8.15 show the variation of measured rock mass permeability coefficient with depth.

Based on field measurements, Louis (1974) found that the rock mass permeability coefficient K decreases with depth z by a negative exponential formula:

$$K = K_0 e^{-Az} \tag{8.30}$$

Permeability coefficient K (m/s)

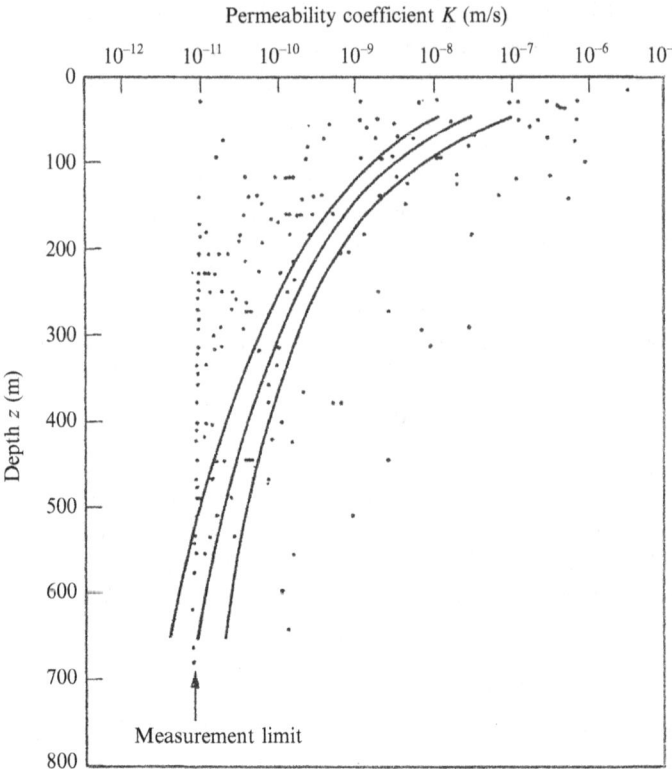

FIG. 8.14 Variation of measured permeability coefficient with depth in granitic rock mass, Sweden. *(From Carlsson, A., Olsson, T., 1993. The analysis of fractures, stress and water flow for rock engineering projects. In: Hudson, J.A. (Ed.), Comprehensive Rock Engineering—Principle, Practice & Projects, vol. 2. Pergamon, Oxford, pp. 415–437.)*

where K_0 is the surface rock permeability coefficient and A is an empirical coefficient. At the Grand Maison dam site, he observed that K_0 varied between 10^{-7} and 10^{-6} m/s and A between 7.8 and 3.4×10^{-3}/m. Meng et al. (2011) applied the same form of expression as Eq. (8.30) to describe the variation of intrinsic permeability with depth in the Southern Qinshui coalbed reservoir in China:

$$k = 11.642e^{-0.0061z} \tag{8.31}$$

where k is the permeability in md and z is the depth in m.

Based on the data given by Snow (1968a,b) about the variation of the permeability coefficient of fractured crystalline rocks with depth, Carlsson and Ollsson (1993) proposed the following relation between permeability coefficient K and depth z:

$$K = 10^{-(1.6 \log z + 4)} \tag{8.32}$$

Permeability coefficient K (m/s)

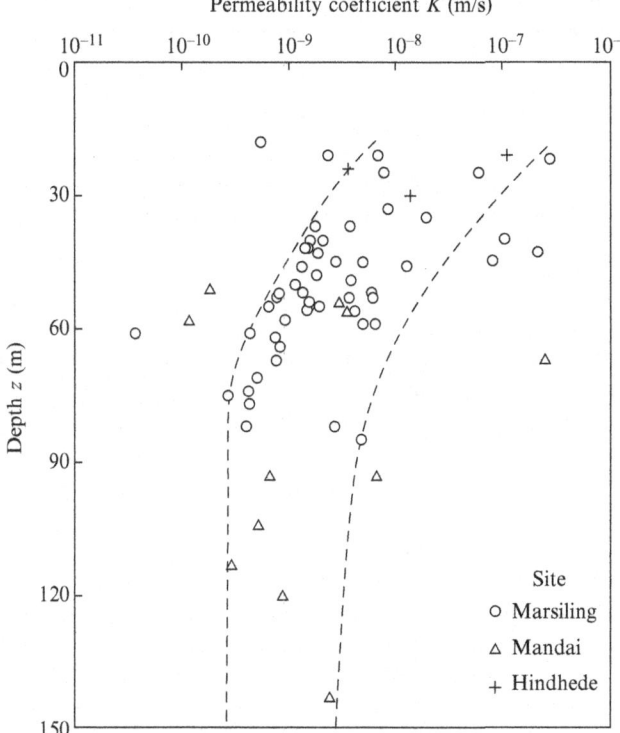

FIG. 8.15 Variation of rock mass permeability coefficient of the Bukit Timah granite with depth at three different sites. *(From Zhao, J., 1998. Rock mass hydraulic conductivity of the Bukit Timah granite, Singapore. Eng. Geol. 50, 211–216.)*

where K and z are in the units of m/s and m, respectively.

Strack (1989) proposed the following relation between permeability coefficient K and depth z for modeling purposes in crystalline rock masses:

$$K = K_0 \left(1 - \frac{z}{\mu} \right)^{\beta} \tag{8.33}$$

where K_0 is the initial rock mass permeability coefficient at the surface and β and μ are constants.

A numerical study conducted by Wei and others based on rock discontinuity network simulation (Wei and Hudson, 1988; Wei et al., 1995) suggested the following relation between rock mass permeability coefficient K and depth z:

$$K = K_0 \left(1 - \frac{z}{58.0 + 1.02z} \right)^{3} \tag{8.34}$$

where K_0 is the rock mass permeability coefficient at initial stage where the normal stress approaches to zero.

Based on the measurements in Sweden, Burgess (1977) presented the following empirical relation between the mean horizontal permeability coefficient K and depth z:

$$\log K = 5.57 + 0.352 \, \log z - 0.978 \, (\log z)^2 + 0.167 \, (\log z)^3 \qquad (8.35)$$

where K and z are in the units of m/s and m, respectively.

The effective vertical in situ rock stress due to the weight of the overburden can be simply estimated by:

$$\sigma = \gamma' z \qquad (8.36)$$

where γ' is the effective unit weight of the overlying rock mass and z is the depth below ground surface. Combining Eqs. (8.27), (8.36) yield the following general relation between rock mass permeability coefficient K and depth z:

$$K = A + Cz^{-m} \qquad (8.37)$$

where A, C $(=B\gamma'^{-m})$, and m are constants.

The decrease of rock mass permeability with depth is mainly due to the decrease of discontinuity aperture with depth. Fig. 8.16 shows the variation of discontinuity aperture with depth based on the data of Snow (1968a,b).

8.7 EFFECT OF TEMPERATURE ON ROCK PERMEABILITY

Changes in temperature also affect the rock permeability. An increase in temperature will cause a volumetric expansion of the rock material leading to reduction in discontinuity aperture and thus an overall reduction in the rock mass permeability. Mineral dissolution and precipitation due to increased temperature will also cause redistribution of minerals in the rock, such that asperities are chemically removed while pores and discontinuities are filled leading to reduction of the rock mass permeability (Moore et al., 1994; Polak et al., 2003; Rosenbrand et al., 2012). Fig. 8.17 shows the reduction of hydraulic aperture of a natural discontinuity in novaculite due to temperature increase, under a constant effective stress of 3.5 MPa. The hydraulic aperture decreased from above 12 to 2.7 μm as temperature was increased from 20°C to 150°C over a period of 900 h. Fig. 8.18 shows the variation of the permeability of tuff as temperature was increased from below 30°C to 150°C and then decreased back to below 30°C. The permeability decreased with higher temperature and then increased with lower temperature (Lin et al., 1997).

Morrow et al. (1981) measured the changes in permeability of intact and fractured granite samples between 200°C and 310°C at confining pressures of 30 and 60 MPa, respectively. In both cases, the permeability decreased between one and two orders of magnitude with the increase in temperature.

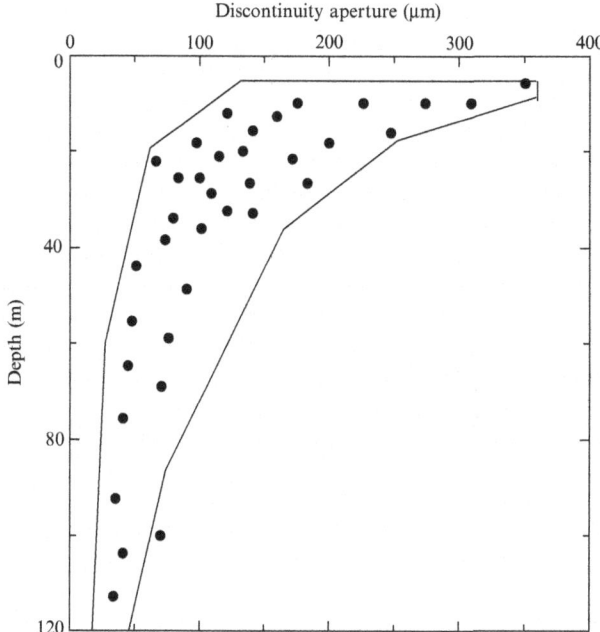

FIG. 8.16 Variation of discontinuity aperture with depth. *(Data from Snow, D.T., 1968a. Anisotropic permeability of fractured media. Water Resour. Res. 5, 1273–1289; Snow, D.T., 1968b. Rock fracture spacings, openings and porosities. J. Soil. Mech. Found. Div., ASCE 94, 73–91.)*

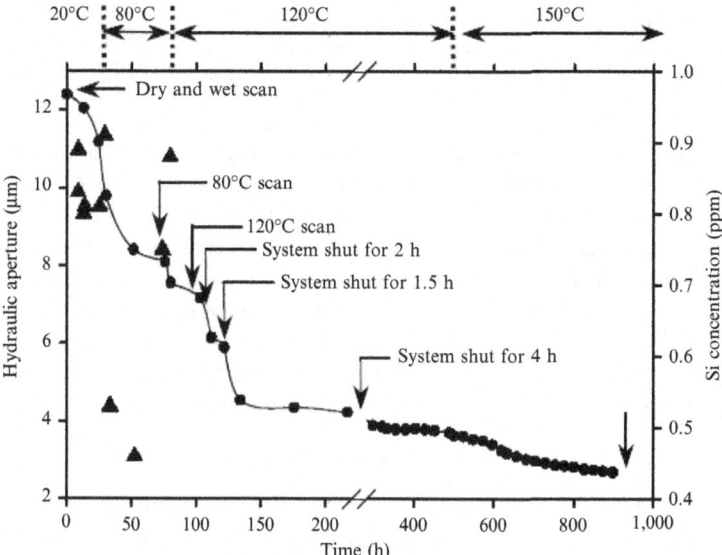

FIG. 8.17 Effect of temperature on hydraulic aperture of a natural discontinuity in novaculite. *(From Polak, A., Elsworth, D., Yahuhara, H., Grader, A.S., Halleck, P.M., 2003. Permeability reduction of a natural fracture under net dissolution by hydrothermal fluids. Geophy. Res. Lett. 30, 2020.)*

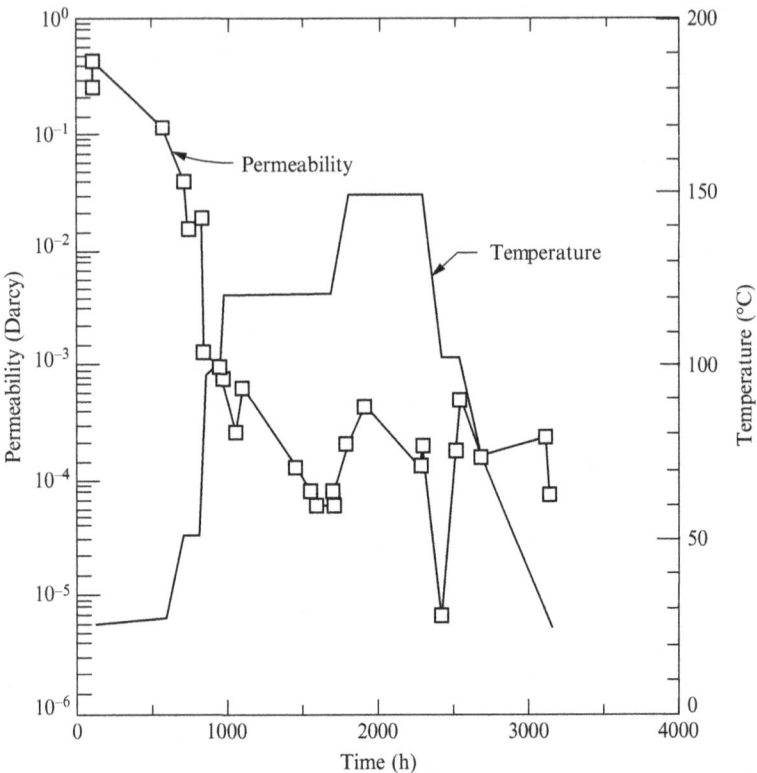

FIG. 8.18   Effect of temperature on the permeability of tuff. *(From Lin, W., Roberts, J., Glassley, W., Ruddle, D., 1997. Fracture and matrix permeability at elevated temperatures. In: Workshop on Significant Issues and Available Data. Near-Field/Altered-Zone Coupled Effects Expert Elicitation Project, San Francisco, CA, November.)*

The permeability decreasing rate was higher at the higher temperature. At 200°C, the permeability dropped by one order of magnitude per month, whereas at 310°C, the permeability dropped to 5% of the initial value within a few days. The dissolution of quartz and feldspar (under pressure) and re-deposition of these minerals within cracks was found to be the major cause for the reduction of permeability. The same authors (Morrow et al., 2001) also conducted permeability measurements on both intact and fractured Westerly granite samples at an effective pressure of 50 MPa and at temperatures from 250°C to 500°C and found that the permeability decreases with time following an exponential relation:

$$k = k_0 e^{-2.303rt} \tag{8.38}$$

where k is the permeability in m^2; k_0 is the initial permeability; r is the rate of permeability decrease; and t is the time in days. The r varies from 0.001 per day at 250°C to 0.1 per day at 500°C for unfractured samples and shows a greater value for prefractured samples at a given temperature.

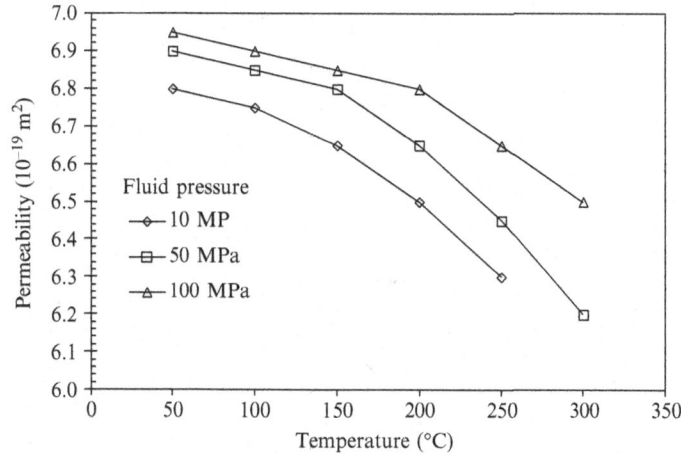

FIG. 8.19 Variation of permeability with temperature for Westerly granite at three different fluid pressures. *(Data from Zharikov, A.V., Shmonov, V.M., Vitotova, V.M., 2005. Experimental study of rock permeability at high temperature and pressure: implication to the continental crust. J. Geophys. Res. 7, 04318.)*

Studies in the Stripa Iron Ore Mine, Sweden demonstrated a decrease of permeability coefficient for granites from $4 \pm 0.8 \times 10^{-11}$ m/s to $1.8 \pm 0.3 \times 10^{-11}$ m/s when temperature was increased by 25°C by circulating warm water. Considering the change in viscosity of the permeant at the higher temperatures, the intrinsic permeability of the rock mass had been reduced by a factor of approximately 4 (Lee and Farmer, 1993).

Barton and Lingle (1982) presented the results of tests conducted in situ on fractured gneiss. The permeability of fractured gneiss was decreased 10-fold with a temperature increase of 74°C. A 10-fold decrease in permeability was also observed based on the in situ test on the Idaho Springs granite when the temperature was increased from 41°C to 73°C (Voegele et al., 1981).

Fig. 8.19 shows the variation of permeability of Westerly granite with temperature at three different fluid pressures based on the study by Artemieva (1997). The permeability obviously decreases with higher temperature although the amount of decreases is not as large as those described above. The study by Zharikov et al. (2005) on granite even observed that the permeability initially decreases up to a minimum value and then increases with subsequent heating. Such permeability behavior is governed by rock microstructure transformations due to the effect of temperature and pressure.

8.8 SCALE EFFECT ON ROCK PERMEABILITY

Research results have shown that the rock mass permeability is strongly scale-dependent (Kunkel et al., 1988; Hudson and Harrison, 1997; Singhal and Gupta, 1999; Wyllie and Mah, 2004; Matsuki et al., 2006). As illustrated in

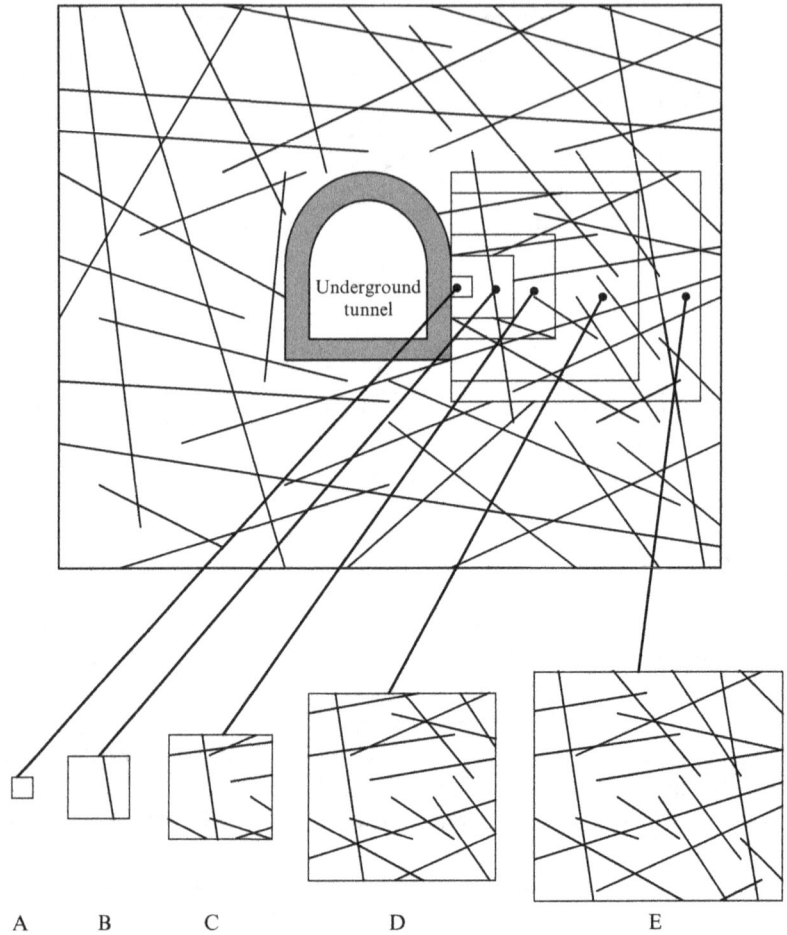

FIG. 8.20 Simplified representation of scale effect on rock mass permeability. *(Modified from Brady, B.H.G., Brown, E.T., 1985. Rock Mechanics for Underground Mining. George Allen & Unwin, London.)*

Fig. 8.20, the permeability of rock will vary as the problem domain enlarges. For domain A, water can flow only through the intact rock and the rock mass permeability is simply the intact rock permeability. For domain B, water can flow vertically through the intact rock and along a single discontinuity and thus the rock mass permeability in the vertical direction is the sum of the intact rock permeability and the permeability of that single discontinuity. In the lateral direction, however, the water can flow only through the intact rock and thus the rock mass permeability is simply the intact rock permeability. As the domain enlarges to C, water will flow through the intact rock and along discontinuities in both the vertical and lateral directions. Therefore, the rock mass permeability in both the vertical and lateral directions will be the sum of the intact rock permeability and the permeability of the corresponding discontinuities. As the domain further enlarges and thus the number of discontinuities in it increases, water will flow along more discontinuities in both the vertical and lateral directions. When the domain enlarges to a certain volume, called "representative elementary volume" (REV), the rock mass permeability will reach a steady magnitude.

The concept of REV is illustrated in Fig. 8.21. The rock mass permeability will become constant at some REV if the discontinuity occurrence is statistically homogeneous in the region considered. If the discontinuity occurrence is inhomogeneous, the permeability may show further oscillations in the trace or steady increases or decreases.

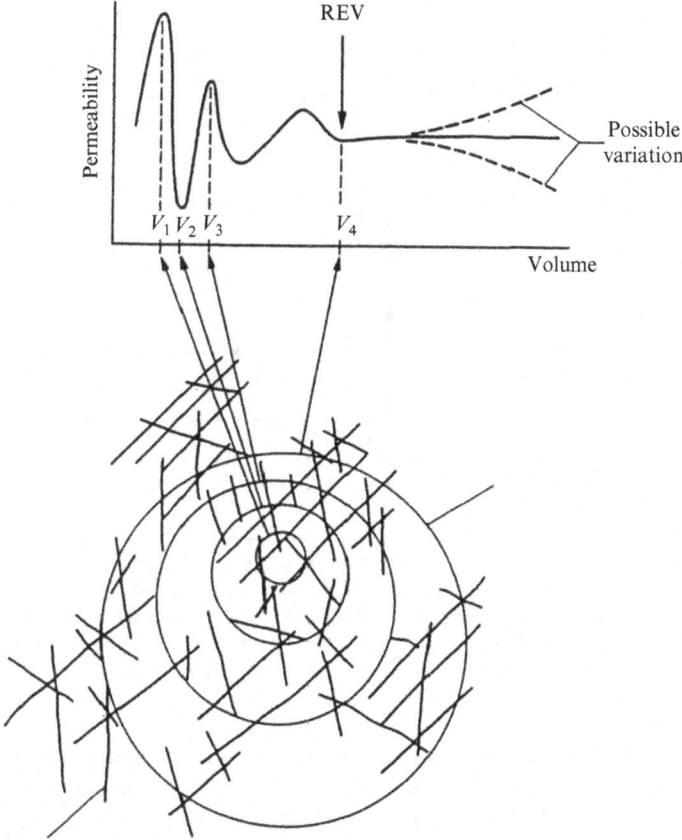

FIG. 8.21   Representative elemental volume (REV) for rock mass permeability. *(From Elsworth, D., Mase, C.R., 1993. Groundwater in rock engineering. In: Hudson, J.A. (Ed.), Comprehensive Rock Engineering—Principle, Practice & Projects, vol. 2. Pergamon, Oxford, pp. 201–226.)*

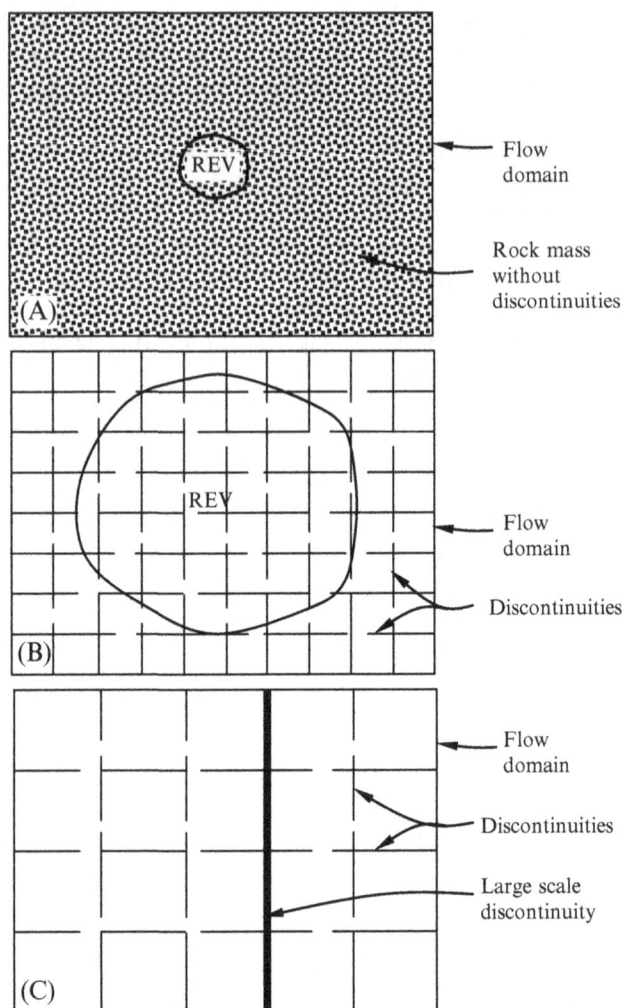

FIG. 8.22 Representative elemental volume (REV) in different rock conditions: (A) rock mass without discontinuities; (B) rock mass containing discontinuities where REV includes sufficient discontinuity interactions; and (C) rock mass containing large-scale discontinuities where REV is either very large or nonexistent. *(Modified from Kunkel, J.R., Way, S.C., McKee, C.R., 1988. Comparative Evaluation of Selected Continuum and Discrete-Fracture Models. US Nuclear Regulatory Commission, NUREG/CR-5240.)*

REV increases in size with larger discontinuity spacing (Kunkel et al., 1988). Fig. 8.22 illustrates how discontinuities affect REV. In rocks without discontinuities, small REV can be representative of the rock mass (Fig. 8.22A). In rock masses containing discontinuities, REV should be large enough to include sufficient discontinuity intersections to represent the flow domain (Fig. 8.22B). The size of REV will be large compared to the discontinuity lengths in order to provide a good statistical sample of the discontinuity population. In the case of large-scale discontinuities, such as faults and dykes, REV may not be feasible as it will be too large (Fig. 8.22C). So the REV concept may not be satisfied for every rock mass. The only way to define REV for a rock mass is to investigate in detail the discontinuity geometry. For detailed characterization of discontinuity geometry, the reader can refer to Chapter 4.

8.9   INTERCONNECTIVITY OF DISCONTINUITIES

Another important point that can be seen from Fig. 8.20 is the interconnection of discontinuities. For example, there are eight discontinuities included in domain C; but only five of them are interconnected and may act as flow path for the lateral and vertical water flow (Fig. 8.23). Interconnectivity of discontinuities is one of the most important factors affecting the permeability of rock masses. Since all discontinuities are of finite length, a discontinuity can act as a flow path only when it extends completely across the zone interested or is connected to other conductive discontinuities. McCrae (1982) estimated that only about 20% of discontinuities encountered during construction of the Muna highway tunnels, Saudi Arabia, were potential water conduits, of which only about 25% had positive evidence of being so. Andersson et al. (1988) found that only 10–40% of discontinuities in the Brandan area, Finnsjon, Sweden were conductive. Of the 11,000 discontinuities documented throughout the Äspö Hard Rock Laboratory in Sweden, only 8% were wet when they were excavated (Talbot and Sirat, 2001).

8.10   ANISOTROPY OF ROCK PERMEABILITY

Like mechanical properties, the permeability of rocks also shows appreciable anisotropy (Zoback and Byerlee, 1976; Pratt et al., 1977; Ayan et al., 1994; Bieber et al., 1996; Renard et al., 2001; Benson et al., 2005; Louis et al., 2005; Meyer and Krause, 2006; Clavaud et al., 2008). The anisotropy of intact rock permeability is primarily a function of the preferred orientation of mineral particles and micro-discontinuities. The permeability of intact rock parallel to

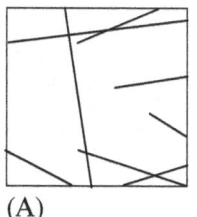
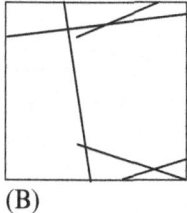

(A)   (B)

FIG. 8.23   (A) Domain C in Fig. 8.20 containing eight discontinuities and (B) five discontinuities in domain C are interconnected.

the bedding (k_{par}) is usually larger than that perpendicular to it (k_{per}). Table 8.1 lists the values of k_{par}/k_{per} for different rocks.

In many cases, the bedding plane may not be horizontal but inclined (Fig. 8.24). Meyer and Krause (2006) measured the permeability values of a sandstone reservoir in three orientations: horizontal and parallel to the strike of the inclined cross-bedding or cross-laminate (k_{hpar}), horizontal and perpendicular to the cross-bedding or cross-laminate at a shallow angle (k_{hper}), and

TABLE 8.1 Ratio of Permeability Parallel to Bedding k_{par} to Permeability Perpendicular to Bedding k_{per} for Different Rocks

Rock	k_{par}/k_{per}	Reference
Rothbach sandstone	7.1	Louis et al. (2005)
Andesite	4.0	Clavaud et al. (2008)
Limestone	1.6–8.3	Clavaud et al. (2008)
Sandstone	1.3–5.9	Clavaud et al. (2008)
Berea sandstone	4.0	Zoback and Byerlee (1976)
Triassic Sherwood sandstone	2.0–3.3	Ayan et al. (1994)
Granite	2.5	Pratt et al. (1977)
Crab Orchard sandstone	2.2	Benson et al. (2005)
Sandstone	1.4	Meyer (2002)
Bentheim sandstone	1.2	Louis et al. (2005)

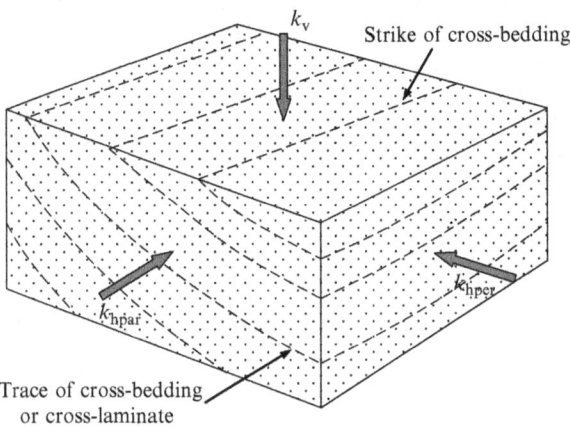

FIG. 8.24 Schematic of inclined cross-bedding or cross-laminate and permeability in three different orientations.

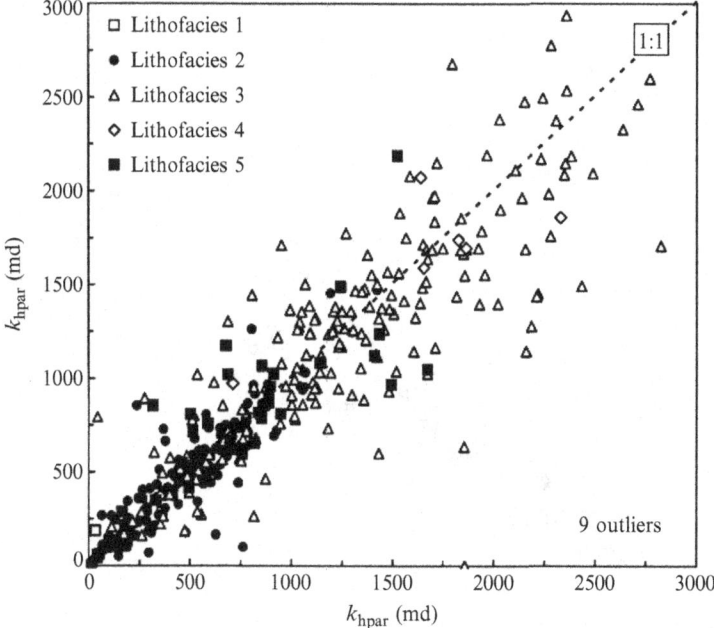

FIG. 8.25 Plot of permeability in the direction horizontal and perpendicular to the strike of the inclined cross-bedding or cross-laminate at a shallow angle (k_{hper}) versus permeability in the direction horizontal and parallel to the cross-bedding or cross-laminate (k_{hpar}). *(From Meyer, R., Krause, F.F., 2006. Permeability anisotropy and heterogeneity of a sandstone reservoir analogue: an estuarine to shoreface depositional system in the Virgelle Member, Milk River Formation, Writing-on-Stone Provincial Park, southern Alberta. Bull. Can. Petro. Geol. 54, 301–318.)*

vertical, at a high angle to the cross-bedding or cross-laminate (k_v) (see Fig. 8.24). Fig. 8.25 plots k_{hper} versus k_{hpar}. The mean values of the ratio k_{hpar}/k_{hper} for all lithofacies are between 1.0 and 1.2, indicating the slight anisotropy of permeability on the horizontal plane. Fig. 8.26 plots k_v versus $k_h = (k_{hpar}+k_{hper})/2$. Most of the k_v/k_h ratios (69%) are between 0.5 and 1.

Because of the discontinuities, the degree of permeability anisotropy for jointed rock masses may be much higher than that for intact rock. The ratio k_{par}/k_{per} may vary from 10^{-2} for rock masses whose discontinuities are mainly vertical to 10^3 for rock masses containing bedding planes. The contribution of discontinuities to the permeability of a rock mass can be estimated using the methods presented in Section 8.4.

Changes in pore pressure can affect the degree of permeability anisotropy. For example, in cases where there is only one dominant discontinuity set, an increase in pore pressure will lower the effective stress. This leads to an increase in anisotropy by opening the discontinuities, thus increasing the permeability parallel to the discontinuity orientation. Where there is more than

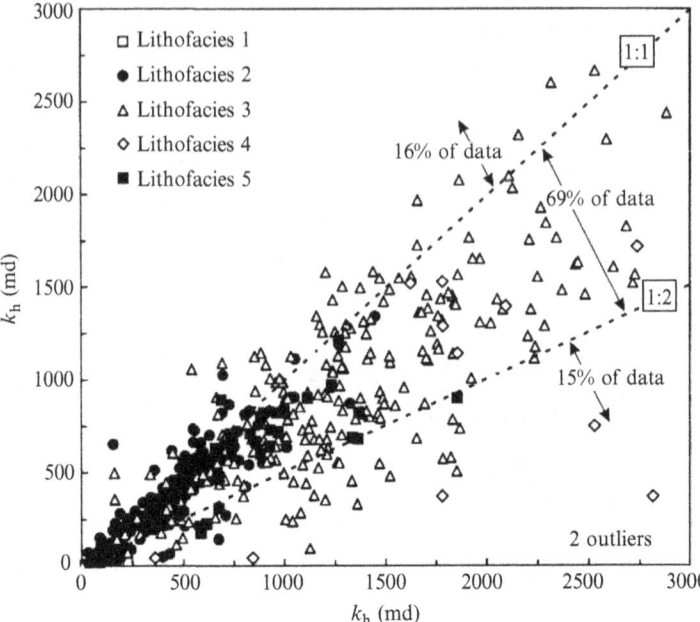

FIG. 8.26 Plot of permeability in the direction vertical, at a high angle to the inclined cross-bedding or cross-laminate (k_v) versus $k_h = (k_{hpar} + k_{hper})/2$, where k_{hpar} is the permeability in the direction horizontal and parallel to the cross-bedding or cross-laminate and k_{hper} is the permeability in the direction horizontal and perpendicular to the strike of the cross-bedding or cross-laminate at a shallow angle. *(From Meyer, R., Krause, F.F., 2006. Permeability anisotropy and heterogeneity of a sandstone reservoir analogue: an estuarine to shoreface depositional system in the Virgelle Member, Milk River Formation, Writing-on-Stone Provincial Park, southern Alberta. Bull. Can. Petro. Geol. 54, 301–318.)*

one discontinuity set, the nature of anisotropy change with stress would depend on which of the sets are more deformable. In a poorly connected discontinuity network, a decrease in pore pressure could theoretically make the rock more isotropic.

REFERENCES

Andersson, J.-E., Ekman, L., Winberg, A., 1988. Detailed hydraulic characterization of a fracture zone in the Brandan area, Finnsjon, Sweden. In: Hitchon, B., Bachu, S. (Eds.), Proc. 4th Canadian/American Conf. on Hydrogeol. Natl. Water Well Assoc., Dublin, OH, pp. 32–39.

Artemieva, I., 1997. In situ permeability of hot dry rock. In: Fractured reservoirs. Nordic Petroleum Technology Series. Middleton, M.F. (Ed.), vol. 1. Goteborg, pp. 99–124.

Ayan, C., Colley, N., Cowan, G., Ezekwe, E., Wannell, M., Goode, P., Halford, F., Joseph, J., Mongini, A., Pop, J., 1994. Measuring permeability anisotropy: the latest approach. Oilfield Rev., 6 (4), 24–27.

Azeemuddin, M., Roegiers, J.C., Suir, P., Zaman, M., Kukreti, A.R., 1995. Stress-dependent permeability measurement of rocks in a triaxial cell. In: Daemen, J.J.K., Schultz, R.A. (Eds.), Proc. 35th US Symp. on Rock Mech., Reno, Nevada, pp. 645–650.

Baghbanan, A., Jing, L., 2008. Stress effects on permeability in a fractured rock mass with correlated fracture length and aperture. Int. J. Rock Mech. Min. Sci. 45, 1320–1334.

Barton, N., 2008. Training Course in Rock Engineering. Organized by ISRMTT & CSMRS, New Delhi. December 10–12.

Barton, N., Lingle, R., 1982. Rock mss characterization methods for nuclear waste repositories in jointed rock. In: Rock Mech.: Caverns and Pressure Shafts, ISRM Symp., Aachen, pp. 3–18.

Barton, N., Bandis, S., Bakhtar, K., 1985. Strength, deformation and conductivity coupling of rock joints. Int. J. Rock Mech. Min. Sci. Geomech. Abstr. 22, 121–140.

Benson, P.M., Meredith, P.G., Platzman, E.S., White, R.E., 2005. Pore fabric shape anisotropy in porous sandstones and its relation to elastic wave velocity and permeability anisotropy under hydrostatic pressure. Int. J. Rock Mech. Min. Sci. 42, 890–899.

Bieber, M.T., Rasolofosaon, P., Zinszner, B., Zamora, M., 1996. Measurements and overall characterization of permeability anisotropy by tracer injection. Rev. Inst. Fr. Pétrol. 51, 333–347.

Brace, W.F., 1978. A note on permeability changes in geological material due to stress. Pure Appl. Geophys. 116, 627–633.

Brace, W.F., Walsh, J.B., Frangos, W.T., 1968. Permeability of granite under high pressure. J. Geophys. Res. 73, 2225–2236.

Burgess, A., 1977. Groundwater Movements Around a Repository—Regional Groundwater Flow Analysis. Kaernbraenslesaekerhet, Stockholm. KBS 54:03.

Cade, C.A., Evans, I.J., Bryant, S.L., 1994. Analysis of permeability controls—a new approach. Clay Miner. 29, 491–501.

Carlsson, A., Olsson, T., 1993. The analysis of fractures, stress and water flow for rock engineering projects. In: Hudson JA, editor, Comprehensive Rock Engineering – Principle, Practice & Projects, Pergamon, Oxford, UK, 2, 415–437.

Cha, S.S., Lee, J.Y., Lee, D.H., Amantini, E., Lee, K.K., 2006. Engineering characterization of hydraulic properties in a pilot rock cavern for underground LNG storage. Eng. Geol. 84, 229–243.

Chen, Z., Pan, Z., Liu, J., Connell, L.D., Elsworth, D., 2011. Effect of the effective stress coefficient and sorption-induced strain on the evolution of coal permeability: experimental observations. Int. J. Greenhouse Gas Control 5, 1284–1293.

Clavaud, J.-B., Maineult, A., Zamora, M., Rasolofosaon, P., Schlitter, C., 2008. Permeability anisotropy and its relations with porous medium structure. J. Geophys. Res. 113, B01202.

David, C., Wong, T.F., Zhu, W., Zhang, J., 1994. Laboratory measurement of compaction-induced permeability change in porous rocks: implications for the generation and maintenance of pressure excess in the crust. Pure Appl. Geophys. 143, 425–456.

El-Naqa, A., 2001. The hydraulic conductivity of the fractures intersecting Cambrian sandstone rock masses, central Jordan. Environ. Geol. 40, 973–982.

Fernandez, G., Moon, J., 2010. Excavation-induced hydraulic conductivity reduction around a tunnel—part 1: guideline for estimate of ground water inflow rate. Tunn. Undergr. Space Technol. 25, 560–566.

Gangi, A.F., 1978. Variation of whole and fractured porous rocks permeability with confining pressure. Int. J. Rock Mech. Min. Sci. Geomech. Abstr. 15, 335–354.

Ghabezloo, S., Sulem, J., Guedon, S., Martineau, F., 2009. Effective stress law for the permeability of a limestone. Int. J. Rock Mech. Min. Sci. 46, 297–306.

Hakami, E., 1995. Aperture distribution of rock fractures. Doctoral thesis, Department of Civil and Environmental Engineering, KTH.

Hoek, E., Bray, J.W., 1981. Rock Slope Engineering. Institution of Mining and Metallurgy, London.

Hudson, J.A., Harrison, J.P., 1997. Engineering Rock Mechanics: An Introduction to the Principles. Elsevier, Amsterdam.

Huitt, J.L., 1956. Fluid flow in simulated fracture. Am. Inst. Chem. Eng. J. 2, 259–264.

Indraratna, B., Ranjith, P., Gale, W., 1999. Deformation and permeability characteristics of rocks with interconnected fractures. In: 9th Int. Cong. on Rock Mech. vol. 2, Paris, France, pp. 755–760.

Jiang, X.-W., Wan, L., Wang, X.-S., Wu, X., Zhang, X., 2009. Estimation of rock mass deformation modulus using variations in transmissivity and RQD with depth. Int. J. Rock Mech. Min. Sci. 46, 1370–1377.

Jiang, T., Shao, J.F., Xu, W.Y., Zhou, C.B., 2010. Experimental investigation and micromechanical analysis of damage and permeability variation in brittle rocks. Int. J. Rock Mech. Min. Sci. 47, 703–713.

Jones, F.O., 1975. A laboratory study of the effects of confining pressure on fracture flow and storage capacity in carbonate rocks. J. Petro. Tech. 21, 21–27.

Jouanna, P., 1993. A summary of field test methods in fractured rocks. In: Bear, J., Tsang, C.F., de Marsily, D. (Eds.), Flow and Contaminant Transport in Fractured Rock. Academic Press, San Diego, CA, pp. 437–543.

Kawano, S., Katayama, I., Okazaki, K., 2011. Permeability anisotropy of serpentinite and fluid pathways in a subduction zone. Geology 39, 939–942.

Kishida, K., Kawaguchi, Y., Nakashima, S., Yasuhara, H., 2011. Estimation of shear strength recovery and permeability of single rock fractures in shear-hold-shear type direct shear tests. Int. J. Rock Mech. Min. Sci. 48, 782–793.

Knill, J.L., 1969. The application of seismic methods in the prediction of grout taken in rock. In: Proc. Conf. In-Situ Investigations in Soils and Rocks, London, pp. 93–99.

Konecny, P., Kozusnikova, A., 2011. Influence of stress on the permeability of coal and sedimentary rocks of the upper silesian basin. Int. J. Rock Mech. Min. Sci. 48, 347–352.

Kranz, R.L., Frankel, A.D., Engelder, T., Scholz, C.H., 1979. The permeability of whole and jointed barre granite. Int. J. Rock Mech. Min. Sci. Geomech. Abstr. 16, 225–234.

Kunkel, J.R., Way, S.C., McKee, C.R., 1988. Comparative Evaluation of Selected Continuum and Discrete-Fracture Models. US Nuclear Regulatory Commission. Washington, DC. NUREG/CR-5240.

Lee, C.-H., Farmer, I., 1993. Fluid Flow in Discontinuous Rocks. Chapman & Hall, London.

Lin, W., Roberts, J., Glassley, W., Ruddle, D., 1997. Fracture and matrix permeability at elevated temperatures. In: Workshop on Significant Issues and Available Data, Near-Field/Altered-Zone Coupled Effects Expert Elicitation Project, San Francisco, CA, November.

Liu, H.-H., Rutqvist, J., 2010. A new coal-permeability model: internal swelling stress and fracture–matrix interaction. Transp. Porous Media 82, 157–171.

Lomize, G., 1951. Fluid Flow in Fissured Formation. Moskva, Leningrad (in Russian).

Louis, C.A., 1969. A study of groundwater flow in jointed rock and its influence on the stability of rock masses. Rock Mechanics Research Report No. 10, Imperial College, London, England, Technical Information Center, US Army Engineer Waterways Experiment Station, Vicksburg, MS.

Louis, C.A., 1974. Rock hydraulics. In: Muller, L. (Ed.), Rock Mechanics. Springer Verlag, Vienna, pp. 299–382.

Louis, C.A., Dessenne, J.-L., Feuga, B., 1977. Interaction between water flow phenomena and the mechanical behavior of soil or rock masses. In: Gudehus, G. (Ed.), Finite Elements in Geomechanics. Wiley, New York, NY, pp. 479–511.

Louis, L., David, C., Metz, V., Robion, P., Menéndez, B., Kissel, C., 2005. Microstructural control on the anisotropy of elastic and transport properties in undeformed sandstones. Int. J. Rock Mech. Min. Sci. 42, 911–923.

Matsuki, K., Chida, Y., Sakaguchi, K., Glover, P.W.J., 2006. Size effect on aperture and permeability of a fracture as estimated in large synthetic fractures. Int. J. Rock Mech. Min. Sci. 43, 726–755.

McCrae, R.W., 1982. Geotechnical factors affecting the design of highway tunnels in hard rock, PhD thesis, University of Newcastle, Newcastle, NSW.

Meng, Z., Zhang, J., Wang, R., 2011. In-situ stress, pore pressure and stress-dependent permeability in the Southern Qinshui Basin. Int. J. Rock Mech. Min. Sci. 48, 122–131.

Metwally, Y.M., Sondergeld, C.H., 2011. Measuring low permeabilities of gas-sands and shales using a pressure transmission technique. Int. J. Rock Mech. Min. Sci. 48, 1135–1144.

Meyer, R., 2002. Anisotropy of sandstone permeability. CREWES Research Report, vol. 14, pp. 1–12. https://www.crewes.org/ForOurSponsors/ResearchReports/2002/2002-05.pdf

Meyer, R., Krause, F.F., 2006. Permeability anisotropy and heterogeneity of a sandstone reservoir analogue: an estuarine to shoreface depositional system in the Virgelle Member, Milk River Formation, Writing-on-Stone Provincial Park, southern Alberta. Bull. Can. Petrol. Geol. 54, 301–318.

Moore, D.E., Lockner, D.A., Byerlee, J.A., 1994. Reduction of permeability in granite at elevated temperatures. Science 265, 1558–1561. New Series.

Morrow, C., Lockner, D., Moore, D., Byerlee, J., 1981. Permeability of granite in a temperature gradient. J. Geophys. Res. 86, 3002–3008.

Morrow, C.A., Moore, D.E., Lockner, D.A., 2001. Permeability reduction in granite under hydrothermal conditions. J. Geophys. Res. 106, 30551–30560.

Nelson, R., 1975. Fracture permeability in porous reservoirs: experimental and field approach. PhD dissertation, Dept. of Geology, Texas A&M University, College Station, TX.

Oda, M., Saitoo, T., Kamemura, K., 1989. Permeability of rock mass at great depth. In: Maury, V., Fourmaintraux, D. (Eds.), Proc. Symp. Rock at Great Depth. Balkema, Rotterdam, pp. 449–456.

Olsson, R., Barton, N., 2001. An improved model for hydromechanical coupling during shear of rock joints. Int. J. Rock Mech. Min. Sci. 38, 317–329.

Polak, A., Elsworth, D., Yahuhara, H., Grader, A.S., Halleck, P.M., 2003. Permeability reduction of a natural fracture under net dissolution by hydrothermal fluids. Geophys. Res. Lett. 30, 2020.

Pratt, H.R., Schrauf, T.A., Bills, L.A., Hustrulid, W.A., 1977. Thermal and mechanical properties of granite: Stripa, Sweden. TerraTek Summary Report TR-77-92, Salt Lake City, Utah.

Quadros, E.F., 1982. Determinacao das caracteristicas do fluxo de agua em fracturas de rochas. Dissert. de Mestrado, Dept. of Civil Eng., Polytech School, University of Sao Paulo, Sao Paulo.

Ranjith, P.G., 2000. Analytical and numerical investigation of water and air flow through rock media. PhD thesis, Dept. of Civil Eng., Univ. of Wollongong, Wollongong.

Read, M.D., Meredith, P.G., Murrell, S.A.F., 1989. Permeability measurement techniques under hydrostatic and deviatoric stress conditions. In: Maury, V., Fourmaintraux, D. (Eds.), Proc. Symp. Rock at Great Depth. Balkema, Rotterdam, pp. 211–217.

Renard, P., Genty, A., Stauffer, F., 2001. Laboratory determination of the full permeability tensor. J. Geophys. Res. 106, 26443–26452.

Rosenbrand, E., Fabricius, I.L., Yuan, H., 2012. Thermally induced permeability reduction due to particle migration in sandstones: the effect of temperature on kaolinite mobilization and aggregation. In: Proc. 37th Workshop on Geothermal Reservoir Eng., January 30–February 1, SGP-TR-194. Stanford University, Stanford, CA.

Schopper, J.R., 1982. Porosity and permeability. Numerical Data and Functional Relationships in Science and Technology, New Series; Group V Geophysics and Space Research, Vol 1 Physical properties of Rocks. Subvol A. Springer-Verlag Berlin.

Seidle, J.P., Jeansonne, M.W., Erickson, D.J., 1992. Application of matchstick geometry to stress dependent permeability in coals. SPE Rocky Mountain Regional Meeting, Casper, Wyoming. Paper SPE 24361.

Sen, Z., 1996. Theoretical RQD-porosity-conductivity-aperture charts. Int. J. Rock Mech. Min. Sci. Geomech. Abstr. 33, 173–177.

Sharp, J.C., 1970. Fluid flow through fissured media. Ph.D. thesis, University of London (Imperial College).

Singhal, B.B.S., Gupta, R.P., 1999. Applied Hydrology of Fractured Rocks. Kluwer Academic Publishers, Dordrecht.

Snow, D.T., 1968a. Anisotropic permeability of fractured media. Water Resour. Res. 5, 1273–1289.

Snow, D.T., 1968b. Rock fracture spacings, openings and porosities. J. Soil. Mech. Found. Div., ASCE 94, 73–91.

Strack, O.D.L., 1989. Groundwater Mechanics. Prentice Hall, Englewood Cliffs, NJ.

Talbot, C.J., Sirat, M., 2001. Stress control of hydraulic conductivity in fractured-saturated Swedish bedrock. Eng. Geol. 61, 145–153.

Tiller, F.M., 1953. The role of porosity in filtration: numerical methods for constant rate and constant pressure filtration based on Kozeny's law. Chem. Eng. Process. 49, 467–479.

Van-Golf Racht, T.D., 1982. Fundamentals of Fractured Reservoir Engineering. Elsevier, Amsterdam.

Voegele, M., Hardin, E., Lingle, D., Board, M., Barton, N., 1981. Site characterisation of joint permeability using the heated block test. In: Proc. 22nd Symp. on Rock Mech., Cambridge, MA, pp. 120–127.

Walsh, J.B., 1981. Effects of pore pressure and confining pressure on fracture permeability. Int. J. Rock Mech. Min. Sci. Geomech. Abstr. 18, 429–434.

Walsh, J.B., Grosenbaugh, M.A., 1979. A new model for analyzing the effect of fractures on compressibility. J. Geophys. Res. 84, 3532–3536.

Wei, Z.Q., Hudson, J.A., 1988. Permeability of jointed rock masses. In: Proc. Int. Symp. Rock Mech. and Power Plants. Balkema, Rotterdam, pp. 613–625.

Wei, Z.Q., Egger, P., Descoeudres, F., 1995. Permeability predictions for jointed rock masses. Int. J. Rock Mech. Min. Sci. Geomech. Abstr. 32, 251–261.

Wong, L.G.Y., Li, D., Liu, G., 2013. Experimental studies on permeability of intact and singly jointed meta-sedimentary rocks under confining pressure. Rock Mech. Rock Eng. 46, 107–121.

Wyllie, D.C., Mah, C.W., 2004. Rock Slope Engineering—Civil and Mining, fourth ed. Spon Press, London.

Yang, Y.L., Aplin, A.C., 2007. Permeability and petrophysical properties of 30 natural mudstones. J. Geophys. Res. 112.

Yang, Y.L., Aplin, A.C., 2010. A permeability–porosity relationship for mudstones. Mar. Pet. Geol. 27, 1692–1697.

Yang, D., Billiotte, J., Su, K., 2010. Characterization of the hydromechanical behavior of argillaceous rocks with effective gas permeability under deviatoric stress. Eng. Geol. 114, 116–122.

Zhang, J., Bai, M., Roegiers, J.C., Wang, J., Liu, T., 2000. Experimental determination of stress–permeability relationship. In: Proc. Fourth North Amer. Rock Mech. Symp., Seattle, pp. 817–822.

Zhang, J., Standifird, W.B., Roegiers, J.-C., Zhang, Y., 2007. Stress-dependent fluid flow and permeability in fractured media: from lab experiments to engineering applications. Rock Mech. Rock Eng. 40, 3–21.

Zharikov, A.V., Shmonov, V.M., Vitotova, V.M., 2005. Experimental study of rock permeability at high temperature and pressure: implication to the continental crust. J. Geophys. Res. 7, 04318.

Zimmerman, R.W., Bodvarsson, G.S., 1996. Hydraulic conductivity of rock fractures. Transp. Porous Media 23, 1–30.

Zoback, M.D., Byerlee, J.D., 1976. Effect of high-pressure deformation on the permeability of Ottawa Sand. Bull. Am. Assoc. Petrol. Geol. 60, 1531.

Zou, L., Tarasov, B.G., Dyskin, A.V., Adhikary, D.P., Pasternak, E., Xu, W., 2013. Physical modelling of stress-dependent permeability in fractured rocks. Rock Mech. Rock Eng. 46, 67–81.

Index

Note: Page numbers followed by *f* indicate figures, and *t* indicate tables.

Printed in the United States
By Bookmasters